EXPERIMENTAL PHYSICS

EXPERIMENTAL PHYSICS

A SELECTION OF EXPERIMENTS

BY

G. F. C. SEARLE, Sc.D., F.R.S.

UNIVERSITY LECTURER IN EXPERIMENTAL PHYSICS
AND
DEMONSTRATOR OF EXPERIMENTAL PHYSICS
AT THE CAVENDISH LABORATORY
SOMETIME FELLOW OF PETERHOUSE

CAMBRIDGE
AT THE UNIVERSITY PRESS
1934

CAMBRIDGE
UNIVERSITY PRESS

University Printing House, Cambridge CB2 8BS, United Kingdom

Published in the United States of America by Cambridge University Press, New York

Cambridge University Press is part of the University of Cambridge.

It furthers the University's mission by disseminating knowledge in the pursuit of education, learning and research at the highest international levels of excellence.

www.cambridge.org
Information on this title: www.cambridge.org/9781107698307

First published 1934
First paperback edition 2014

A catalogue record for this publication is available from the British Library

ISBN 978-1-107-69830-7 Paperback

PREFACE

In the course of my work as a Demonstrator of Experimental Physics, which began in October 1888, many experiments have been devised for the instruction of my students at the Cavendish Laboratory. The three manuals on *Experimental Elasticity* (1908), on *Experimental Harmonic Motion* (1915) and on *Experimental Optics* (1925) describe courses of experiments in those subjects. Many experiments done in my Class had, by their nature, no place in these three volumes. Accounts of some have appeared in the *Proceedings of the Cambridge Philosophical Society* and elsewhere; others* have not, up to the present, been described by me, except in the manuscripts I have written for use in my Class.

The present volume contains accounts of experiments in Mechanics, Elasticity, Surface Tension, Viscosity, Heat and Sound. These experiments are not described in the three manuals already published. Prof. E. V. Appleton has called the new volume "The Odds and Ends Book"; I do not quarrel with his description.

My work as Demonstrator ceases on 30 September, 1935. It seemed right to prepare for that event by writing this book. I hope it may be of service to those who succeed me, and that it may be useful to students and teachers at the Cavendish Laboratory and elsewhere.

The book is, in one aspect, love's thank-offering for the many happy years spent in the service of students. The work has often been hard, sometimes very hard, but hard work need not be misery.

Mathematical discussions of some problems in Surface Tension, Conduction of Heat and Sound occupy Chapters v, viii and x. The Chapter (x) on Sound may seem over-long for the small

* In his *Practical Physics* (1926) my colleague and former assistant demonstrator, Mr T. G. Bedford, gave, by my permission, brief accounts of a few of these experiments.

number of acoustical experiments described in Chapter XI, but I was anxious to bring out the character of the motion which the air has when a resonator is sounding.

The increase in the number of students led, some years ago, to the organisation of a new Practical Class in Electricity and Magnetism. The removal of these subjects from my Practical Course allowed them to be taught more fully than was previously possible, and compelled me to extend my course in Optics. The only, but very real, disadvantage is that students are now offered more than they can absorb in the time at their disposal.

I have abandoned a former project of writing a manual on *Experimental Electricity and Magnetism.*

To avoid the monotony of repetition, I record here my cordial thanks to each of the many friends who gave me help at any stage of the work.

Prof. Lord Rutherford, F.R.S., has seconded my endeavour to secure efficiency in my Class; he has been indulgent to me.

I have often sought the advice of Dr G. T. Bennett, F.R.S. By his help, many statements, particularly in the dynamical sections of the book, were put into precise forms.

The theory of the Sessile Drop in Chapter V was suggested by the interesting, but mathematically invalid, work of Mathieu. By using an elliptic integral method, to the same degree of approximation as I had adopted, Prof. R. H. Fowler, F.R.S., verified equation (25) of § 102 and equation (30) of § 104.

Several students gave their help during vacations. Messrs T. B. Rymer, C. L. Cook, J. G. P. Baker, R. Stone, K. G. Tupling, W. A. B. Carter, J. W. Jeffery and J. L. Roberts assisted in the preparation of the manuscript; Mr C. H. Garrett and Mr D. A. Crooks gave effective help in the correction of the proof sheets. My chief debt is to Miss F. W. Stubbins, of Girton College, who gave her help in the summers of 1929 and 1930. Besides transcribing for the press much of my rough manuscript, she made a large number of the drawings for the figures.

For some years I have had the help of Mr J. A. Ratcliffe, Fellow of Sidney Sussex College, and of Dr N. Feather, Fellow of Trinity College, as demonstrators. Their experience gave weight to their comments on the proofs of the first six chapters. It is usually taught that the stable length of a cylindrical soap film, of radius a, is $2\pi a$, and this is true under suitable conditions. Dr Feather's criticism led me to the result that, if the pressure excess be maintained at the *constant* value $2T/a$, where T is the surface tension, the maximum stable length is only πa.

Dr Guy Barr, of the Department of Metallurgy and Metallurgical Chemistry, National Physical Laboratory, read the proofs of the Chapter (VII) on Viscosity. His expert knowledge of Viscometry made his help of great value.

Dr Barr, and Dr Allan Ferguson, East London College, made some useful remarks on the Chapters (V and VI) on Surface Tension.

Dr J. K. Roberts, sometime Assistant in the Heat Department of the National Physical Laboratory, made a careful revision of the Chapters on Heat. The Chapters on Sound were read by Dr J. E. R. Constable, of the Physics Department of the National Physical Laboratory, and also by Mr W. R. Dean, Fellow of Trinity College. Mr Dean's detection of a mathematical error led me to set out in full the *exact* equation for spherical waves of sound.

The Cambridge Philosophical Society and Messrs W. G. Pye & Co. have lent blocks for several figures.

The work of the Staff of the Cambridge University Press has won my admiration and my gratitude.

I owe very much to my assistant, Mr C. G. Tilley, for his faithful help for over fifteen years, and for his unfailing kindness and thought for me. If the apparatus described in this book be efficient for its purpose, it is, in many cases, due to Mr Tilley's resourcefulness, to his knowledge of workshop methods, and to his skilled craftsmanship. In former years, I received much help from Mr F. Lincoln and Mr H. D. Roff, instrument makers at the Laboratory.

I have authorised Messrs W. G. Pye & Co., of Cambridge, to supply apparatus made to my designs.

I dare not conclude without a word of grateful testimony. For about seven years, I went through a dark time of nervous and physical weakness, with a complete breakdown in 1910. At the end of 1914, there came into my hands Miss Dorothy Kerin's little book, *The Living Touch*, a record of her miraculous healing. It opened my eyes, as they had never been opened before, to the present-day Power of the Living God, the Creator of the universe. As an immediate result, all the old weakness left me, and I was well. In recent years, at Miss Kerin's hostel, Chapel House, Mattock Lane, Ealing, London, many have realised the greatness of the Lord's Power and Love, and have learned to trust Him.

G. F. C. SEARLE

Cavendish Laboratory
Cambridge
17 *September* 1934

CONTENTS

CHAPTER I

EXPERIMENTS IN DYNAMICS

CHAPTER II

THE STROBOSCOPE

CHAPTER III

EXPERIMENTS IN ELASTICITY

CHAPTER VII

EXPERIMENTS ON VISCOSITY

CHAPTER VIII

MATHEMATICAL DISCUSSIONS OF PROBLEMS IN CONDUCTION OF HEAT

CHAPTER IX

EXPERIMENTS IN HEAT

CHAPTER X

MATHEMATICAL DISCUSSIONS OF PROBLEMS IN SOUND

Spherical Waves and Radiation of Energy

Theory of Resonators

CHAPTER XI

EXPERIMENTS IN SOUND

CHAPTER I

EXPERIMENTS IN DYNAMICS

EXPERIMENT 1. **An example of conservation of angular momentum.***

1. Angular momentum of a particle about an axis. If at any instant a particle of mass m be moving with velocity v, it has momentum mv. The momentum is a vector lying along the tangent to the path of the particle at the point where the particle is at the instant considered.

Just as a force has a moment about any straight line or axis A unless its line of action either intersect A or be parallel to it, so momentum P along a given straight line has a moment about an axis A unless the straight line either intersect A or be parallel to it. To find the value of the moment of P about A we take a plane cutting A at right angles at O and project P upon this plane. If the perpendicular from O upon the projection of P be ON, and if the angle between P and the axis be θ, the moment of the projection about OA is $ON \,.\, P \sin\theta$. The momentum P has the component $P\cos\theta$ parallel to A, but this has no moment about A. Hence $ON \,.\, P \sin\theta$ is the moment of the momentum P about A.

The moment about A of the momentum mv of the particle m is often called the angular momentum of m about A.

2. Angular momentum of a system about an axis. For the purpose of the experiment we may confine ourselves to the simple case in which every particle of the system moves parallel to a fixed plane, which we take as the plane of Oxy (Fig. 1). The axis of z is perpendicular to Oxy and its positive direction is towards the reader. The three axes then form a right-handed system. In the experiment the plane Oxy is horizontal. Let P be the projection on Oxy of a particle of mass m. Let the coordinates of P relative to the fixed

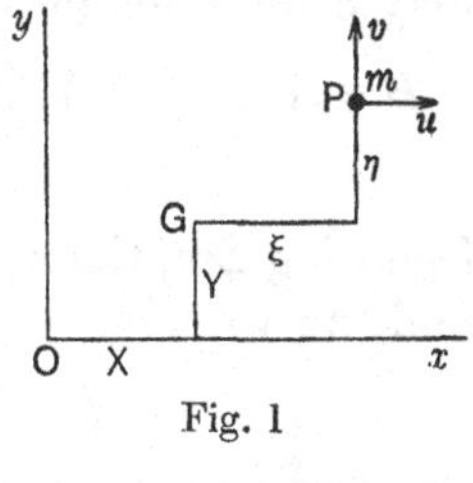

Fig. 1

* "An experiment illustrating the conservation of angular momentum," *Proc. Camb. Phil. Soc.* Vol. XXI, p. 75 (1922).

axes Ox, Oy be x, y, and let the components of the velocity of P be u, v. Then the components of the momentum of the particle are mu, mv. Hence, when counter-clockwise rotation about Oz is counted as positive, the sum of the moments about Oz of the momenta mu, mv, or the angular momentum of the particle about Oz, is $m(vx-uy)$. If H be the angular momentum about Oz of the whole system of particles,

$$H=\Sigma m(vx-uy). \quad \text{.....................(1)}$$

Let G be the projection on Oxy of the centre of gravity of the system. Let its coordinates relative to the fixed axes Ox, Oy be X, Y, and let the components of its velocity be U, V. Let $x=X+\xi$, $y=Y+\eta$, so that ξ, η are the coordinates of P relative to axes through G parallel to the fixed axes Ox, Oy. Then, since G is the centre of gravity,

$$\Sigma m\xi=0, \qquad \Sigma m\eta=0. \quad \text{..................(2)}$$

Let $u=U+\alpha$, $v=V+\beta$, so that α, β are the components of the velocity of P relative to G. Then, since $u=dx/dt$, $v=dy/dt$, $U=dX/dt$, $V=dY/dt$, we have $\alpha=d\xi/dt$, $\beta=d\eta/dt$. But, by (2),

$$\Sigma m\, d\xi/dt=0, \qquad \Sigma m\, d\eta/dt=0,$$

and hence $\qquad \Sigma m\alpha=0, \qquad \Sigma m\beta=0. \quad$ ....(3)

Now, by (1),

$$H=\Sigma m\{(V+\beta)(X+\xi)-(U+\alpha)(Y+\eta)\}$$
$$=\Sigma m\{VX-UY+V\xi-U\eta+\beta X-\alpha Y+\beta\xi-\alpha\eta\}.$$

Since U, V and X, Y do not change from particle to particle, we may bring them outside the sign of summation. Thus

$$H=(VX-UY)\Sigma m+V\Sigma m\xi-U\Sigma m\eta$$
$$+X\Sigma m\beta-Y\Sigma m\alpha+\Sigma m(\beta\xi-\alpha\eta).$$

Denoting Σm, the mass of the whole system, by M and using (2) and (3), we have

$$H=M(VX-UY)+\Sigma m(\beta\xi-\alpha\eta). \quad \text{...........(4)}$$

The first term in (4) is the angular momentum about Oz of a particle of mass M placed at G and moving with G. The second term is the angular momentum of the system, for the motion relative to G, about an axis through G parallel to Oz.

3. Method. A board D (Fig. 2) is suspended by a thread which we suppose to exert no torsional control. The thread is attached to a torsion head, which is carried by a fixed support. The plane of the board is horizontal and the axis of suspension cuts the board in O. The vertical through O is the axis Oz of § 2. The inertia bar AB turns about a vertical shaft fixed to the board, the axis of the shaft passing through G, the centre of gravity of AB. A second bar C, suitably fixed to the board, acts as a counterpoise to AB. By adjusting C, the plane of D is made horizontal. By a light spring E attached to the board and operating by a string wound round a drum carried by AB, this bar can be set in motion about G relative to the board, when a thread attached to the board and holding AB in its initial position is burned. Before the thread is burned, the system is at rest. At any later time let the axis OG of the board make an angle θ with OF, its initial direction, and let the axis GA of the bar make an angle $\phi+\epsilon$ with GO, where ϵ is the angle between GA and GO before the thread is burned. Let ϕ be measured in the opposite direction to θ.

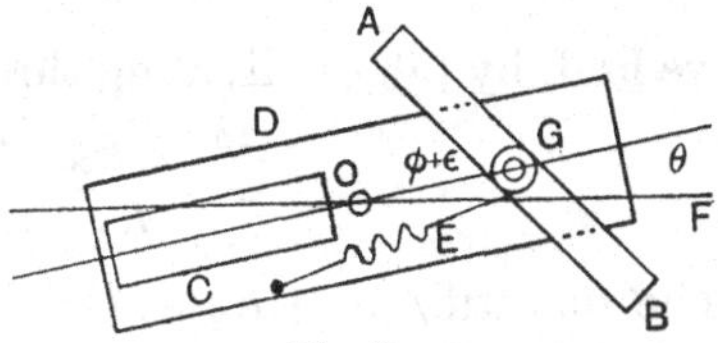

Fig. 2

With the exception of gravity and the tension of the suspension, no external forces act on the system at any time. Before the thread is burned, the system has no angular momentum about Oz. The burning of the thread does not call any external force into action and hence the angular momentum of the system about the vertical axis Oz remains zero. If the moment of inertia about Oz of the board, the counterpoise and all the other fittings *except* the bar AB be K_1, the angular momentum about Oz of this part of the system is $K_1 d\theta/dt$.

Let M be the mass of AB and let $OG=a$. Since the linear velocity of G is $a d\theta/dt$, the momentum of M at G is $Ma d\theta/dt$, and the moment of this momentum about Oz is $Ma^2 d\theta/dt$. This angular momentum is in the same direction as that of the board.

The angular velocity of AB, in the same direction as that of the board, is $d\theta/dt - d\phi/dt$, and hence, if the moment of inertia of AB about a vertical axis through G be K_2, its angular momentum about that axis is $K_2(d\theta/dt - d\phi/dt)$. Hence, by (4), the angular momentum of AB about Oz is

$$Ma^2 d\theta/dt + K_2(d\theta/dt - d\phi/dt).$$

The angular momentum of the whole system is zero, and hence

$$K_1 d\theta/dt + Ma^2 d\theta/dt + K_2(d\theta/dt - d\phi/dt) = 0. \quad \text{......(5)}$$

Since
$$\frac{d\theta}{dt}\bigg/\frac{d\phi}{dt} = \frac{d\theta}{d\phi},$$

we find, by (5), on dividing throughout by $d\phi/dt$,

$$\frac{d\theta}{d\phi} = \frac{K_2}{K_1 + K_2 + Ma^2} = \frac{K_2}{K_3}. \quad \text{..............(6)}$$

The quantity $K_1 + K_2 + Ma^2$, or K_3, is the moment of inertia about Oz of the whole system, when the bar AB is fixed relative to the board.

Since (6) holds at every instant, we have

$$\theta/\phi = K_2/K_3, \quad \text{.......................(7)}$$

where θ and ϕ are, respectively, the angles turned through by the board relative to the initial line OF and by the bar relative to the board in any time, each measured from the corresponding zero.

By (5), $d\theta/dt = 0$ when $d\phi/dt = 0$. Hence if, at any instant, the motion of the bar relative to the board be arrested by interaction between them, the motion of the board will cease at the same instant.

Let the bar AB be removed from the board and be attached to a vertical torsion wire and be made to execute torsional vibrations, the axis of suspension agreeing with that about which AB turns when on the board. Let T_2 be the periodic time. Let T_3 be the periodic time when the *complete* system is suspended from the same wire; in this measurement AB is fixed relative to the board. Then $K_2/K_3 = T_2^2/T_3^2$ and thus, by (7),

$$\theta/\phi = T_2^2/T_3^2. \quad \text{........................(8)}$$

In the experiment θ/ϕ is compared with T_2^2/T_3^2.

4. Experimental details. Fig. 3 shows some details of the apparatus. The counterpoise does not appear, as it is fixed to the part of the board which is omitted. The counterpoise and other fittings should be designed so that the axis of suspension is as nearly as possible a "principal axis." If this condition be not secured, the motion of the board will be unsteady. The ends of the torsion wire W, used in comparing K_2 with K_3, are soldered into two short cylinders 0·5 cm. in diameter. At the centre of the board is a socket H, provided with a set-screw. Into this socket fits either the rod N, by which the system is attached to the thread, or the cylinder at the end of the torsion wire. The hole in the bar AB at G is of the same diameter as that in H. By a set-screw, the bar can be secured to the torsion wire.

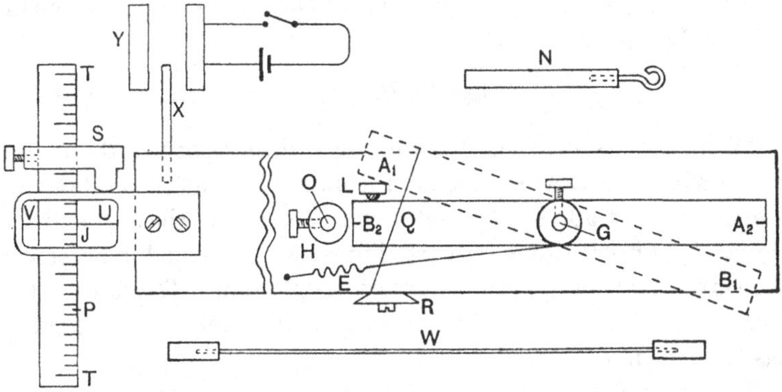

Fig. 3

The bar is held in its initial position $A_1 B_1$ against the pull of the spring E* by a thread Q secured under the button R; this position is defined by the stop L. When the thread is burned, the bar turns round until it hits the other side of L and is arrested in the position $A_2 B_2$. To prevent rebound, a small pellet of plasticene is placed on L as shown. Unless the plasticene be re-moulded into pellet form after each impact, the bar will rebound. Though the rebound does not vitiate the experiment, it makes the observations more difficult. The angle ϕ turned through by AB relative to the board is $\pi - A_1 G B_2$. If $A_1 G = r$, we have $\sin \frac{1}{2} A_1 G B_2 = \frac{1}{2} A_1 B_2 / r$. Then

$$\phi = \pi - A_1 G B_2 = \pi - 2 \sin^{-1} (\tfrac{1}{2} A_1 B_2 / r) \text{ radians.} \quad \ldots(9)$$

* In order that the spring may have the length needed to allow the extension required, the anchorage of E is much farther from G than is suggested in Fig. 3. The thread Q may pass through an eye in the end of a short horizontal arm so fixed to the board that the part of Q between the eye and the button R is held away from the side of the board. The flame is then easily applied to the thread.

For the determination of $A_1 B_2$, the board is placed on the table and a vertical pin is held in a clip so that its tip is just above the point A_1 on the end of the bar. The bar is then turned round into the second position shown in Fig. 3 and the distance between the tip of the pin and the point B_2 is measured.

The angle θ turned through by the board is measured by the fine wire UV, which is stretched in a metal frame carried by the board and intersects the axis Oz at right angles; it is convenient if the edges of the board be parallel to OUV. The wire moves over a horizontal scale whose edge is TT. A zero J on TT is chosen, and TT is set perpendicular to OJ. If UV cut the edge in J initially and in P finally,

$$\tan\theta = PJ/JO. \quad \text{.....................(10)}$$

After the observations for the angle θ have been taken, the board with its rod N is unhooked from the loop at the end of the suspending thread. A plummet is then hung from the suspending thread by a hook and is immersed in water contained in a vessel; the water damps any oscillations of the plummet. The horizontal distance between J and the axis of the plummet's thread is now easily measured.

For success, the system must be *at rest* when the thread Q is burned. The silk thread supporting the system is attached to a torsion head, which is adjusted so that UV cuts TT in some point very near J when the system is at rest. A stop S is fixed to the scale so that, when the frame touches S, UV cuts TT in J. A current sent through the coil Y attracts the small magnet X attached to the board with a force which should be only just great enough to keep the frame against the stop. The centre O of the board is then reduced to rest. A flame or a small gas jet is prepared and, when the system is at rest, the thread Q is burned. The current is stopped just before the flame is applied. The board moves round, and the reading of UV in its new position of rest at P is taken. The wire may subsequently drift very slowly from P on account of slight torsion of the silk thread or on account of draughts, and thus the reading should be taken without delay.

The supporting thread should be of *plaited* silk; *fine* fishing line may be used. If a *twisted* thread be used, it will be difficult to obtain anything like a steady zero position, since a twisted thread exerts a couple approximately proportional to the load, unless the thread be practically "unwound."

5. Practical example.

In the present (1934) apparatus the board is 56 cm. long, 8 cm. wide and 1·5 cm. thick. The inertia bar is 26 cm. long, 1·6 cm. wide and 1·6 cm. thick. The mass of the whole system is 1770 grm.

Mr J. A. Pattern, using an earlier form of apparatus, obtained the following results. Values found for JP were 7·50, 7·80, 7·50, 7·70. Mean 7·625 cm. Distance JO was 38·5 cm. Hence $\tan\theta = 7{\cdot}625/38{\cdot}5 = 0{\cdot}19805$, and

$$\theta = 11^\circ\, 12'\, 9'' = 0{\cdot}19552 \text{ radian}.$$

Distances were, $A_1G = 15{\cdot}0$ cm., $A_1B_2 = 2{\cdot}42$ cm. Hence

$$\phi = \pi - 2\sin^{-1}(1{\cdot}21/15) = 180^\circ - 9^\circ\, 15'\, 14'' = \pi - 0{\cdot}1615 \text{ radian}.$$

Thus $$\phi = 2{\cdot}9801 \text{ radians}.$$

Inertia bar was suspended by torsion wire and transits were observed.

Transit	Time		Transit	Time		$50T_2$	
	mins.	secs.		mins.	secs.	mins.	secs.
0		59	50	3	43	2	44
10	1	32	60	4	16	2	44
20	2	5	70	4	48	2	43
30	2	37	80	5	21	2	44
40	3	10	90	5	53	2	43

Mean value of $50T_2$ is 163·6 secs. Hence $T_2 = 3{\cdot}272$ secs.

When whole system, including inertia bar, was suspended by the same wire, similar observations gave $50T_3 = 638{\cdot}9$ secs. Hence $T_3 = 12{\cdot}778$ secs.

For the times, we have $T_2^2/T_3^2 = 3{\cdot}272^2/12{\cdot}778^2 = 0{\cdot}06557.$

For the angles, we have $\theta/\phi = 0{\cdot}19552/2{\cdot}9801 = 0{\cdot}06561.$

Hence equation (8) is closely verified.

Experiment 2. Kater's pendulum.

6. Introduction. Since "g" is an acceleration, its determination involves the measurement of a time and a length. The "simple" pendulum of theory, with its formula $t = 2\pi\sqrt{(l/g)}$, involving a single time and a single length, would be an ideal means of finding "g." But such a pendulum cannot be made. If we use a sphere of finite size as bob, we must use a wire of sufficient strength to carry the bob and must fix some sort of knife edge to the wire. A little consideration will show that the method is beset with such difficulties that it cannot be used in an accurate determination of "g."

We therefore abandon the "simple" pendulum and use a rigid pendulum. It is impossible to determine the moment of inertia of the pendulum about any given axis by a calculation depending upon a knowledge of the geometrical form of the pendulum, for we cannot tell how the density of the material varies from point to point. We can, however, use the theorem of parallel axes* in such a way as to eliminate any uncertainty.

If we use *two* parallel knife edges *fixed* to the pendulum in such positions that the centre of gravity, G, lies *between* the knife edges in the plane containing them, we can, in theory, arrange matters so that (1) the periodic time about either knife edge is the same and (2) the distances of G from the two knife edges are different. In each case, of course, the knife edge rests on a horizontal plane. The length of the corresponding "simple equivalent pendulum" is then equal to the shortest distance between the knife edges. We cannot attain this ideal, but, if the times be very nearly equal, the necessary correction is found with ample accuracy.

7. Periodic time of a rigid pendulum. Let the mass of the pendulum be M and let G be its centre of gravity. Through G we take an axis in a direction fixed relative to the pendulum; this we call the G-axis. Let the moment of inertia of the pendulum about the G-axis be Mk^2. Let the pendulum have a knife edge parallel to the G-axis and at distance h from it, and let the pendulum oscillate through a small arc while the knife edge rests upon a horizontal plane. By the theorem of parallel axes, the moment of inertia about the knife edge is $M(k^2+h^2)$. If t be the periodic time,

$$t^2 = 4\pi^2(k^2+h^2)/gh, \quad \ldots\ldots\ldots\ldots(1)$$

or

$$h^2 - ght^2/4\pi^2 + k^2 = 0. \quad \ldots\ldots\ldots\ldots(2)$$

If l be the length of the simple equivalent pendulum, $l = gt^2/4\pi^2$, and thus

$$h^2 - lh + k^2 = 0. \quad \ldots\ldots\ldots\ldots(3)$$

We are not restricted to a single position for the knife edge, for if the edge lie along *any* generating line of a cylinder of radius h described about the G-axis, t^2 will have the value given by (1).

* *Experimental Elasticity*, Note IV.

Solving (2) or (3) for h, we have

$$h = \frac{gt^2}{8\pi^2} \pm \sqrt{\left\{\left(\frac{gt^2}{8\pi^2}\right)^2 - k^2\right\}} = \frac{l}{2} \pm \sqrt{\left(\frac{l^2}{4} - k^2\right)}. \quad \text{......(4)}$$

For any value of t greater than $\sqrt{(8\pi^2 k/g)}$, or for any value of l greater than $2k$, there are two values of h. Since h is real, the least value of l is $2k$, and, when l has this minimum value, the distance of the knife edge from G is k, the radius of gyration of the pendulum about the G-axis.

8. Pendulum with two parallel knife edges. Let the knife edges be E, F; each is parallel to the G-axis. Let the distances of E, F from G be h_1, h_2, let t_1, t_2 be the periodic times for those knife edges and l_1, l_2 the lengths of the simple equivalent pendulums. Then

$$h_1 t_1^2 = 4\pi^2 (k^2 + h_1^2)/g, \qquad h_2 t_2^2 = 4\pi^2 (k^2 + h_2^2)/g.$$

By subtraction we eliminate k and obtain

$$h_1 t_1^2 - h_2 t_2^2 = 4\pi^2 (h_1^2 - h_2^2)/g.$$

Hence

$$\frac{4\pi^2}{g} = \frac{h_1 t_1^2 - h_2 t_2^2}{h_1^2 - h_2^2} = A + B, \quad \text{...............(5)}$$

where

$$A = \frac{t_1^2 + t_2^2}{2(h_1 + h_2)}, \qquad B = \frac{t_1^2 - t_2^2}{2(h_1 - h_2)}.$$

By the use of *two* knife edges instead of one we gain the great advantage that k, whose value we cannot calculate, does not appear in (5). To find g we now have to measure the two times t_1, t_2 and the two distances h_1, h_2 of E, F from G. The times present no difficulty, but we cannot identify the position of G with any accuracy. If, however, E, F be so placed that (1) the plane through them passes also through G and (2) G lies between them, $h_1 + h_2$ is simply the distance between the knife edges, and this can be measured accurately.

If we can make $t_2 = t_1$ *without making* $h_2 = h_1$, the term B will vanish and then we have

$$4\pi^2/g = t^2/(h_1 + h_2),$$

where t is the common value of t_1 and t_2. We have thus recovered the simplicity of the theoretical "simple" pendulum with a single time and a single distance. In practice we cannot quite attain this ideal.

To determine the difference $h_1 - h_2$ which occurs in B we must measure h_1 and h_2 separately, since E and F are on opposite sides of G, and these measurements cannot be made with accuracy. The numerator in B is small if t_2 nearly equals t_1. If h_1 and h_2 differ widely, B will be so small compared with A, that the error due to the uncertainty in $h_1 - h_2$ will be negligible. We have therefore to secure that t_1 and t_2 are nearly equal while h_1 and h_2 are widely different. How to do this we can discover by aid of equation (4).

For a given value of l_1 we have a choice of two values of h_1 and similarly for l_2 and h_2. For h_1 we take the positive sign before the square root and for h_2 the negative sign. Then

$$h_1 - h_2 = \tfrac{1}{2}(l_1 - l_2) + \surd(\tfrac{1}{4}l_1^2 - k^2) + \surd(\tfrac{1}{4}l_2^2 - k^2),$$

and so $h_1 - h_2$ will be large even though l_1 and l_2 be nearly equal provided that both l_1 and l_2 be large compared with the minimum $2k$ or that the periodic times be much greater than the minimum $2\pi\surd(2k/g)$.

We shall fail in our purpose if, in (4), we take the same sign before the square root for both h_1 and h_2. For then we should have

$$h_1 - h_2 = \tfrac{1}{2}(l_1 - l_2) \pm \{\surd(\tfrac{1}{4}l_1^2 - k^2) - \surd(\tfrac{1}{4}l_2^2 - k^2)\},$$

and $|h_1 - h_2|$ will be small when l_1 and l_2 are nearly equal.

We can, of course, make $t_1 = t_2$ by making $h_1 = h_2 = h$, where h has any arbitrary value, but we must avoid the error of supposing that this will necessarily make $l = 2h$. We shall not have $l = 2h$ unless h_1 and h_2 be not only roots of (3) but also be equal. In this case, as shown in § 7, $l = 2k$ and $h_1 = h_2 = k$. These conditions could be secured by adjusting two massless knife edges until the periodic time about each reached a minimum value, which would be the same for each, but this process would be difficult and without any advantage.

An approximate knowledge of masses and dimensions will enable us to design a pendulum with two knife edges in such positions that h_1, h_2 are widely different although the periodic times are approximately equal. The knife edges being fixed,

h_1+h_2 is determinate. To allow an easy comparison with a clock, h_1+h_2 would be made nearly equal to the length of a simple pendulum making a complete vibration in two seconds. By small masses sliding on the rod, we can change not only k^2 but also the position of the centre of gravity and thus, without changing h_1+h_2, we can vary k^2 and the product $h_1 h_2$ until the two periodic times be very nearly equal (see § 13). The value of g is then deduced from the observations by aid of (5).

It is only when G lies on the straight line EF that h_1+h_2 equals EF; here E, F denote the points in which the knife edges are cut by a plane through G at right angles to them. Since, for given values of h_1 and h_2, EF is a maximum when G lies on EF, any first order perpendicular distance x of G from EF will cause only a second order difference between EF and h_1+h_2. Thus, as far as x^2,

$$EF = \sqrt{(h_1{}^2 - x^2)} + \sqrt{(h_2{}^2 - x^2)} = h_1 + h_2 - \tfrac{1}{2}x^2(1/h_1 + 1/h_2)$$
$$= (h_1+h_2)(1 - \tfrac{1}{2}x^2/h_1 h_2),$$

or
$$h_1 + h_2 = EF(1 + \tfrac{1}{2}x^2/h_1 h_2). \quad \text{..............(6)}$$

In the example of § 12, $h_1 = 70$, $h_2 = 30$ cm. approximately, and then the correcting factor is $1 - x^2/4200$. The correction is less than one in 10^4 when $x < 0{\cdot}648$ cm., and is less than one in 10^5 when $x < 0{\cdot}205$ cm.

9. Buoyancy and inertia of air. The resultant force exerted by the air on the pendulum when at rest is $V\rho g$ upwards, where V is the volume of the pendulum and ρ the density of the air. The pendulum is symmetrical in form, though not in mass, about its mid-point O, and hence $V\rho g$ acts upwards through O. The buoyancy effect is little changed by the slow motion of the pendulum. As the pendulum swings, the air near it is set into motion. On account of the symmetry of form, the increase in the moment of inertia of the system about each of the knife edges E, F will be the same and may be allowed for by writing k'^2 for k^2, where k' is slightly greater than k. The slight damping will not appreciably affect the periodic time.

Neither the force $V\rho g$ nor the increase of moment of inertia has any effect if the pendulum be so adjusted that $t_1 = t_2 = t$. For then since $OE = OF$,

$$gt^2 (Mh_1 - \tfrac{1}{2}V\rho \, . \, EF) = 4\pi^2 M (k'^2 + h_1{}^2),$$
$$gt^2 (Mh_2 - \tfrac{1}{2}V\rho \, . \, EF) = 4\pi^2 M (k'^2 + h_2{}^2).$$

By subtraction, $gt^2 = 4\pi^2 (h_1 + h_2)$, just as when there is no air.

10. The pendulum. Two knife edges E, F (Fig. 4) are fixed to a straight rod which carries a large brass bob P at one end and a wooden bob Q at the other. These bobs are of equal dimensions and are similarly placed with respect to E and F. The mass of P is so much greater than that of Q that G, the centre of gravity of the pendulum, is much farther from E than from F. Thus $GE > GF$, or $h_1 > h_2$. A small sliding bob R of brass and a correponding slider S of wood are so placed between E and F that, as nearly as may be, $ER = FS$, and R and S are adjusted so that the periodic times t_1 and t_2 about E and F respectively are nearly equal. The brass slider R is much nearer to E than to F. The adjustment of R and S necessarily occupies much time and hence they should not be disturbed without good reason.

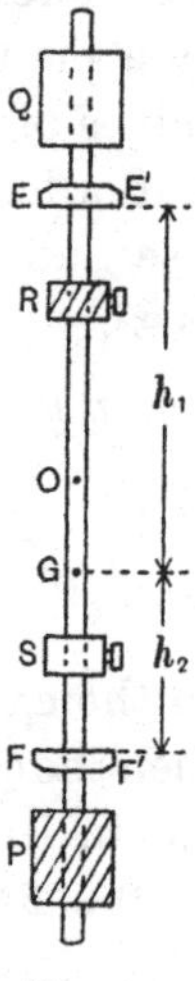

Fig. 4

The distances $GE = h_1$ and $GF = h_2$ are found approximately by balancing the pendulum on a knife-edge fulcrum at G and then measuring GE or GF.

The distance between the knife edges EE' and FF' (Fig. 4) may be measured by aid of an optical bench. A carriage bearing a stiff vertical rod can slide along the bench. The pendulum rests on two supports, with its length parallel to the bench, in such a position that the vertical rod can be brought into contact first with E and then with F. The pendulum is so secured as to prevent any motion relative to the bench during the measurements. If the bench slides be *well* oiled so that the carriage moves freely, the resistance due to contact between rod and knife edge will be perceived instantly. If the diameter of the rod be d and if the carriage move through distance c, then $EF = c + d$.

The pendulum is then turned on its axis and the distance $E'F'$

between the other ends of the knife edges is measured. The mean value is used in the calculations.

11. Observation of periodic time. The determination of the periods of the vibrations about the two knife edges is carried out by the method given in § 27 of *Experimental Harmonic Motion*. The times at which the transits 0, 10, 20, ... 190, 200, 210, ... 390 occur are observed. In this way, 20 independent observations of the time occupied by 200 vibrations are obtained. Unless the timepiece employed be known to be keeping correct time, or its rate be known, it is compared with the laboratory clock at the beginning of the experiment and again after the timings have been made; the interval should be as long as is available. The timepiece should be allowed to run continuously and should not be stopped.

Care must be taken that the periodic time denoted by t_1 really corresponds to h_1 and not to h_2. In the notation adopted, t_1 is the periodic time when the large heavy bob hangs below the knife edge EE', and h_1 is the distance EG from the centre of gravity to this knife edge. Of necessity $h_1 > h_2$, but t_1 may be larger or smaller than t_2 according to the adjustment of the sliders on the pendulum.

By *Experimental Harmonic Motion*, Note II, if t_0 be the periodic time for infinitely small arcs,

$$t_0 = t(1 + \alpha^2/16)^{-1} = t(1 - \alpha^2/16),$$

where t is the periodic time when the pendulum swings through a small angle of α radians on either side of its position of rest.

The comparison of the pendulum with the laboratory clock may also be made by the method of coincidences. The simplest method is to hang the Kater's pendulum in front of the clock and to observe the coincidences between the two pendulums. If, however, a drop of mercury be so placed that a wire forming the tip of the clock pendulum makes contact with the mercury when that pendulum is nearly vertical, a neon lamp can be caused to flash at each contact, and, of course, the lamp can be placed in any part of the laboratory.

12. Practical example.

In an experiment by Mr W. E. Miller, three measurements of movement of carriage on bench gave 98·60, 98·57, 98·60 cm. Mean 98·59 cm. Diameter of rod on carriage was 0·32 cm. Hence

$$h_1 + h_2 = 98{\cdot}59 + 0{\cdot}32 = 98{\cdot}91 \text{ cm.}$$

For distance GF from C.G. (point of balance) to nearer knife edge

$$h_2 = 26{\cdot}90 \text{ cm.}$$

Hence $$h_1 = 98{\cdot}91 - 26{\cdot}90 = 72{\cdot}01 \text{ cm.}$$

Thus $$h_1 - h_2 = 45{\cdot}11 \text{ cm.}$$

Mr Miller observed 390 vibrations about each knife edge; to save space only 190 vibrations are recorded here. The times are those read on the timepiece.

Observations for t_1. Large heavy bob P (Fig. 4) below knife edge EE' employed.

Transit	Time		Transit	Time		$100t_1$	
	mins.	secs.		mins.	secs.	mins.	secs.
0	5	11·0	100	8	29·0	3	18·0
10	5	31·0	110	8	49·0	3	18·0
20	5	51·0	120	9	9·0	3	18·0
30	6	10·5	130	9	28·5	3	18·0
40	6	30·5	140	9	48·5	3	18·0
50	6	50·5	150	10	8·0	3	17·5
60	7	10·0	160	10	28·0	3	18·0
70	7	29·5	170	10	48·0	3	18·5
80	7	49·5	180	11	7·5	3	18·0
90	8	9·5	190	11	27·0	3	17·5

Mean $t_1 = \frac{1}{100}$ (3 mins. 17·95 secs.) = 1·9795 timepiece secs.

Observations for t_2, with large heavy bob P above knife edge FF' employed, gave $100t_2$ = 3 mins. 16·65 secs. and t_2 = 1·9665 timepiece secs.

In 46 mins. timepiece lost 10·5 secs. Hence 46 × 60 or 2760 true secs. are indicated as 2760 − 10·5 or 2749·5 secs. on timepiece. Thus

1 sec. on timepiece = 2760·0/2749·5 = 1·00382 true secs.

Hence $t_1 = 1{\cdot}9795 \times 1{\cdot}00382 = 1{\cdot}98706$ secs., $t_1^2 = 3{\cdot}948407$ sec.^{2},

$t_2 = 1{\cdot}9665 \times 1{\cdot}00382 = 1{\cdot}97401$ secs., $t_2^2 = 3{\cdot}896715$ sec.^{2},

and $\frac{1}{2}(t_1^2 + t_2^2) = 3{\cdot}922561$, $\frac{1}{2}(t_1^2 - t_2^2) = 0{\cdot}025846$ sec.^{2}.

Then, by (5), § 8,

$$\frac{4\pi^2}{g} = \frac{t_1^2 + t_2^2}{2(h_1 + h_2)} + \frac{t_1^2 - t_2^2}{2(h_1 - h_2)} = \frac{3{\cdot}922561}{98{\cdot}91} + \frac{0{\cdot}025846}{45{\cdot}11}$$

$$= 0{\cdot}0396579 + 0{\cdot}0005730 = 0{\cdot}0402309.$$

Hence*

$$g = 4\pi^2/0{\cdot}0402309 = 981{\cdot}30 \text{ cm. sec.}^{-2} \qquad [4\pi^2 = 39{\cdot}4784\ldots].$$

* If, through confusion, t_1 and t_2 had been interchanged so that t_1 was made to correspond to h_2 and t_2 to h_1, we should have had

$$4\pi^2/g = 0{\cdot}0396579 - 0{\cdot}0005730 = 0{\cdot}0390849,$$

and $$g = 1010{\cdot}07 \text{ cm. sec.}^{-2}$$

13. Theory of adjustment by sliding masses.*

The pendulum carries two sliders R, S (Fig. 4). The remaining part of the system we shall call the "pendle" to distinguish it from the complete pendulum. Let M_0 be the mass, and $M_0 k_0^2$ the moment of inertia, of the pendle about an axis parallel to the knife edges through G_0, its centre of gravity. Let the distances of G_0 from the knife edges E, F be a, b. The masses of R, S are m, n and their moments of inertia about axes through their centres of gravity parallel to the knife edges are mr^2, ns^2. Let the distances of the centre of gravity of R from E, F be x, y. Then, because R, S are symmetrically arranged, the distances of the centre of gravity of S from E, F are y, x. Since the mass of the whole pendulum is M and the distances of its centre of gravity G from E, F are h_1, h_2, we have

$$Mh_1 = M_0 a + mx + ny, \qquad Mh_2 = M_0 b + my + nx.$$

If the lengths of the simple equivalent pendulum for oscillations about E and F be l_1 and l_2, we have

$$Mh_1 l_1 = M_0 (a^2 + k_0^2) + m(x^2 + r^2) + n(y^2 + s^2),$$
$$Mh_2 l_2 = M_0 (b^2 + k_0^2) + m(y^2 + r^2) + n(x^2 + s^2).$$

Since $x + y = EF$, we have

$$dy/dx = -1, \qquad M\,dh_1/dx = m - n, \qquad M\,dh_2/dx = n - m,$$

and thus $$Mh_1\,dl_1/dx = 2mx - 2ny - l_1(m-n)$$

and $$Mh_2\,dl_2/dx = 2nx - 2my - l_2(n-m).$$

Hence $$\frac{d}{dx}(l_1 - l_2) = \frac{m}{M}\left\{\frac{2x - l_1}{h_1} + \frac{2y - l_2}{h_2}\right\} - \frac{n}{M}\left\{\frac{2x - l_2}{h_2} + \frac{2y - l_1}{h_1}\right\}. \quad \text{...(7)}$$

All the quantities on the right side can be found. If $d(l_1 - l_2)/dx$ be positive, a small increase in x will increase $l_1 - l_2$. If $l_1 > l_2$, the increase in x will increase the difference between the lengths of the simple equivalent pendulums and increase the difference between the periods. In this case x must be diminished to bring the periods into closer agreement. It is necessary to consider a small change in x, for a large change might alter the right side from being a positive to being a negative quantity.

When l_1 and l_2 are nearly equal, we may, as a first approximation, write $l_1 = l_2 = l$ on the right side of (7), where l is the actual distance between the knife edges. Then $y = l - x$. We thus obtain

$$\frac{d}{dx}(l_1 - l_2) = \frac{m+n}{M}(l - 2x)\left(\frac{1}{h_2} - \frac{1}{h_1}\right).$$

In the pendulum, as described in § 10, h_1 is much larger than h_2 and thus $1/h_2 - 1/h_1$ is positive. Further, the distance x of the slider R from E is much less than $\tfrac{1}{2}l$ and hence $l - 2x$ is positive. Thus, when the sliders R, S are nearly in correct adjustment, $d(l_1 - l_2)/dx$ is positive. If $l_1 > l_2$, we must move R and S so as to diminish $l_1 - l_2$, i.e. we must diminish x by moving R towards the knife edge E through a small distance. Since S is symmetrical with R, we must at the same time move S towards F through an

* I owe the method of this section to Dr G. T. Bennett, F.R.S.

equal distance. A new comparison of l_1 and l_2 is now made by timing and, if necessary, a second adjustment of R and S is made.

EXPERIMENT 3. Moment of inertia of a body about a non-principal axis.

14. Introduction. Let O be a point fixed in a rigid body and OR an axis through O. Let M be the mass of the body and Mk^2 its moment of inertia about OR. Then Mk^2 depends, in general, upon the direction of OR. For instance, if O be the centre of a thin uniform disk of radius a, the moments of inertia about axes through O in the plane of the disk and perpendicular to that plane are $\frac{1}{4}Ma^2$ and $\frac{1}{2}Ma^2$ respectively. In some special cases Mk^2 is independent of the direction of OR. Thus Mk^2 has the same value for all axes passing through the centre of a cube of uniform material.

We shall show that, corresponding to any given length λ, there is a geometrical ellipsoid which has its centre at O, is fixed relative to the body and is such that, if OR cut the ellipsoid in R, Mk^2 for OR equals $M\lambda^4/\rho^2$, where $\rho = OR$. Thus Mk^2 is proportional to OR^{-2}, the reciprocal of the square of the radius. The *form* of the ellipsoid depends upon the position of O and upon the distribution of matter in the body; the *magnitude* depends upon λ.

15. The momental ellipsoid. Let O be the point for which the ellipsoid is to be found. Through O we take rectangular axes $Oxyz$ fixed relative to the body. The mass of any particle of the body is m and its coordinates are x, y, z. Then

$$A = \Sigma m(y^2 + z^2), \quad B = \Sigma m(z^2 + x^2), \quad C = \Sigma m(x^2 + y^2) \quad \text{...(1)}$$

are the "moments of inertia" of the body about the axes Ox, Oy, Oz respectively. Further,

$$D = \Sigma myz, \quad E = \Sigma mzx, \quad F = \Sigma mxy \quad \text{.........(2)}$$

are the "products of inertia" of the body with respect to the three pairs of axes Oyz, Ozx and Oxy respectively. The moments A, B, C are necessarily positive, since mass is positive, but the products D, E, F may be positive or negative. Since $x^2 + y^2 \geqslant |2xy|$, it follows that $|D| \leqslant \frac{1}{2}A$. Similarly $|E| \leqslant \frac{1}{2}B$, $|F| \leqslant \frac{1}{2}C$.

We will now find, in terms of A, B, C, D, E, F, the moment of inertia about an axis OR (Fig. 5) whose forward direction makes with the forward directions of Ox, Oy, Oz angles whose cosines are α, β, γ. These cosines are called the "direction cosines" of OR. Let there be a particle of mass m at the point P whose coordinates are x, y, z. Let M be the mass of the body and let Mk^2 be its moment of inertia about OR. Then

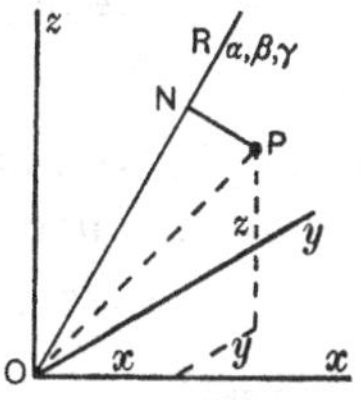

Fig. 5

$$Mk^2 = \Sigma m \,.\, PN^2, \quad \text{.....................(3)}$$

where PN is the perpendicular distance of P from OR. Since PNO is a right angle, $PN^2 = OP^2 - ON^2$. But ON is the projection of OP on OR and is therefore the sum of the projections on OR of x, y and z. The distances x, y, z make with OR angles whose cosines are α, β, γ and hence, taking the sum of their projections on OR, we have

$$ON = x\alpha + y\beta + z\gamma.$$

Since $OP^2 = x^2 + y^2 + z^2$,

$$PN^2 = OP^2 - ON^2 = x^2 + y^2 + z^2 - (x\alpha + y\beta + z\gamma)^2$$
$$= x^2(1-\alpha^2) + y^2(1-\beta^2) + z^2(1-\gamma^2) - 2yz\beta\gamma - 2zx\gamma\alpha - 2xy\alpha\beta.$$

A simple relation exists between the direction cosines. Let Q be a point on OR, let $OQ = q$ and let the coordinates of Q be f, g, h. Then $f^2 + g^2 + h^2 = q^2$. Since $\alpha = f/q$, $\beta = g/q$, $\gamma = h/q$,

$$\alpha^2 + \beta^2 + \gamma^2 = (f^2 + g^2 + h^2)/q^2 = 1. \quad \text{............(4)}$$

Using this result, we obtain

$$PN^2$$
$$= x^2(\beta^2 + \gamma^2) + y^2(\gamma^2 + \alpha^2) + z^2(\alpha^2 + \beta^2) - 2yz\beta\gamma - 2zx\gamma\alpha - 2xy\alpha\beta$$
$$= \alpha^2(y^2 + z^2) + \beta^2(z^2 + x^2) + \gamma^2(x^2 + y^2) - 2\beta\gamma yz - 2\gamma\alpha zx - 2\alpha\beta xy.$$
$$\text{......(5)}$$

Hence, on summation,

$$Mk^2 = \Sigma m \,.\, PN^2 = \alpha^2 \Sigma m(y^2 + z^2) + \ldots - 2\beta\gamma \Sigma myz - \ldots$$
$$= A\alpha^2 + B\beta^2 + C\gamma^2 - 2D\beta\gamma - 2E\gamma\alpha - 2F\alpha\beta, \quad \text{.........(6)}$$

where A, B, C and D, E, F are the moments and products of inertia of the body with respect to the axes of coordinates through O, as defined by (1) and (2).

If $OR=\rho$, and if the coordinates of R be ξ, η, ζ, we have $\alpha=\xi/\rho$, $\beta=\eta/\rho$, $\gamma=\zeta/\rho$. Then, by (6),

$$\rho^2Mk^2=A\xi^2+B\eta^2+C\zeta^2-2D\eta\zeta-2E\zeta\xi-2F\xi\eta.$$

Hence, if we take OR or ρ so that $\rho^2Mk^2=M\lambda^4$, the point R will lie on the quadric surface

$$A\xi^2+B\eta^2+C\zeta^2-2D\eta\zeta-2E\zeta\xi-2F\xi\eta=M\lambda^4 \text{.......(7)}$$

This surface has its centre at O. Since Mk^2 is positive and neither zero† nor infinite for any direction of R, ρ^2 is positive and ρ is real and never infinite or zero. Hence the surface is an ellipsoid. It is called the "momental ellipsoid" for the point O and has the property that the moment of inertia about any radius vector from the centre to the surface is proportional to the inverse square OR^{-2} of that radius. Thus the form and size of the momental ellipsoid depend upon the distribution of mass in the body and upon λ, but not upon the directions of the coordinate axes.

If we take through O a new set of rectangular axes which coincide with the axes of the momental ellipsoid, the equation to the ellipsoid referred to the new axes will be

$$A_0\xi^2+B_0\eta^2+C_0\zeta^2=M\lambda^4. \text{..................(8)}$$

Since there are no terms in (8) which involve the products $\eta\zeta$, $\zeta\xi$, $\xi\eta$, it follows, by comparison with (7), that, if D_0, E_0, F_0 be the products of inertia with respect to the new axes,

$$D_0=0, \qquad E_0=0, \qquad F_0=0. \text{ ....(8a)}$$

The three mutually perpendicular directions of the axes of the momental ellipsoid are called the "principal axes" of the body at the point O. If the momental ellipsoid for O have an axis of revolution, *any* axis through O perpendicular to that axis is a principal axis. If $A_0=B_0=C_0$, the momental ellipsoid for the point O is a sphere, and then every axis through O is a principal axis.

When the axes of coordinates are the principal axes at O, the moment about an axis OR, whose direction cosines are α, β, γ, is Mk^2, where, by (6) and (8a),

$$Mk^2=A_0\alpha^2+B_0\beta^2+C_0\gamma^2. \text{ ....(9)}$$

† A material rod cannot be of *zero* thickness.

The moments about the new axes of x, y, z are A_0, B_0, C_0 respectively.

Let the principal axes through G, the centre of gravity, be the axes of coordinates and let A_0, B_0, C_0 be the moments of inertia about them. If the moment about an axis GR with direction cosines α, β, γ be Mk^2, and if Mk'^2 be the moment about a parallel axis PQ through the point P whose coordinates are f, g, h, then $Mk'^2 = Mk^2 + M.PN^2$, where PN is the perpendicular distance from P to GR. By (9) and (5) we find

$$Mk'^2 = \alpha^2\{A_0 + M(g^2+h^2)\} + \ldots - 2Mgh\beta\gamma - \ldots$$
$$= A_0\alpha^2 + B_0\beta^2 + C_0\gamma^2 + M\{(g\gamma - h\beta)^2 + (h\alpha - f\gamma)^2 + (f\beta - g\alpha)^2\}.$$

Thus the moment of inertia about any axis can be found in terms of A_0, B_0, C_0.

Since Mk'^2 contains terms such as $-2Mgh\beta\gamma$, involving *products* of two direction cosines, the axes through P parallel to the principal axes through G are not principal axes at P unless two of the coordinates f, g, h be zero, in which case P lies upon one of the principal axes Gx, Gy, Gz.

16. Apparatus. The apparatus is designed to test relation (9) for the simple case when $\gamma = 0$, i.e. when the axis OR is perpendicular to the axis Oz. As (9) implies, the principal axes of the body at O are taken as axes of coordinates.

If a rigid pendulum oscillating about a horizontal knife edge be built up of two parts of masses M_0 and M, its periodic time is T, where

$$T = 2\pi\sqrt{\left\{\frac{M_0k_0^2 + Mk_1^2}{(M_0h_0 + Mh_1)g}\right\}}. \qquad \ldots\ldots\ldots\ldots(9\,a)$$

Here $M_0k_0^2$, Mk_1^2 are the moments of the two parts about the knife edge and h_0, h_1 are the distances of their centres of gravity from it. It is supposed that the two centres of gravity lie in a plane containing the knife edge. If h_0, h_1 and k_0^2 be kept constant, variations in Mk_1^2 will show themselves as variations in T. In the compound pendulum, described below, the radius of gyration of one part about the knife edge can be varied without changing the position of the centre of gravity of the whole.

The axle UV (Fig. 6) which bears the knife edges is fixed at right angles to the rod HJ. This rod passes through a "clearing" hole in the inertia bar PQ, which can be clamped to UV at any angle θ by a suitable clamping nut. The rod HJ bears a mass L which hangs below the knife edges when the system is at rest, and L is so adjusted on the rod that convenient periodic times are obtained. The knife edges may be formed of two pieces of hack-saw blade soldered into saw cuts in the axle. They rest on two horizontal co-planar plates of glass attached to two arms projecting from a board fixed to a shelf. The pendulum hangs between the arms.

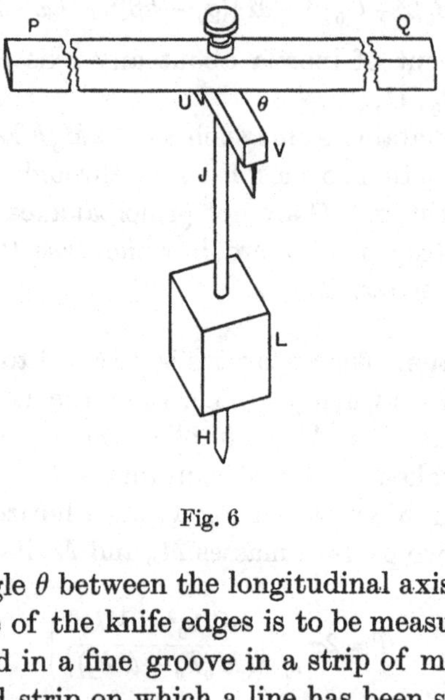

Fig. 6

When the angle θ between the longitudinal axis of the inertia bar and the line of the knife edges is to be measured, the knife edges are placed in a fine groove in a strip of metal fixed to a board. A second strip on which a line has been scribed is fixed near the first strip and is adjusted by aid of a scribing block and a surface plate so that the line is parallel to the groove. The angle between a long edge of the bar and the line is measured by a suitable protractor; the inertia bar should be as straight as possible.

17. Method. We take the axis of the vertical rod as axis of z and draw Ox through O, the centre of gravity of the inertia bar PQ, parallel to the length of the bar, and Oy perpendicular to

Ox and Oz. These are principal axes of the bar at O, since, by symmetry, the products of inertia (§ 15) with respect to them are zero. If the edges of the bar parallel to Ox, Oy, Oz be $2a$, $2b$, $2c$, and if the mass of the bar be M, the moments of inertia of the bar about the axes are

$$A_0 = \tfrac{1}{3}M(b^2+c^2), \qquad B_0 = \tfrac{1}{3}M(c^2+a^2), \qquad C_0 = \tfrac{1}{3}M(a^2+b^2).$$

By (9), the moment of inertia of the bar about an axis through O at right angles to Oz and making angle θ with Ox is

$$Mk^2 = A_0 \cos^2\theta + B_0 \sin^2\theta = (B_0 - A_0)\sin^2\theta + A_0, \quad \ldots(10)$$

since $\alpha = \cos\theta$, $\beta = \sin\theta$, $\gamma = 0$.

This result may also be obtained by integration. The square of the perpendicular from the point x, y, z upon the axis considered in (10) is

$$(x\sin\theta - y\cos\theta)^2 + z^2,$$

if θ be measured from Ox towards Oy.

If the density of the bar be σ, we have

$$\begin{aligned} Mk^2 &= \int_{-a}^{a}\int_{-b}^{b}\int_{-c}^{c} \{(x\sin\theta - y\cos\theta)^2 + z^2\}\,dx\,dy\,dz \\ &= 8\sigma abc\{\tfrac{1}{3}a^2\sin^2\theta + \tfrac{1}{3}b^2\cos^2\theta + \tfrac{1}{3}c^2\} \\ &= M\{\tfrac{1}{3}(b^2+c^2)\cos^2\theta + \tfrac{1}{3}(c^2+a^2)\sin^2\theta\} \\ &= A_0\cos^2\theta + B_0\sin^2\theta. \end{aligned}$$

Let the distance of O from the line of the knife edges be h_1. Then, by the theorem of parallel axes, the moment of the bar PQ about the knife edges is $M(k^2+h_1^2)$; this we have denoted by Mk_1^2 in (9 a). The distance h_1 we count positive when O is below the knife edges; it remains constant when θ is varied.

The bar PQ forms one part of the pendulum. The remaining part has mass M_0, its centre of gravity is at distance h_0 from the knife edges and its moment about them is $M_0k_0^2$. The moment of inertia of the whole is $M_0k_0^2 + Mk_1^2$, or

$$M_0k_0^2 + M(k^2+h_1^2).$$

The periodic time T is given by

$$T^2(M_0h_0 + Mh_1)g/4\pi^2 = M_0k_0^2 + M(k^2+h_1^2).$$

Hence, by (10), T^2 is a linear function of $\sin^2\theta$, or

$$T^2 = U\sin^2\theta + V,$$

where U, V are independent of θ. If we plot T^2 against $\sin^2\theta$, we shall obtain a straight line not passing through the origin. If $T=T_0$ when $\theta=0°$ and $T=T_{90}$ when $\theta=90°$,

$$V=T_0^2, \qquad U=T^2_{90}-V=T^2_{90}-T_0^2$$

and thus

$$T^2=(T^2_{90}-T_0^2)\sin^2\theta+T_0^2. \qquad (11)$$

The angles 0°, 30°, 45°, 60°, 90° are convenient, since they give $0, \frac{1}{4}, \frac{1}{2}, \frac{3}{4}, 1$ for $\sin^2\theta$.

The number of oscillations observed for each value of θ should be great enough to ensure some accuracy in the periodic times.

18. Practical example.

Mr L. Bairstow obtained the following results. Values * are necessarily identical with corresponding values of T^2.

θ	T	T^2 observed	$(T^2_{90}-T_0^2)\sin^2\theta+T_0^2$ calculated
0°	0·923 secs.	0·852 secs.^{2}	0·852* secs.^{2}
30	1·191	1·418	1·422
45	1·413	1·997	1·992
60	1·603	2·570	2·563
90	1·770	3·133	3·133*

EXPERIMENT 4. **Recording gyroscope.**†

19. Introduction. Gyroscopic effects furnish important instances of dynamical actions, but their general discussion is outside the range of this book. If, however, we limit ourselves to the case in which the axis of the wheel makes a constant angle with the vertical, the theory becomes elementary and simple devices suffice for recording the movements of the revolving wheel. The recording gyroscope gives students an opportunity of making practical measurements to verify the theoretical results.

20. Relation between couple and precession in simplest case. The axle of a cycle wheel W (Fig. 7) is rigidly attached to a frame UU which can turn about a horizontal axle or hinge D supported by a block E carried by a vertical shaft S. This shaft turns freely in suitable bearings. The axis of the hinge D does not intersect that of the shaft S. The shortest distance between these axes is l. Fig. 7 is diagrammatic; the details of the apparatus are better seen in Figs. 10, 11.

† "A recording gyroscope," *Proc. Camb. Phil. Soc.* Vol. XXIV, p. 236 (1928).

The axis of W intersects that of D at right angles and also intersects the axis Oz of the shaft S in O. The plane containing the axes of D and of W is a plane of symmetry of the frame as is also the plane which is at right angles to the axis of D and contains the axis of W. The centre of gravity of the wheel and frame will, therefore, be at a point G on OC, the axis of W.

In the present *simple* case, the axis OC of the wheel is horizontal and moves as if rigidly connected to the block E.

The mass of the whole system of wheel and frame is M grm., and the moment of inertia of the wheel alone about its axis is C grm. cm.^{2}

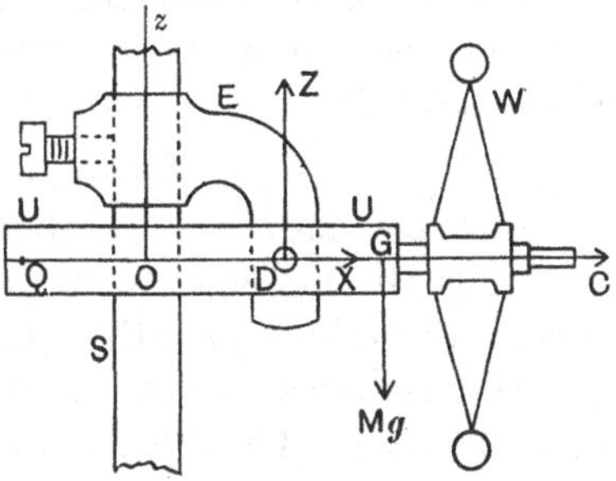

Fig. 7

The block E turns with angular velocity Ω radians per second about the fixed vertical axis Oz.† When Ω is positive, the relation of the common motion of the block E and the axis OC to the upward direction of Oz is right-handed, i.e. is that of the rotation to the translation of a right-handed screw working in a fixed nut. In the horizontal plane in which OC moves, we take a moving line OB always at right angles to OC and such that right-handed rotation about Oz, through $\frac{1}{2}\pi$, turns OC to OB. (See Fig. 8.)

The wheel spins about its axle p times per second relative to the frame UU and thus its angular velocity relative to the frame is $2\pi p$ radians per second. We count p positive when the angular velocity relative to the frame is related right-handedly to the

† When, as at present, we neglect the angular velocity of the laboratory relative to the stars, we equate the angular velocity of the block E relative to the laboratory to its true angular velocity relative to the stars.

direction OC. The electromagnetic gear (§ 22) records p, whether the axle of the wheel be horizontal or not. We shall write

$$2\pi p = \eta. \quad \text{..........................(1)}$$

The frame UU is hinged to the block E, but, in the assumed conditions, the two move as a rigid body and UU has no angular velocity relative to E. The block has angular velocity about Oz. The angular velocity of the wheel is compounded of (1) the angular velocity of E relative to the stars, (2) the angular velocity (zero, in our case) of UU relative to E, and (3) the angular velocity of the wheel relative to UU.

In the present case, OC is perpendicular to Oz, and hence the block E has no angular velocity about OC. Thus, if ω be the angular velocity of the wheel about OC, we have in this case (but *not* in the general case of § 21)

$$\omega = \eta. \quad \text{..............................(2)}$$

The momentum of any particle of the wheel has at any instant two components. One is the momentum it would have if the wheel were at rest relative to the frame; the other is the momentum it would have if the frame were at rest and the wheel turned on its axle relative to the frame. Thus the angular momentum of the wheel about any axis is the sum of the angular momenta about that axis due (1) to the wheel moving with the frame but at rest relative to it and (2) to the wheel spinning about its axle while the frame is at rest.

The angular velocity Ω of the frame and wheel about Oz gives rise to no angular momentum about OC, i.e. about a fixed axis with which OC instantaneously coincides, since (1) the mass of the frame and wheel is symmetrically distributed about OC† and (2) $COz = \frac{1}{2}\pi$. Hence the angular momentum about OC is simply $C\eta$.

Again, neither the angular velocity Ω of the frame and wheel nor the angular velocity η of the wheel relative to the frame gives rise to any angular momentum about the line OB, which is perpendicular to Oz and to the plane COz, since (1) the mass is symmetrically distributed with respect to the plane BOC and (2) COz has the constant value $\frac{1}{2}\pi$.

† The small unsymmetrical parts at V and Q (Fig. 10) are too small to have any sensible effect.

Let the moving axes OC, OB coincide with the fixed axes Ox, Oy (Fig. 8) at time t_0. When, at time t, the axle has reached OC', by turning through ϕ about Oz, and OB has reached OB', the angular momentum about Ox is $C\eta \cos \phi$, since there is no angular momentum about OB'. Hence, if the rate of increase of angular momentum about Ox be S_ϕ when $C'OC = \phi$, we have, since $d\phi/dt = \Omega$,

$$S_\phi = \frac{d}{dt}(C\eta \cos \phi) = C\left(-\eta\Omega \sin \phi + \cos \phi \frac{d\eta}{dt}\right). \quad \text{......(3)}$$

If the rate of increase of angular momentum about Ox, when the axle has the direction Ox and ϕ is zero, be S, we have

$$S = C\, d\eta/dt.$$

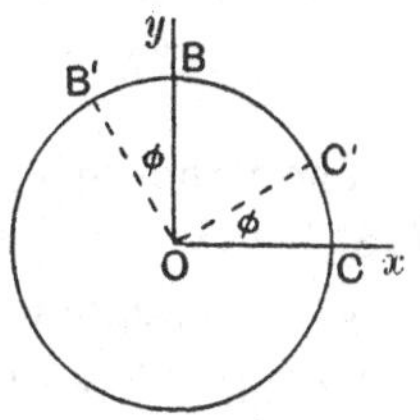

Fig. 8

If the forces applied to the wheel have no moment about OC, as is the case when there is no friction, $S = 0$, and η is constant.

At time t the angular momentum about Oy is $C\eta \sin \phi$. Hence, if the rate of increase of angular momentum about Oy be R_ϕ when $C'OC = \phi$, we have

$$R_\phi = \frac{d}{dt}(C\eta \sin \phi) = C\left(\eta\Omega \cos \phi + \sin \phi \frac{d\eta}{dt}\right), \quad \text{......(4)}$$

since $d\phi/dt = \Omega$. If the rate of increase of angular momentum about Oy, when the axle has the direction Ox and ϕ is zero, be R, we have

$$R = C\eta\Omega. \quad \text{......................(5)}$$

It thus appears that when, as in this case, there is no angular momentum about OB' for any value of ϕ, the value of R depends only upon η and Ω and not upon their rates of change.

The attraction of the earth on the frame and wheel is Mg dynes and acts at G. The effect of the hinge may be analysed into three forces X, Y, Z acting at D, the point in which OC intersects the axis of the hinge, and three couples Γ_1, Γ_2, Γ_3 about the directions of X, Y, Z. The forces X, Z are shown in Fig. 7; they have no moment about the hinge axis. If the hinge be frictionless, $\Gamma_2 = 0$. In practice, the slight vibrations, due to lack of perfect balance of the wheel, cause the hinge friction to be very small. If the resistance of the air be neglected, the force Y and the couple Γ_3 vanish for steady motion. On account of the symmetry noted above, $\Gamma_1=0$.

The centre of gravity has no vertical acceleration and hence $Z=Mg$. The two vertical forces thus constitute a couple Mgh, where h is the distance of G from the axis of the hinge. The moment of this couple about Oy (or about any line parallel to Oy) is Mgh when $\phi=0$, and, with the arrangement of Fig. 7, is positive. The couple generates angular momentum about the line Oy at the rate Mgh when $\phi=0$. Hence

$$C\eta\Omega=R=Mgh. \quad\text{.......................(6)}$$

The value of $\eta\Omega$ in terms of measured quantities is given by (18) below.

The preponderance Mgh is found by hanging such a mass m from a point Q of the frame, on the straight line CGO, as will produce equilibrium, the wheel and frame being at rest. If $QD=q$, we have $Mgh=mgq$, and thus

$$C\eta\Omega=mgq. \quad\text{.........................(7)}$$

We are not here concerned with the force X, but

$$X=-M(h+l)\Omega^2.$$

If the wheel be spun about its axis, which is horizontal, and if the frame and wheel be then given such a precession about Oz that $\Omega=Mgh/C\eta$, the state of motion so established will be steady and, in the absence of frictional forces, will continue indefinitely. The method of starting the precession is described in § 22.

The effect of the earth's rotation is considered in § 27 and is shown to be inappreciable.

In practice, η diminishes while a record is being taken. But the observer gradually increases Ω by applying such a couple to the shaft S that the axle of the wheel remains horizontal; if the axle remain horizontal, the product $\eta\Omega$ remains constant. This couple has, of course, no component about any horizontal axis. Axle friction does not concern us, for it is internal to the system of frame and wheel. The resistance of the air has, by symmetry, no moment about OB. Hence (6) remains true at each instant.

21. General case of motion with constant slope of axle. In general, the axle GC (Fig. 9) is not horizontal but makes angle $CGL = \theta$ with GL, which is horizontal and in the plane COz. We now take as axes for frame and wheel (1) GC, (2) GA in the plane COz, as shown, and (3) GB perpendicular to the plane CGA and pointing away from the reader. These right-handed axes are principal axes at G of the system of frame and wheel. The moments of inertia of the wheel and of the frame about GC are C, C' and about GA are A, A'.

The angular velocity Ω of the block E about Oz (or about any parallel axis) may be resolved into $\Omega \sin\theta$ about GC and $-\Omega\cos\theta$ about GA. The frame UU moves as if rigidly connected to E

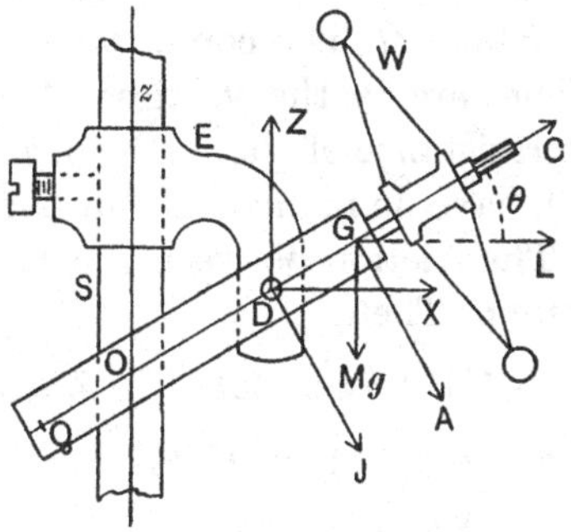

Fig. 9

and therefore has the same angular velocity, viz. $\Omega \sin\theta$, about GC.

The wheel makes p revs. per sec. relative to the frame and has angular velocity $2\pi p$ relative to the frame. As in § 20, we denote $2\pi p$ by η. Hence, if the resultant angular velocity of the wheel about GC be ω, we have

$$\omega = \eta + \Omega \sin\theta. \quad \text{.......................(8)}$$

The angular velocity of the wheel about GA is $-\Omega\cos\theta$.

By symmetry, neither the angular velocity $\Omega\sin\theta$ of the frame nor the angular velocity $\eta + \Omega\sin\theta$ of the wheel about GC gives rise to any angular momentum about either GA or GB. The corresponding angular momenta about GC are $C'\Omega\sin\theta$ and $C(\eta + \Omega\sin\theta)$. By symmetry, the angular velocity $-\Omega\cos\theta$ of frame and wheel about GA gives rise to no angular momentum about either GB or GC. The corresponding angular momentum about GA is $-(A+A')\Omega\cos\theta$. Hence, if the angular momentum of the wheel and frame about the horizontal line GL in the plane COz be F, we have

$$F = C\eta\cos\theta + (C+C'-A-A')\Omega\cos\theta\sin\theta. \quad \text{......(9)}$$

Provided that θ be constant, neither frame nor wheel has angular momentum about GB and this is true whether η and Ω be constant or variable.

By the method of § 20, if the rate of increase of angular momentum about an axis with which GB is instantaneously coincident be R, we have

$$R = F\Omega. \quad \text{..........................(10)}$$

The weight Mg acts at G. The components X, Z of the force exerted by the hinge are in the vertical plane COz, and Y is horizontal. If the resistance of the air be neglected, $Y=0$. The couples Γ_1, Γ_3 (§ 20) have no moment about GB. With a frictionless hinge, $\Gamma_2=0$. The friction between the wheel and its axle is internal to the system. Then

$$X = -M(l+h\cos\theta)\Omega^2, \qquad Z = Mg.$$

The moment of these forces about GB is R, and thus

$$F\Omega = R = -Xh\sin\theta + Zh\cos\theta$$
$$= Mgh\cos\theta + M(l+h\cos\theta)\Omega^2 h\sin\theta. \quad \text{...(11)}$$

Hence, by (9), on dividing by $\cos\theta$,

$$Mgh = C\eta\Omega + (C+C'-A-A'-Mh^2)\Omega^2\sin\theta - Mlh\Omega^2\tan\theta.$$

If the moments of inertia of the system of wheel and frame about the axes DC and DJ (parallel to GA) be K and H, then

$$C+C' = K, \qquad A+A'+Mh^2 = H.$$

Thus $$Mgh = C\eta\Omega + Q\Omega^2, \quad \text{....................(12)}$$

where $$Q = (K-H)\sin\theta - Mlh\tan\theta.$$

If $Q=0$, we have, as in § 20,

$$Mgh = C\eta\Omega. \quad \text{....................(12a)}$$

We can make $Q=0$ by making $\theta=0$. We can also make $Q=0$ for finite values of θ, if we so design the apparatus that both $K=H$ and

$l=0$. In this latter case, the product $\eta\Omega$ for a given steady value of θ would be independent of θ.

When Q is not zero, Ω has the two values given by

$$\Omega=\frac{C\eta}{2Q}\left\{-1\pm\sqrt{\left(1+\frac{4QMgh}{C^2\eta^2}\right)}\right\}. \quad \text{.........(13)}$$

If Qh be positive or if, when Qh is negative, $|4QMgh|<C^2\eta^2$, the two values of Ω are real. In the experiment Q is small, since θ is small. In this case we expand the square root and obtain

$$\Omega=\frac{C\eta}{2Q}\left\{-1\pm\left(1+\frac{2QMgh}{C^2\eta^2}\right)\right\}. \quad \text{............(14)}$$

The two, approximate, values of Ω given by (14) are

$$\Omega_1=Mgh/C\eta \quad \text{and} \quad \Omega_2=-\frac{C\eta}{Q}\left(1+\frac{QMgh}{C^2\eta^2}\right).$$

Since Q is small, we may write $\Omega_2=-C\eta/Q$ as a sufficient approximation. When $\Omega=-C\eta/Q$, $|\Omega|$ is very large when θ is very small. It will be impracticable to make $|\Omega|$ very large, and hence the value of Ω which occurs in the experiment will be that one of the two given by (13) which is nearly equal to $Mgh/C\eta$.

In the experiment, Ω/η is small and θ is small, and hence the term in (12) involving Ω^2 has little effect. When θ is small, $\sin\theta=\tan\theta=\theta$, and then

$$Mgh=C\eta\Omega\{1+f\theta/n\}, \quad \text{..................(15)}$$

where

$$f=(K-H-Mhl)/C. \quad \text{..................(15a)}$$

Here $n=\eta/\Omega$, so that the wheel makes n turns relative to the frame while the frame turns once about the vertical shaft. Thus (12a) needs a correction of the first order in θ, but, with the apparatus used in § 26, f/n is so small that moderate accuracy in adjusting the axle to the horizontal suffices.

22. Apparatus. The apparatus is shown diagrammatically in Fig. 10. The axle of the cycle wheel W is attached to the frame UU. This frame is supported by a horizontal axle D passing through the block E carried by the vertical steel shaft SS. The axle D is clamped by set-screws to the frame UU and turns in the long plain hole in the block E with little friction.

The shaft is supported by a conical bearing R, fixed to the floor, and by a plain bearing J, fixed to a table projecting from the wall of the laboratory. When the wheel is at rest, its axle can be made horizontal by suspending a suitable load, by means of a knife-edged hanger, from the groove Q. The moment of the weight of the wheel and frame about D is then equal to that of the mass hanging from Q. If this mass be m, the latter moment is $mg \, . \, QD$ or mgq (§ 20).

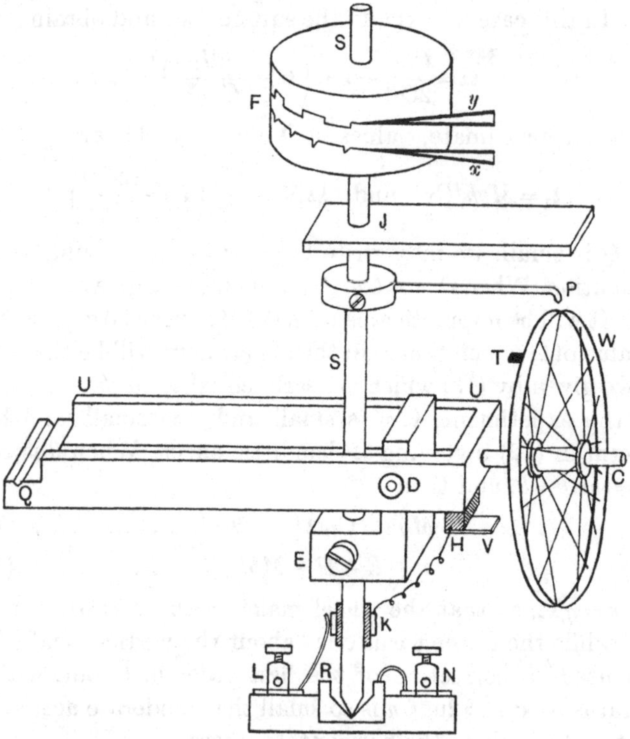

Fig. 10

The upper end of the shaft carries a drum F, on which the records are taken automatically.

To operate the apparatus, the end C of the axle is held in one hand, the axle being horizontal, and the wheel W is given a vigorous spin with the other. If the operator now move the axle in a horizontal plane, he will be able, after a little practice, to give the system such an angular velocity about the shaft S that, when he releases the grip of his hand on the axle, the wheel will precess about S with the axle practically horizontal. Success depends upon so adjusting this angular velocity about S that the force exerted by the axle on the hand vanishes. When this occurs, the grip is released and the hand is removed. Any small oscillations soon disappear under the influence of the slight frictional forces. If the axle be not quite horizontal, an adjustment is made by applying—steadily, not in jerks—a small couple to the vertical shaft S by the fingers. If the couple assist the precession, the end C of the axle will rise; a reverse couple will cause C to fall.

Automatic devices are used in recording the two angular velocities η and Ω. The frame carries an insulating block H from which projects a suitable spring, and the wheel carries a projecting tooth; the spring and the tooth are represented diagrammatically by V and T in Fig. 10. Actually the tooth projects from the hub. The tooth makes contact with the spring once in each revolution of the wheel relative to the frame. This contact puts the spring into electrical connexion with the axle of the wheel and so with the shaft S. A thin, flexible wire from V passes to a collar K (Fig. 10) fixed to the shaft by an insulating bush. A brush pressing on K connects it with the terminal L. The terminal N is in electrical connexion with the shaft. If a battery be connected to L, N, a current will pass whenever the tooth touches the spring.

The battery circuit includes the windings of an electromagnet X (Fig. 11). When the current passes, the armature is attracted and moves a style x. The electromagnet Y, with its style y, is operated by a pendulum which makes and breaks a second electrical circuit.

The two electromagnets are mounted on a sliding piece which has, when released, a slow vertical motion controlled by a piston working in a cylinder containing very viscous oil.

By a simple geometrical slide, the styles can be brought into contact with the drum F. The drum carries smoked paper, and thus the movements of the styles are recorded. By aid of a line AB ruled on the drum parallel to the shaft, the number of revolutions made by the wheel relative to the frame and also the number of beats made by the pendulum during one revolution of the drum are determined. Fractional parts are, in each case, found by interpolation.

Let the periodic time of the pendulum be T seconds, and in one particular revolution of the drum let there be N periods. Then, if the time of that one revolution of the drum be D seconds, $D = NT$. Hence, if the mean angular velocity of the drum during the interval D be Ω_{av}, we have

$$\Omega_{av} = 2\pi/D = 2\pi/(NT). \quad \text{.................(16)}$$

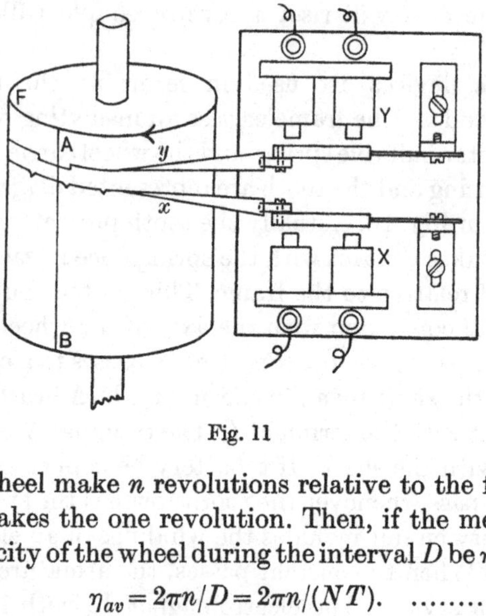

Fig. 11

Let the wheel make n revolutions relative to the frame while the drum makes the one revolution. Then, if the mean relative angular velocity of the wheel during the interval D be η_{av}, we have

$$\eta_{av} = 2\pi n/D = 2\pi n/(NT). \quad \text{..............(17)}$$

It is shown in § 25 that, if η, Ω be the values at any instant during the interval D, we may write

$$\eta\Omega = \eta_{av}\,\Omega_{av} = 4\pi^2 n/D^2 = 4\pi^2 n/(N^2 T^2), \text{.........(18)}$$

with only a second-order error when the friction is small.

Hence, by (6), we may write

$$Mgh = C\eta\Omega = 4\pi^2 Cn/(N^2 T^2). \quad \text{...........(19)}$$

23. Experimental details. The shaft S must be vertical. If, when the wheel is not spinning, the system turn about the axis of S to take up a definite position, the shaft is not vertical. The error is corrected by adjusting one of the bearings. The adjustment may be tested by a level resting on an adjustable table fixed to the shaft.

The shaft carries a pointer, which is set so that its tip, P, is vertically above one edge of the rim of the wheel when the axle is horizontal. If the highest and lowest points of that edge be equidistant from the (vertical) shaft when P is vertically above the selected edge, the adjustment is correct.

A slip of paper, with a hard smooth surface, is wrapped round the drum. The overlapping, ready-gummed, end is stuck to the slip itself. The drum is held over a smoky paraffin flame to smoke the paper lightly, and the drum is then fixed to the shaft.

The pendulum is started and the battery switch is closed. One observer then starts the wheel and gives it approximately the precessional motion which will cause the axle to remain horizontal as the wheel spins. A second observer applies, by his fingers, a small couple to the vertical shaft so as to keep the axle horizontal. The couple is applied continuously while the record is taken and not spasmodically; small errors of level as indicated by the pointer P are of less consequence than sudden changes of precessional motion. When the wheel is spinning correctly, the first observer releases the slider carrying the electromagnets and brings the styles against the drum. If the speed of the slider be suitably adjusted, the drum will make 5 or 6 revolutions while the styles move from the top to the bottom of the drum. The drum is removed from the shaft and is mounted on a short rod which can rest, with its axis horizontal, in V's in a cradle standing on a flat horizontal plate. By means of a scribing block sliding on the plate, a line parallel to the axis of the rod is drawn upon the smoked paper. A second line, to furnish an independent set of readings, may be drawn approximately diametrically opposite the first line. The record is "fixed" by a weak solution of shellac in methylated spirit. When the paper is dry, it is cut along a generating line of the cylinder and is then ready for measurement.

The general appearance of the record, when removed from the drum, is shown in Fig. 12. The pendulum completes its circuit for *about* half a complete vibration during each complete vibration. The contacts are made at the times indicated by the points marked

0, 1, 2, ... The contact made by the wheel is of short duration. The contacts are marked 0, 1, 2, ...

Interpolation is used to find the position of the vertical line AB, which was marked by the scribing block, relative both to the pendulum signals and to the wheel signals. Thus, for the pendulum signals on the original record, which is reproduced on a smaller scale as Fig. 12, the distance from 0 to 1 is 18·5 mm. and that from 0 to AB is 7·2 mm. Hence AB corresponds to $0 + 7{\cdot}2/18{\cdot}5$ or 0·389. Similarly, from 9 to 10 is 21·0 mm. and from 9 to AB is 14·9 mm. Hence AB now corresponds to $9 + 14{\cdot}9/21{\cdot}0$ or 9·710. Thus the revolution has occupied $(9{\cdot}710 - 0{\cdot}389)\,T$ or $9{\cdot}321T$, where T is the period of the pendulum.

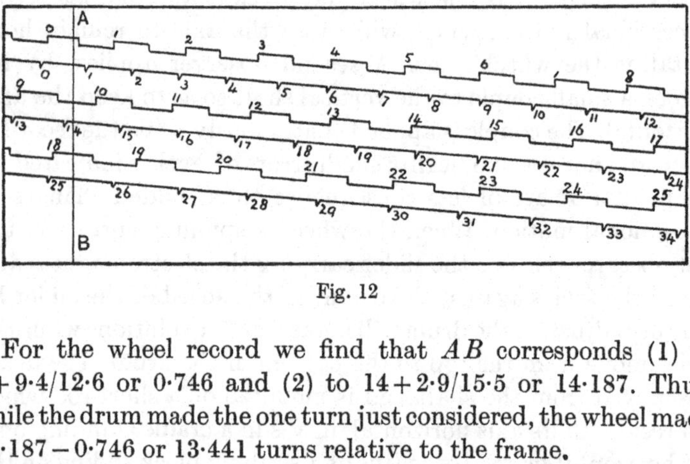

Fig. 12

For the wheel record we find that AB corresponds (1) to $0 + 9{\cdot}4/12{\cdot}6$ or 0·746 and (2) to $14 + 2{\cdot}9/15{\cdot}5$ or 14·187. Thus, while the drum made the one turn just considered, the wheel made $14{\cdot}187 - 0{\cdot}746$ or 13·441 turns relative to the frame.

The mean angular velocity of the drum during this turn is thus $2\pi/(9{\cdot}321T)$. The mean angular velocity of the wheel relative to the frame is 13·441 times that of the drum and is thus $2\pi \times 13{\cdot}441/(9{\cdot}321T)$.

24. The moments of inertia.* In order to find the moment of inertia C of the wheel about its axis, the wheel is removed from the axle and a suitable fitting, of small and known moment of inertia, is secured in the hub. By this fitting, the wheel is fixed,

* The determination of the moments of inertia is not convenient for class work; it involves the dismantling of the apparatus.

with its axis vertical, to a *straight* vertical torsion wire.* An inertia bar can be substituted for the wheel. The moment of inertia C of the wheel is deduced from the periods of torsional vibrations of the wheel and the bar. The moment of inertia C' of the frame and axle about the axis of the latter is found in a similar manner, a light fitting, of known moment of inertia, being used to fix the frame to the torsion wire.

The wheel is now replaced on its axle and the frame UU is disconnected from the block E (Fig. 10). A small fitting provided with a knife edge is fixed to the frame. The line of the knife edge intersects the axis of the hinge and is perpendicular to that axis and to the axis of the wheel. The line of the knife edge is represented by DJ in Fig. 9. The system of wheel and frame is then swung as a pendulum about the knife edge. The fitting is adjusted so that its period about the knife edge is approximately equal to that of the system of fitting, frame and wheel. Then the period T_0, which the frame and wheel would have if the fitting were massless, may be taken as equal to that found for the system of the actual fitting, the frame and the wheel. The moment of inertia H (§ 21) of the system about the knife edge, i.e. about DJ, is given by

$$H = T_0^2 Mgh/4\pi^2. \quad \text{.....................(20)}$$

By § 20, $Mgh = mgq$, where m is the mass which balances the system when hung from the groove Q, and $q = QD$.

25. Relation of average to actual angular velocity. Since it is impossible to observe the actual values of η and Ω at any instant, we use a number of *average* values of η and Ω derived from successive *complete* revolutions of the drum.

In the case of the general equation (12), if we maintain θ constant by adjusting Ω to suit η at each instant, a diminution of η will involve a change in Ω. Whether this will be an increase or a decrease will depend upon the numerical values of the quantities

* The two ends of the wire should be soldered into two rods about $\frac{1}{2}$ cm. in diameter and 3 cm. in length, so as to allow adequate clamping. See *Experimental Elasticity*, § 62 or *Experimental Harmonic Motion*, § 34.

If the wire be not straight when it carries no load, a considerable error may occur in the comparison of the moments of inertia. A steel wire straightened by being raised to a red heat while under tension may be used.

appearing in (12). For convenience, we will rewrite (12) in the form

$$G = \eta\Omega + J\Omega^2. \qquad \text{.....................(21)}$$

When the axis of the wheel is horizontal, $J=0$. For the interval of time, D, occupied by one particular revolution of the drum, it will suffice, when the effects of friction are small, to write

$$\Omega = [\Omega](1+\beta\tau). \qquad \text{.....................(21}a\text{)}$$

Here Ω is the angular velocity at time τ after the beginning of the particular revolution. When $\tau=0$, $\Omega=[\Omega]$. Then, by (21),

$$\eta = \frac{G}{[\Omega](1+\beta\tau)} - J[\Omega](1+\beta\tau).$$

Since the wheel makes n revolutions relative to the frame in time D,

$$2\pi n = \int_0^D \eta\, d\tau = \frac{G}{\beta[\Omega]}\log(1+\beta D) - J[\Omega](D+\tfrac{1}{2}\beta D^2).$$

The drum turns through the one revolution in time D. Hence

$$2\pi = \int_0^D \Omega\, d\tau = [\Omega](D+\tfrac{1}{2}\beta D^2),$$

and $$[\Omega] = 2\pi/(D+\tfrac{1}{2}\beta D^2).$$

Hence $$2\pi n = \frac{GD}{2\pi\beta}(1+\tfrac{1}{2}\beta D)\log(1+\beta D) - 2\pi J.$$

Thus, when βD is small,

$$2\pi n = \frac{GD^2}{2\pi}(1+\tfrac{1}{12}\beta^2 D^2) - 2\pi J. \qquad \text{...........(22)}$$

Since (21) holds at each instant,

$$G = [\eta][\Omega] + J[\Omega]^2, \qquad \text{..................(23)}$$

where $[\eta]$ is the value of η at the beginning of the turn, when $\tau=0$. If we write

$$G^* = \frac{2\pi n}{D}\frac{2\pi}{D} + J\frac{4\pi^2}{D^2}, \qquad \text{..................(24)}$$

then G^* is the value of G *computed* by using the *average* values $\eta_{av} = 2\pi n/D$ and $\Omega_{av} = 2\pi/D$ in place of $[\eta]$ and $[\Omega]$. By (22),

$$G^* = G(1+\tfrac{1}{12}\beta^2 D^2), \qquad \text{..................(25)}$$

or $$G = G^*(1-\tfrac{1}{12}\beta^2 D^2). \qquad \text{..................(26)}$$

With the apparatus at the Cavendish Laboratory, β is less than 0·02, when a second is the unit of time. When the wheel is spun by hand, the greatest value of D is less than 10 seconds, and hence $\beta D < 0{\cdot}2$. Thus G does not differ from G^* by as much as one part in 300.

When the axis of the wheel is horizontal and, in consequence, $J=0$, we have, by (23), (26) and (24), since $D=NT$,

$$[\eta][\Omega]=G=G^*(1-\tfrac{1}{12}\beta^2D^2)=\frac{4\pi^2 n}{N^2T^2}(1-\tfrac{1}{12}\beta^2D^2). \quad ...(27)$$

26. Practical example.

The wheel had mass 1515 grm. and diameter 64·9 cm. The shortest distance between the centre of its hub and the axis of the hinge was 9·3 cm.

Moments of inertia. The moments of inertia with respect to axis OC (Fig. 9) were found as in § 24.

For wheel, $C=1{\cdot}11013\times10^6$ grm. cm.^{2};

for frame, $C'=4{\cdot}488\times10^4$ grm. cm.^{2}

Hence $K=C+C'=1{\cdot}1550\times10^6$ grm. cm.^{2}

The moment of inertia, H, of system of wheel and frame about axis DJ was found as in § 24. Period about knife edge, $T_0=1{\cdot}462$ secs.; as below, $Mgh=1{\cdot}3402\times10^7$ dyne cm.

By (20), $H=7{\cdot}256\times10^5$ grm. cm.^{2}

The shortest distance, l, between axes of vertical shaft and hinge was a little uncertain as shaft was not quite straight and was not very stiff. Approximately, $l=1{\cdot}5$ cm. Thus

$$Mhl=Mgh\,.\,l/g=1{\cdot}3402\times10^7\times1{\cdot}5/981=2{\cdot}05\times10^4 \text{ grm. cm.}^2$$

Hence, as in (15a),

$$f=(K-H-Mhl)/C=4{\cdot}089\times10^5/(1{\cdot}11013\times10^6)=0{\cdot}368.$$

Thus, by (15),

$$Mgh=C\eta\Omega(1+f\theta/n)=C\eta\Omega(1+0{\cdot}368\theta/n) \text{ dyne cm.}$$

The radius of the wheel was 32·5 cm. An error of one cm. horizontally in adjusting rim to pointer P (Fig. 10) made $\theta=1/32{\cdot}5$ radians. In measurements recorded below, least value of n was 7·342. With $\theta=1/32{\cdot}5$, and $n=7{\cdot}342$, $0{\cdot}368\theta/n=0{\cdot}0015$. After practice, it was easy to maintain adjustment to within 0·5 cm.; thus error due to obliquity was less than one per thousand.

Precession measurements. These were made with the help of Mr Leslie Bairstow, of King's College, Cambridge. The periodic time of the pendulum was $T=0{\cdot}71206$ sec.

The second and third columns of Table give respectively swings of pendulum and turns of wheel corresponding to line AB (Fig. 12) ruled on record; fractions were obtained by interpolation (§ 23). Columns "N"

and "n" give swings of pendulum and turns of wheel during successive turns of drum.

Turns of drum	Swings of pendulum	Turns of wheel	Per turn of drum		$\frac{n}{N^2}$	$\frac{\beta^2 D^2}{12}$
			Swings, N, of pendulum	Turns, n, of wheel		
0	0·817	0·658	—	—	—	—
1	12·331	21·260	11·514	20·602	0·15540	0·00196
2	22·569	37·544	10·238	16·284	0·15536	0·00116
3	31·816	50·848	9·247	13·304	0·15559	0·00076
4	40·301	62·023	8·485	11·175	0·15522	0·00053
5	48·163	71·602	7·862	9·579	0·15497	0·00039
6	55·488	79·946	7·325	8·344	0·15551	0·00030
7	62·368	87·288	6·880	7·342	0·15511	0·00023

Mean value of $n/N^2 = 0{\cdot}15531$.

Two other records gave 0·15558 and 0·15551 for n/N^2. The mean of the three values is 0·15547.

Hence, by (18), which omits the correction due to (27),

$$C\eta\Omega = C \times 4\pi^2 (n/N^2)/T^2 = 4\pi^2 \times 1{\cdot}11013 \times 10^6 \times 0{\cdot}15547/(0{\cdot}71206)^2$$
$$= 1{\cdot}3438 \times 10^7 \text{ dyne cm.}$$

The correction due to friction can be found by plotting θ/t against t; here θ is the angle turned through by the drum in time t, counted from the beginning of the first turn of the drum shown in the Table. It was more convenient to plot d/p against p, where d is the number of turns of the drum and p is the number of swings of the pendulum. Thus, when $d = 0, 1, 2, \ldots, p = 0, 11{\cdot}514, 21{\cdot}752, \ldots$. For the whole range of the Table, i.e. up to $d = 7$, and $p = 61{\cdot}551$, the points lay very nearly on a straight line. In terms of θ and t, the line was $\theta/t = a(1 + \frac{1}{2}\gamma t)$, where

$$a = 9{\cdot}65\pi/(60T), \qquad \gamma T = 1/75{\cdot}1;$$

here T secs. is the pendulum period. Thus

$$\theta = a(t + \tfrac{1}{2}\gamma t^2), \qquad \Omega = d\theta/dt = a(1 + \gamma t).$$

Awbery's method (Note I) gave 9·644 and 74·91 in place of 9·65 and 75·1.

If the angular velocity be $[\Omega]$ when $t = rT$, where r is an integer, and Ω_τ when $t = rT + \tau$, we have

$$[\Omega] = a\{1 + \gamma rT\}, \qquad \Omega_\tau = a\{1 + \gamma(rT + \tau)\},$$

and hence
$$\Omega_\tau = [\Omega]\left(1 + \frac{\gamma\tau}{1 + \gamma rT}\right).$$

By comparison with (21a), we have $\beta = \gamma/(1 + \gamma rT)$. If time occupied by the single turn, which begins when $t = rT$, be sT, we have $D = sT$; then

$$\beta D = \gamma sT/(1 + \gamma rT) = s/\{r + (\gamma T)^{-1}\}.$$

In the experiment, $(\gamma T)^{-1} = 75{\cdot}1$, and then

$$\beta D = s/\{r + 75{\cdot}1\}.$$

For the first turn, $r=0$, $s=11{\cdot}514$, $\beta D=0{\cdot}153$;
for the last turn,

$$r=55{\cdot}488-0{\cdot}817=54{\cdot}671, \qquad s=6{\cdot}88, \qquad \beta D=0{\cdot}053.$$

The values of $\frac{1}{12}\beta^2D^2$ are given in Table; their mean is 0·00076. The three records were so nearly alike and the values of n/N^2 were so nearly equal that the mean correction was applied to the mean of n/N^2.

Since the axle of the wheel is horizontal, $J=0$, and then, by (21), (23),

$$\eta\Omega=G=[\eta][\Omega].$$

Hence, by (27),

$$C\eta\Omega=C\times 4\pi^2(n/N^2)(1-0{\cdot}00076)/T^2=1{\cdot}3438\times 0{\cdot}99924\times 10^7$$
$$=1{\cdot}3428\times 10^7 \text{ dyne cm.}$$

The distance, q, between groove Q (Fig. 10) and axis of hinge was 14·709 cm. On account of hinge friction, the balancing mass was a little uncertain. It was therefore measured ten times. One out of ten different objects of masses unknown to the observer was included in the load, the remainder consisting of known masses. After the balancing values of the latter had been found, the ten objects were weighed. The ten values of the load, m, varied from 927·45 to 929·75 grm. The mean gave $m=928{\cdot}54$ grm. with a probable error of 0·15 grm. By § 20, Mgh for frame and wheel equals mgq. At Cavendish Laboratory, $g=981{\cdot}27$ cm. sec.$^{-2}$. Hence

$$Mgh=928{\cdot}54\times 981{\cdot}27\times 14{\cdot}709=1{\cdot}3402\times 10^7 \text{ dyne cm.}$$

This value differs from the corrected value of $C\eta\Omega$ by 2·0 per thousand.

27. Effect of the earth's rotation.

The earth turns about its axis once in a sidereal day of 86,164 secs. Its angular velocity, Θ, is $2\pi/86164$ or $7{\cdot}292\times 10^{-5}$ radians per sec. We suppose the precession of the gyroscope about the vertical shaft S (Fig. 7) to be so adjusted by a couple applied to the shaft S that the axis of the wheel is always perpendicular to that shaft and so is horizontal.

For the moving origin we take D (Fig. 7), the point in which the axis of the wheel intersects the axis of the hinge. We take three moving axes (1), (3) and (2), of which (1) is the upward line parallel to the shaft through D, (3) is the axis $ODGC$ of the wheel, and (2) is a line through D perpendicular to (1) and (3) to complete a right-handed system. At any time the axis (3) makes an angle ψ, measured in the positive direction, with the southward meridian through O. The north latitude of O is λ.

If $\theta_1, \theta_2, \theta_3$ be the angular velocities of the axes about themselves,

$$\theta_1=\Theta\sin\lambda+\dot{\psi}, \qquad \theta_2=-\Theta\cos\lambda\sin\psi, \qquad \theta_3=-\Theta\cos\lambda\cos\psi.$$

If the components of the velocity of the origin D relative to the centre of the earth be u_1, u_2, u_3; if those of the velocity of G, the centre of gravity, be v_1, v_2, v_3; and if the perpendicular distance of O from the earth's axis be r, we have, since $DG=h$,

$$u_1=l\theta_2, \qquad u_2=-r\Theta\cos\psi-l\theta_1, \qquad u_3=r\Theta\sin\psi,$$
$$v_1=(l+h)\theta_2, \qquad v_2=-r\Theta\cos\psi-(l+h)\theta_1, \qquad v_3=r\Theta\sin\psi.$$

The moment of inertia of the wheel about (3) is C and of the frame is C'.

The moments of inertia of the wheel and frame system, whose mass is M, about axes through G parallel to (1) and (2) respectively are A_0 and B_0. If h_1, h_2, h_3 be the angular momenta about the axes (1), (2), (3),

$$h_1 = A_0(\Theta \sin\lambda + \dot{\psi}) - Mhv_2 = (A_0 + E)(\Theta \sin\lambda + \dot{\psi}) + Mhr\Theta \cos\psi,$$
$$h_2 = -B_0\Theta \cos\lambda \sin\psi + Mhv_1 = -(B_0 + E)\Theta \cos\lambda \sin\psi,$$
$$h_3 = C\omega - C'\Theta \cos\lambda \cos\psi,$$

where $$E = Mh(l+h).$$

Here ω is the angular velocity of the wheel about (3); in the absence of friction, ω remains constant. The angular velocity of the wheel relative to the frame is η and $\eta = \omega - \theta_3$.

In order to find how the motion of the system in azimuth depends upon the applied forces, we find the rate of increase at time t of angular momentum about the *fixed* axis with which the moving axis (2) coincides at time t and equate this rate to the moment about the same axis, at the same instant, of the applied forces.

Let the coordinates of a point D' relative to fixed axes Dx, Dy, Dz through D be ξ, η, ζ; let the coordinates of any particle m relative to *parallel* axes through D' be x', y', z'; and let the velocity of m have components α, β, γ. If k_2 be the angular momentum of the system about Dy,

$$k_2 = \Sigma m\{\alpha(z' + \zeta) - \gamma(x' + \xi)\}.$$

Since ξ, ζ are independent of m, we may put them outside the sign of summation. Then

$$k_2 = \Sigma m(\alpha z' - \gamma x') + \zeta\Sigma m\alpha - \xi\Sigma m\gamma = k_2' + p_1\zeta - p_3\xi, \quad \text{......(28)}$$

where k_2' is the angular momentum about $D'y'$ and p_1, p_3 are the components parallel to $D'x'$, $D'z'$ of the momentum of the system.

Let Dx, Dy, Dz be three axes with which the moving axes (1), (2), (3) coincide at time t. At $t + dt$, D has moved to D' and then, relative to $Dxyz$, has coordinates $LN = u_1 dt$, $DL = u_2 dt$, $ND' = u_3 dt$, as in Fig. 13. The moving axis (1) through D', which at t coincides with Dx, has at $t + dt$ turned through $\theta_2 dt$ about $D'y'$ and through $\theta_3 dt$ about $D'z'$, and similarly for the other two moving axes.

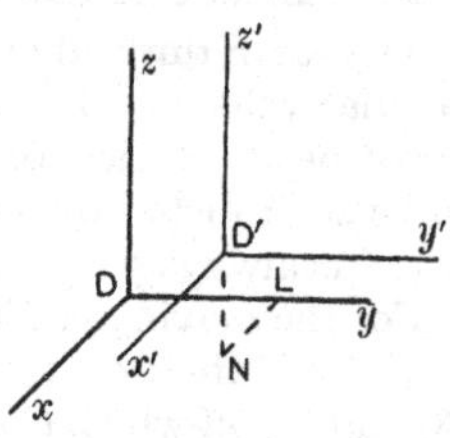

Fig. 13

The angular momenta about the three moving axes (1), (2), (3) in their positions at $t + dt$ are

$$h_1 + \dot{h}_1 dt, \qquad h_2 + \dot{h}_2 dt, \qquad h_3 + \dot{h}_3 dt.$$

Axis (2) has turned in the plane $D'y'z'$ through $\theta_1 dt$ and in $D'x'y'$ through $\theta_3 dt$. The angle between (2) and $D'y'$ is thus of the first order in dt and the difference between its cosine and unity is of the second order. Axis (3) has turned in $D'y'z'$ through $\theta_1 dt$ away from $D'y'$ and in $D'z'x'$ through $\theta_2 dt$. To sufficient accuracy, the angle between (3) and $D'y'$ is $\frac{1}{2}\pi + \theta_1 dt$ and its cosine is $-\theta_1 dt$. Axis (1) has turned in $D'x'y'$ through $\theta_3 dt$ towards $D'y'$ and in $D'z'x'$ through $\theta_2 dt$. The angle between (1) and $D'y'$ is $\frac{1}{2}\pi - \theta_3 dt$ and its cosine is $\theta_3 dt$.

Hence the angular momentum about $D'y'$ at $t+dt$ is

$$h_2+\dot{h}_2 dt+(h_3+\dot{h}_3 dt)(-\theta_1 dt)+(h_1+\dot{h}_1 dt)(\theta_3 dt),$$

or, when terms involving dt^2 are rejected,

$$h_2+(\dot{h}_2-h_3\theta_1+h_1\theta_3)\,dt.$$

The coordinates of D' relative to $Dxyz$ are $\xi=u_1 dt$, $\eta=u_2 dt$, $\zeta=u_3 dt$, and hence, by (28), the angular momentum about Dy at $t+dt$ is

$$h_2+(\dot{h}_2-h_3\theta_1+h_1\theta_3)\,dt+(p_1 u_3-p_3 u_1)\,dt.$$

At t the angular momentum is h_2 and hence the rate of increase about Dy is

$$\dot{h}_2-h_3\theta_1+h_1\theta_3+M(v_1 u_3-v_3 u_1),$$

for $p_1=Mv_1$ and $p_3=Mv_3$, since the mass of the system is M and its centre of gravity has component velocities v_1, v_2, v_3.

The equation of moments about axis (2) is thus

$$\dot{h}_2-h_3\theta_1+h_1\theta_3+M(v_1 u_3-v_3 u_1)=H_2, \quad \text{............(29)}$$

where H_2 is the moment about (2) of the forces acting on the system of wheel and frame. These forces arise from the earth's attraction and the reaction of the hinge; the reaction has no moment about (2). Any small moment due to the air is neglected.

If a particle of one gramme be at rest relative to the earth at O (Fig. 7), the support exerts g dynes along the upward vertical. This force with a force f dynes due to the earth's attraction causes the acceleration $r\Theta^2$ necessary to make the particle move in its circular path of radius r about the earth's axis. The line of action of f intersects the earth's axis in L. If OL make angle ϵ with the vertical through O, we have, by resolving along the vertical,

$$f\cos\epsilon=g+r\Theta^2\cos\lambda.$$

Since the dimensions of the apparatus are very small compared with the earth's radius, we may consider that each gramme of matter in the system of wheel and frame experiences an attraction of f dynes parallel to OL. Hence the resultant force due to attraction passes through G, is parallel to OL and is Mf dynes.

The component $Mf\sin\epsilon$ lies in the horizontal plane of (2) and (3) and so has no moment about (2). The moment of $Mf\cos\epsilon$ about (2) is $-Mfh\cos\epsilon$. Thus

$$H_2=-M(g+r\Theta^2\cos\lambda)\,h.$$

When we substitute in (29) the values of the various quantities and neglect terms involving Θ^2, we obtain

$$\dot{\psi}(C\omega+U\Theta\cos\lambda\cos\psi)=Mgh-C\omega\Theta\sin\lambda, \quad \text{........(30)}$$

where

$$U=A_0+B_0+2E-C'.$$

The apparatus records the time NT (§ 22) occupied by a complete revolution, relative to the earth, of the frame about the vertical shaft. When we integrate (30) with respect to t from 0 to NT, the term involving U disappears, and thus

$$2\pi C\omega=NT(Mgh-C\omega\Theta\sin\lambda).$$

If S be the sidereal day of 86,164 secs., $\Theta=2\pi/S$, and thus

$$Mgh=\frac{2\pi C\omega}{NT}\left\{1+\frac{NT\sin\lambda}{S}\right\}. \quad \text{.................(31)}$$

The angular velocity of the wheel relative to the frame is η, where $\eta=\omega-\theta_3=\omega+\Theta\cos\lambda\cos\psi$. In time NT the wheel makes n revolutions relative to the frame. Hence

$$2\pi n=\eta NT=\omega NT+\Theta\cos\lambda\int_0^{2\pi}\cos\psi\,(dt/d\psi)\,d\psi. \quad \text{............(32)}$$

When we use the value of $dt/d\psi$ given by (30), we find

$$\int_0^{2\pi}\cos\psi\,(dt/d\psi)\,d\psi=\pi U\Theta\cos\lambda/(Mgh-C\omega\Theta\sin\lambda), \quad \text{......(33)}$$

since

$$\int_0^{2\pi}\cos\psi\,d\psi=0, \quad \text{and} \quad \int_0^{2\pi}\cos^2\psi\,d\psi=\pi.$$

When we retain only the first power of Θ, we have, by (32) and (33), $2\pi n=\omega NT$, or $\omega=2\pi n/(NT)$. Thus, by (31), we have, in place of (19),

$$Mgh=\frac{4\pi^2Cn}{N^2T^2}\left\{1+\frac{NT\sin\lambda}{S}\right\}. \quad \text{...............(34)}$$

The greatest value of NT recorded in § 26 is

$$11{\cdot}514\times0{\cdot}71206=8{\cdot}199\text{ secs.}$$

We have $S=86{,}164$ secs., and at Cambridge $\lambda=52°\,13'$. Hence

$$NT\sin\lambda/S=7{\cdot}5\times10^{-5}.$$

The correction due to the earth's rotation, as given by (34), is therefore inappreciable.

In § 26, the rotation of the frame about the vertical was opposite to the component of the earth's rotation about the same line. In this case the correcting factor is $1-NT\sin\lambda/S$.

The precession of the Cavendish Laboratory gyroscope is so rapid in comparison with the earth's rotation that the latter has little effect. But in many important applications it is precisely the earth's rotation which makes the gyrostatic mechanism effective for its purpose.

CHAPTER II

THE STROBOSCOPE

EXPERIMENT 5. **Determination of frequency of alternating current.**

28. Introduction. The comparison of the frequency of rotation of a wheel or other rotating piece with the frequency of a vibrating body or of a periodically intermittent source of light is an operation of much importance in mechanical and electrical work. Very accurate synchronisation can be obtained under suitable conditions. The method depends upon the persistence of vision.

A rotating wheel bearing a suitable pattern and illuminated by periodically intermittent light is called a stroboscope—the word means the viewing of a rotating body.*

Let a wheel, centre O (Fig. 14), which bears a mark A, turn steadily in the direction of the arrow at the rate of f revolutions per second. Let the wheel be illuminated by intermittent light which gives flashes of very brief duration at regular intervals of $1/N$ seconds; the frequency of the flashes is N per second. The intermittent illumination may be furnished by a neon glow lamp (Osglim lamp) excited by an alternating current. In another method, two thin plates of metal, each with a slit in it, are fixed to the two prongs of an electrically maintained tuning fork, so that their planes are nearly coincident, in the manner indicated in Fig. 15. When the fork is not vibrating, the light from a properly placed source, which must be steady, can pass through both slits. When the fork vibrates with sufficient amplitude, the

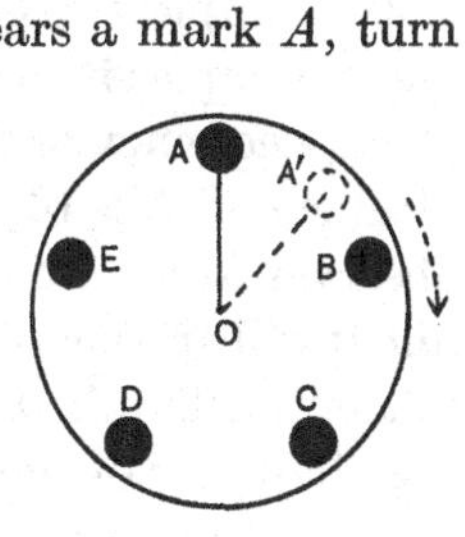

Fig. 14

* Many other methods of comparison, in which intermittent light is used, are, by an extension of the meaning, called stroboscopic. Thus the vibrating balance wheel of a watch may be illuminated by light which flashes periodically. If the frequency of the wheel equal that of the flashes, the wheel appears stationary; if the two frequences differ very slightly, the wheel will appear to vibrate very slowly. If the frequency of the flashes be very steady and be very accurately known, a test of the going of the watch can be made in a few minutes.

light is cut off except for brief intervals when the prongs are near their positions of rest. For each complete vibration of the fork, the light passes through the slits twice. It may be more convenient to illuminate the wheel by a *steady* source of light and to observe it through the slits. For very accurate work the fork is kept at a constant temperature.

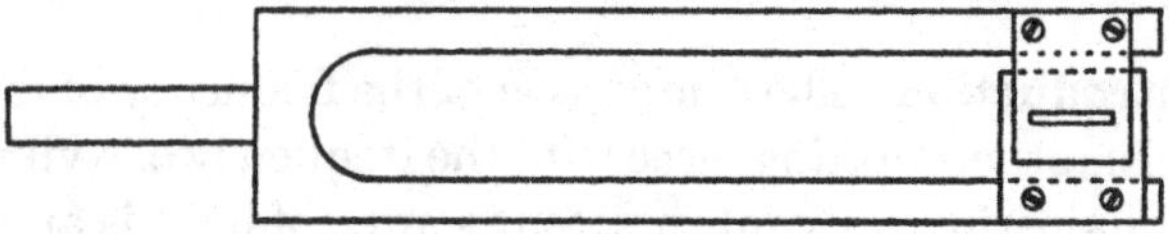

Fig. 15

If the mark be at A at one flash and at A' at the next flash, then AOA' or θ is the angle through which the wheel has turned in $1/N$ seconds. Since the wheel turns through $2\pi f$ radians per sec., we have

$$\theta = 2\pi f/N.$$

The simplest case occurs when $f=N$, for then $\theta = 2\pi$ and the mark occupies its original position when the second flash occurs. If the frequency of the illumination be sufficiently great, which is attained if $N>16$, the persistence of vision will prevent the observer from perceiving that the mark is not always at A, and to him it will appear as a fixed object which is steadily, and not intermittently, illuminated. The definition will, of course, depend upon the character of the illumination. If each flash be of such short duration that it may be described as "instantaneous," the mark will not move appreciably while the flash lasts and, consequently, will appear to be well defined. If, however, at each flash, the illumination last for an appreciable time t seconds, each point of the mark will appear to be drawn out into a circular arc subtending an angle $2\pi ft$ at the centre O. If the rise and fall of the illumination during the time t be gradual, the intensity will not appear uniform along the arc but will fade away towards either end. It will be seen, however, that, provided the duration of each flash be short compared with the interval between one flash and the next, the moderate extent of blurring is of no consequence.

When $f=N$, the mark is always seen in the same position, and the same is true for any other mark or any pattern on the wheel. If the speed of the wheel can be regulated so that $f=N$, we can find N by finding f, which may be done by the aid of suitable counting mechanism and a clock. If N be known, the method may be used to make f have this known value.

When f is not quite equal to N but is a little greater, so that $f=N+e$, where e is small compared with N, the mark will not make exactly one revolution between two successive flashes, but will move through an angle θ, where

$$\theta=2\pi f/N=2\pi(N+e)/N=2\pi+2\pi e/N,$$

and thus will appear to move forward through $2\pi e/N$ in $1/N$ sec.; it will, therefore, appear to make e or $f-N$ turns per sec. We can decide whether the frequency of rotation of the wheel is greater or less than that of the flashes, for the mark appears to rotate in the same direction as the wheel or in the opposite direction according as the wheel is turning too quickly or too slowly.

If $f=mN$, then $\theta=2m\pi$ and thus, when m is an integer, the mark appears stationary. If the wheel be turning a little too fast, so that $f=mN+e$, we have

$$\theta=2\pi(mN+e)/N=2\pi m+2\pi e/N,$$

and thus the mark will now appear to make e or $f-mN$ turns per sec. in the same direction as the wheel.

If the wheel rotate with $f=N/m$, where m is an integer, the mark will appear in m equally spaced positions round the wheel, but the resulting pattern will be weak, for, in each of these m positions, a mark will appear only once in an interval of m/N secs. or N/m times per sec. To make the pattern appear more strongly we must multiply the number of marks.

We will now suppose that there are m marks equally spaced. Let $A, B, C, D, \ldots$ be their positions at a certain flash (1) of the lamp (in Fig. 14, $m=5$). If the wheel rotate with $f=N/m$, then at the next flash (2), i.e. after $1/N$ secs., A will be where B was at flash (1), B will be where C was and so on. Thus, if the eye attend to the position in which A was at flash (1), it will see a mark there m times in m/N secs. or N times per sec. Hence, when $f=N/m$, the pattern—an "m-ring"—will appear much more vigorous than when there is only one mark, particularly when m is large.

If the wheel with m marks rotate with $f=pN/m$, where p is an integer, a strong m-ring will be seen. Hence, for a wheel with m marks, the values of f which give strong m-rings are

$$N/m,\quad 2N/m,\quad 3N/m,\ \dots\ (m-1)\,N/m,\quad N,\quad (m+1)\,N/m,\ \dots.$$

This series includes the values N, $2N$, $3N$, ..., which were found for a single mark, but it also includes other values.

If the wheel have several patterns of—say—m_1, m_2, ... marks, the stroboscopic effect can be obtained for a great number of values of f.

In the experiment the wheel makes less than N revs. per sec. and it is convenient to have m_1, m_2, ... considerable. Fig. 16 shows a stroboscopic disk with four rings having 24, 20, 18 and 16

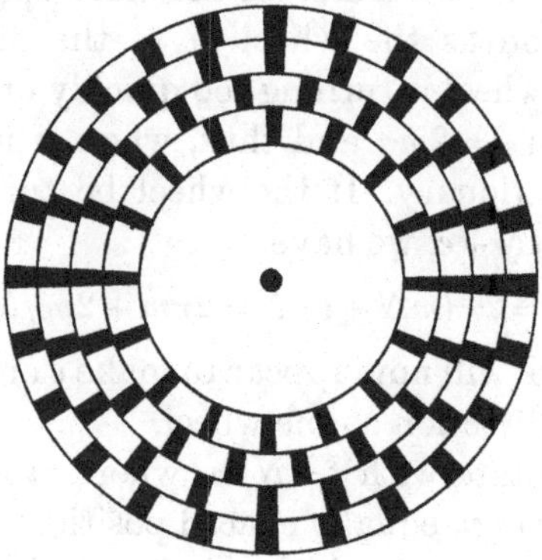

Fig. 16

marks respectively. The pattern may be drawn in Indian ink on a disk of white card which is attached concentrically to the wheel.

The Table gives some values of m/p for this disk. If the wheel make f revs. per sec., we can find N, the frequency of the lamp, for $N=fm/p$. If, on the other hand, N be known, we can find f, for $f=pN/m$. When m/p is not integral, its equivalent in decimals is given, so that the values may be arranged in a series of decreasing magnitude. The *exact* value of m/p, such as 20/3 or 24/5, should be used in the calculation.

In the Table an asterisk (*) in the column for a given value of m indicates that a ring with that number of marks will be seen. Thus, for $m/p=9$, a ring of 18 marks will be clearly seen. For

$m/p=8$, two rings, one of 24 and one of 16 marks, will appear. For $m/p=4$, three rings of 24, 20 and 16 marks will be seen. For these rings the values of p are 6, 5 and 4 respectively.

m/p	Values of m				m/p	Values of m			
	24	20	18	16		24	20	18	16
24	*				$\frac{16}{3}=5\cdot33\ldots$				*
20		*			5		*		
18			*		$\frac{24}{5}=4\cdot80$	*			
16				*	$\frac{9}{2}=4\cdot50$			*	
12	*				4	*	*		*
10		*			$\frac{18}{5}=3\cdot60$			*	
9			*		$\frac{24}{7}=3\cdot42\ldots$	*			
8	*			*	$\frac{10}{3}=3\cdot33\ldots$		*		
$\frac{20}{3}=6\cdot66\ldots$		*			$\frac{16}{5}=3\cdot20$				*
6	*		*		3	*		*	

29. Method. The stroboscopic disk may be attached to the shaft of a small electric motor. A large range of speeds may be obtained if a shunt-wound motor be used. Two suitable resistances, AB and BC, are joined in series and are connected to the direct current mains. The resistance BC is adjustable. The field magnet windings are joined to A and C. One armature brush is connected to A and one to a travelling contact D capable of moving from A to B. By varying the position of D on AB, the E.M.F. applied to the armature can be varied from zero to the maximum available, with a corresponding variation in the speed of the motor. For the lower speeds, the resistance BC may be large. When a suitable rheostat is used for AB, the speed is easily regulated. After a little practice it is easy to keep any desired ring of marks very nearly stationary.

Some counting mechanism is necessary. A toothed wheel of s teeth engages with a screw thread cut on a rod carried by the motor shaft and coaxial with it, and thus makes one turn to s turns of the shaft. It is convenient if s be large, say 100. A pin carried by the toothed wheel completes an electric circuit once in each turn of that wheel and thereby causes a lamp to light or a buzzer to sound.

If the toothed wheel revolve once in T secs., the shaft revolves

s times in T secs. or s/T times per sec., and hence $f=s/T$. Since $f=pN/m$, we have

$$N=\frac{fm}{p}=\frac{s\,(m/p)}{T}\,\text{sec.}^{-1} \qquad \text{..................(1)}$$

A neon lamp is mounted close to the disk. The lamp has two electrodes—one large, the other small. As the direction of the E.M.F. applied to the electrodes alternates, the electrodes light up alternately. The large electrode emits much more light than the small electrode, but the light from the latter is sufficient to give a weak stroboscopic effect. The lamp should be so placed that little light from the small electrode reaches the disk, for that light weakens the contrasts in the pattern seen by the light from the large electrode.

The disparity between the amounts of light emitted by the two electrodes is easily shown. Let the speed be adjusted so that, with the light from the large electrode, a pattern of m marks is seen stroboscopically as a strong m-ring. If, now, the lamp be turned so that more light from the small electrode reaches the disk, a set of m weak marks will be seen, each lying midway between two adjacent strong marks, provided that $p=1, 3, 5, \ldots$.

If a copper oxide rectifier be connected in series with the neon lamp, the current will pass through the rectifier in one direction only. If the lamp be so connected that, when the current flows through the rectifier and lamp, it is the large electrode which glows, the distinctness of the stroboscopic pattern will be somewhat improved, because the weak effect due to the glow of the smaller electrode will be suppressed.

For the lower speeds of the disk, the patterns will be strong and comparatively well defined. As the speed increases, the patterns become blurred and beyond a certain speed are hardly visible. This effect is due to the fact that each "flash" emitted by the lamp is not instantaneous but lasts for a time which is not very small compared with the interval between successive flashes.

Two observers are desirable. One regulates the motor, a task calling for all his attention. The other determines by a timepiece the time nT secs. occupied by a known number, n, of turns of the toothed wheel. Accuracy cannot be expected unless nT be great compared with the errors of reading the timepiece. To gain acquaintance with the method, the students should make observations, each lasting two or three minutes, at various speeds of the disk so as to use the different patterns. When they have gained skill, they may make a careful determination of the frequency using one particular ring of marks and giving ten minutes or more to the observations. A convenient method is to observe the times at which signals 0, 1, 2, ... n, $n+1$, ... $2n-1$ occur and to subtract the time of signal 0 from that of signal n, the time of 1 from that of $n+1$, and so on; the process gives n independent observations of nT in time $(2n-1)T$, where T is the interval between one signal and the next. (See § 31.)

Unless the time-piece be known to indicate true seconds, it should be compared, over a considerable interval, with a reliable clock. (See *Experimental Harmonic Motion*, § 27.)

A ring of m marks will give rise to a stationary m-ring when the disk makes f revs. per sec., where $f = pN/m$ and p is integral. Let the speed be gradually increased from zero. Then, for the first m-ring which appears, $p = 1$, for the second ring, $p = 2$, and so on.

A pattern may be seen when p is less than unity, but it will be one of more than m marks. Thus if $p = \frac{1}{3}$, a $3m$-ring will be seen, and, for a greater speed, when $p = \frac{1}{2}$, a $2m$-ring will appear. Thus, before f reaches N/m, rings of ... $4m$, $3m$, $2m$ marks will have appeared and disappeared.

A *gas-filled* tungsten filament lamp may be used in place of a neon lamp. The filament of the tungsten lamp, when supplied with alternating current of 50 cycles per second, varies periodically in temperature and in luminosity. The frequency of the maxima of brightness is 100 per second. When a *gas-filled* tungsten lamp is used, the contrast between light and shade in the stroboscopic pattern, though less pronounced than when a neon lamp or a fork is used, is still obvious. If a disk of 77 marks, such as is supplied for use with gramophones, be viewed in the light of a tungsten lamp, the stroboscopic effect is plainly visible. With a current of 50 cycles per second, the pattern is stationary when the disk makes 6000/77 or 77·922... revolutions per minute.

Mr Milton Metfessel* has found that stroboscopic effects can be obtained with *steady* illumination, such as daylight, if the eye

* *Science* (New York, U.S.A.), Vol. 78, p. 416 (1933).

be caused to oscillate by the action of a vibrating body upon the head of the observer. If a small electromagnet be provided with an iron armature just out of contact with the poles, the magnet will vibrate vigorously when the alternating current passes. If the magnet be pressed against the observer's cheek-bone, the stroboscopic effect is easily obtained.

The eye is so sensitive to the vibrations of the head that, if the disk have a suitable number of marks, the stroboscopic effect may be seen if the observer merely sing the appropriate musical note.

30. Accuracy of standard frequency. In many parts of England the alternating current supply is regulated to 50 cycles per second with considerable accuracy. This condition is necessary in order that the various power stations connected to the "grid" system may be capable of inter-connexion. The method of regulation is roughly as follows. A synchronous motor driven by the alternating current is connected by gearing to hands moving round a clock dial. The number of revolutions which the motor makes per second bears some simple and *constant* numerical relation to the number of cycles per second made by the alternating current. When the frequency of the current is exactly 50 cycles per sec., the "seconds" hand actuated by the synchronous motor makes exactly one revolution per minute. The power station is provided with a first-rate clock which is accurately regulated to Greenwich mean time. If the hands of the motor clock be suitably adjusted, the two clocks will always show the same time, if the frequency be accurately 50 cycles per sec. There are little discrepancies due to the fact that the seconds hand of an ordinary Greenwich time clock moves on in steps at intervals of one second, whereas the seconds hand of the motor clock moves continuously. But mechanism has been devised by which the seconds hand of the Greenwich time clock is caused to move uniformly. If the motor clock be found to be losing relative to the Greenwich time clock, the supply of steam to the turbines driving the alternating current dynamos is slightly increased. The power taken from the station varies, of course, from time to time and, consequently, the speed of the alternators is liable to more or less sudden changes. Since the sudden variations of load are small compared with the average load, the changes of speed are, as a rule, small; they are nearly corrected by automatic regulating devices. The careful and constant hand-regulation of the turbines at the station ensures that the "motor time" does not differ from Greenwich time over any interval by more than a second or two, however great the interval may be, and thus motor clocks driven by the alternating current are satisfactory for ordinary household or office purposes unless the current supply fail.

Some small irregularities may occur in the results obtained with the stroboscope, but the average of a number of determinations of the frequency will probably differ very little from 50 per sec., if the supply be connected to the "grid" system.

31. Practical example.

G. F. C. Searle and C. G. Tilley used disk with rings of 24, 20, 18, 16 marks. The toothed wheel had 100 teeth and so $s = 100$. They used a neon lamp and measured frequency (N) of Cambridge alternating current supply, which, at the date of the experiment, was about 90 sec.$^{-1}$. For lowest speed for which a 24-ring was visible, observations were as follows:

Signal	Time		Signal	Time		$10T$	
	mins.	secs.		mins.	secs.	mins.	secs.
0	0	45·4	10	5	12·5	4	27·1
1	1	12·5	11	5	39·4	4	26·9
2	1	39·7	12	6	5·8	4	26·1
3	2	6·4	13	6	32·6	4	26·2
4	2	33·2	14	6	59·2	4	26·0
5	2	59·8	15	7	26·3	4	26·5
6	3	26·4	16	7	53·2	4	26·8
7	3	52·7	17	8	20·0	4	27·3
8	4	19·4	18	8	46·4	4	27·0
9	4	46·2	19	9	12·8	4	26·6

Mean value of $10T = 4$ mins. 26·65 secs. $=$ 266·65 secs. and $T = 26{\cdot}665$ secs. Since $p = 1$ and $m = 24$, we have, by (1),

$$N = s\,(m/p)/T = 100 \times 24/26{\cdot}665 = 90{\cdot}006 \text{ sec.}^{-1}.$$

This result and others obtained with circles of 20, 18 and 16 marks are given in the following Table:

m	p	m/p	T	N
24	1	24	26·665 secs.	90·006 sec.$^{-1}$
20	1	20	22·299	89·690
18	1	18	19·962	90·171
16	1	16	17·827	89·752

The frequency N was known from other experiments, e.g. EXPERIMENT 7, to vary from time to time in consequence of changes in the load on the machines at the power station.

CHAPTER III

EXPERIMENTS IN ELASTICITY

32. Transverse vibrations of a thin rod. If a uniform rod of circular section be clamped at one end and be set into transverse vibration, the frequency n will depend upon the radius r, the free length l of the rod, the density ρ and the Young's modulus E for the material, in a manner which we can discover by aid of the method of "dimensions," when r is small compared with l. In this case, any expression such as $(l^p+r^p)^q$ can be replaced with sufficient accuracy by l^{pq}, if $p>1$. We shall, therefore, assume that

$$n=\Omega r^x l^y \rho^z E^u, \quad \text{......................(1)}$$

where Ω is a numerical constant depending upon the number of nodes on the vibrating rod. Since frequency is of dimensions T^{-1} and Young's modulus, being force/area, has dimensions $ML^{-1}T^{-2}$, the method shows that

$$T^{-1}=L^x L^y (ML^{-3})^z (ML^{-1}T^{-2})^u = M^{z+u} L^{x+y-3z-u} T^{-2u}.$$

Hence $\qquad u=\tfrac{1}{2}, \qquad z=-u=-\tfrac{1}{2}, \qquad x+y=-1. \quad$ ......(2)

When the bar vibrates, it is bent and the bending is resisted by the elasticity. When the radius of curvature of the axis of the rod is R, the moment, due to the elastic forces, tending to straighten the rod is $\tfrac{1}{4}\pi E r^4/R$,* and hence E and r will appear together in the form Er^4. The mass of the rod per unit length is $\pi\rho r^2$, and thus ρ and r will appear together in the form ρr^2. To the approximation adopted above, r can enter only through these two associations. Hence we have

$$n=\Omega l^y \sqrt{\left(\frac{Er^4}{\rho r^2}\right)}=\Omega l^y r \sqrt{\left(\frac{E}{\rho}\right)}.$$

Thus, by (1), $x=1$ and therefore, by (2), $y=-2$. The formula then becomes

$$n=\Omega\frac{r}{l^2}\sqrt{\left(\frac{E}{\rho}\right)}. \quad \text{......................(3)}$$

* § 56. See also *Experimental Elasticity*, §§ 27 to 34.

Hence, for a given rod, so long as we keep to the same number of nodes, the frequency is inversely proportional to l^2.

The complete theory shows that $\Omega = m^2/4\pi$, where m is a root of the equation

$$\cosh m \,.\, \cos m = -1.$$

For no node, $m = 1{\cdot}8751$, and for one node, $m = 4{\cdot}6941$.

33. Mathematical theory of vibrations of a rod. We consider a rod OA (Fig. 17) of circular section, of radius r and of free length l. Young's modulus for the material is E and the density is ρ. The rod is held at one end in a massive clamp. The point O in which the plane of the face of the clamp cuts the axis of the rod is taken as origin. The axis of x coincides with the axis of the unbent rod and Oy is perpendicular to Ox. The vibrations take place in the plane Oxy.

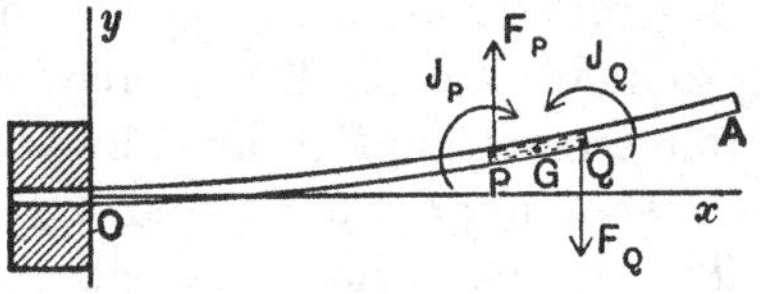

Fig. 17

Let F be the shearing force at G, the centre of gravity of the element PQ, and let J be the bending moment. We count F as positive when it acts in the positive direction of y upon the part of the rod which lies on the positive side, with regard to x, of the section through G, and J as positive when, in the case of Fig. 17, it tends to make the part of the rod which lies on the positive side of G rotate in the direction of the hands of a watch. The angle between the axis of the rod at G and the axis of x is always so small that we may replace it by its tangent dy/dx, and may equate the length of PQ to dx, its projection on Ox.

If F_P, F_Q be the values of F at P and Q, where the abscissae are $x - \frac{1}{2}dx$ and $x + \frac{1}{2}dx$,

$$F_P = F - \tfrac{1}{2}(dF/dx)\,dx, \qquad F_Q = F + \tfrac{1}{2}(dF/dx)\,dx.$$

Similarly

$$J_P = J - \tfrac{1}{2}(dJ/dx)\,dx, \qquad J_Q = J + \tfrac{1}{2}(dJ/dx)\,dx.$$

The mass of PQ is $\pi\rho r^2 dx$ and the product of this mass and the acceleration of G, its centre of gravity, parallel to Oy equals the force acting on it in that direction. Hence

$$\pi\rho r^2 dx\,(d^2y/dt^2) = F_P - F_Q = -(dF/dx)\,dx$$

or

$$\pi\rho r^2 . d^2y/dt^2 = -dF/dx. \quad \text{.................(4)}$$

If θ be the angle between PQ and Ox, the angular acceleration of PQ about the axis of bending through G perpendicular to Oxy, which is the plane of bending, is $d^2\theta/dt^2$; to our approximation $\theta = dy/dx$. Hence

$$\frac{d^2\theta}{dt^2} = \frac{d^3y}{dt^2dx}. \quad \text{......................(5)}$$

For the purpose of finding dK, the moment of inertia of PQ about the axis of bending through G, it is sufficiently accurate to treat PQ as straight. Thus

$$dK = \pi\rho r^2 dx\,\{\tfrac{1}{4}r^2 + \tfrac{1}{12}(dx)^2\}.$$

The product $dK . d^2\theta/dt^2$ equals the moment about the axis of bending, tending to increase θ, of all the forces which act on the element. The portion OP acts on PQ partly by a bending moment J_P and partly by a shearing force F_P, and AQ acts on PQ in a similar manner. The moment about the axis through G of the force F_P is $-\frac{1}{2}F_P dx$. By the principles of dynamics, the angular acceleration of the element about an axis through G, its centre of gravity, is independent of the velocity or acceleration of G. Hence we have

$$dK . d^2\theta/dt^2 = -J_P + J_Q - \tfrac{1}{2}(F_P + F_Q)\,dx,$$

or

$$\pi\rho r^2\{\tfrac{1}{4}r^2 + \tfrac{1}{12}(dx)^2\}\,dx . d^2\theta/dt^2 = (dJ/dx)\,dx - F\,dx.$$

In the limit we have, by (5),

$$\tfrac{1}{4}\pi\rho r^4 \frac{d^3y}{dt^2dx} = \frac{dJ}{dx} - F. \quad \text{...............(6)}$$

By § 56, $J = EI/R$, where E is Young's modulus, $I = \frac{1}{4}\pi r^4$ is the "moment of inertia" of the cross-section of the rod through G about the axis of bending at G, and R^{-1} is the curvature of the axis of the rod at G. When we neglect $(dy/dx)^2$ in comparison with unity, $R^{-1} = d^2y/dx^2$, and then

$$J = E . \tfrac{1}{4}\pi r^4 d^2y/dx^2. \quad \text{...................(7)}$$

If we differentiate (6) with respect to x and use (4), we have

$$\tfrac{1}{4}\pi\rho r^4 \frac{d^4y}{dt^2dx^2} = \frac{d^2J}{dx^2} - \frac{dF}{dx} = E . \tfrac{1}{4}\pi r^4 \frac{d^4y}{dx^4} + \pi\rho r^2 \frac{d^2y}{dt^2}. \quad \text{...(8)}$$

When the motion of each element is simple harmonic, with periodic time $2\pi/p$, $d^2y/dt^2 = -p^2y$ and $d^4y/dt^2dx^2 = -p^2d^2y/dx^2$. Thus (8) becomes

$$\frac{E}{\rho}\frac{d^4y}{dx^4} + p^2\frac{d^2y}{dx^2} - \frac{4p^2y}{r^2} = 0. \qquad (9)$$

Now $y = Ae^{qx}$ will be a solution of (9), and $y = Ae^{qx}\sin pt$ will be a solution of (8), if

$$Eq^4/\rho + p^2q^2 - 4p^2/r^2 = 0,$$

or if

$$q^2 = \{-1 \pm \sqrt{(H^2+1)}\}\rho p^2/2E, \qquad (10)$$

where $H^2 = 16E/\rho p^2 r^2$. When p^2 is small* compared with $16E/\rho r^2$, we can neglect unity in the square root in comparison with H^2. Then

$$q^2 = \{\pm H - 1\}\rho p^2/2E. \qquad (11)$$

We may now neglect unity in comparison with H, and then

$$q^2 = \pm \tfrac{1}{2}H\rho p^2/E = \pm f^2, \qquad (12)$$

where

$$f^2 = \frac{2p}{r}\sqrt{\left(\frac{\rho}{E}\right)}. \qquad (13)$$

Hence

$$q = \pm f, \quad \text{or} \quad q = \pm f\sqrt{(-1)} = \pm if. \qquad (14)$$

If we neglect, from the beginning, the moment about the axis of bending of the forces which give the element angular acceleration, we have, by (6),

$$F = dJ/dx. \qquad (15)$$

Then, since $d^2J/dx^2 = dF/dx$, (8) becomes

$$(E/\rho)\,d^4y/dx^4 + 4r^{-2}d^2y/dt^2 = 0. \qquad (16)$$

Since the motion is harmonic with period $2\pi/p$, $d^2y/dt^2 = -p^2y$, and hence, by (16),

$$(E/\rho)\,d^4y/dx^4 - 4p^2y/r^2 = 0. \qquad (17)$$

If $y = Ae^{qx}$ be a solution of (17), $q^4 = 4p^2\rho/r^2E$, and thus q^2 has the value given by (12). We may, therefore, under appropriate conditions, take (16) as the differential equation satisfied by the vibrating rod.

In the experiment of § 39, the frequency $p/2\pi$ of the vibrating rod was always less than 1000 vibrations per second. Hence $p < 2000\pi$ and $p^2 < 4 \times 10^7$. For the rod $E = 2{\cdot}59 \times 10^{12}$, $\rho = 7{\cdot}8$,

* The velocity, v, of longitudinal waves in the material is $\sqrt{(E/\rho)}$. When p^2 is small compared with $16E/\rho r^2$, the periodic time $2\pi/p$ is large compared with $\pi r/2v$.

$r = 0{\cdot}127$ cm. Thus, since $H^2 = 16E/\rho p^2 r^2$, $H^2 > 8{\cdot}19 \times 10^6$ and $H > 2860$. Hence the four values of q given by (14) may be used in place of the four values

$$q = \pm f\sqrt{(1 - 1/H)}, \qquad q = \pm if\sqrt{(1 + 1/H)}, \quad \text{......(18)}$$

which follow from (11), since $\rho p^2/2E = f^2/H$.

The solution of (16) is

$$y = \{A' e^{fx} + B' e^{-fx} + C' e^{ifx} + D' e^{-ifx}\} \sin pt,$$

or, changing the constants, and using hyperbolic functions,

$$y = \{A \cosh fx + B \sinh fx + C \cos fx + D \sin fx\} \sin pt. \quad \text{...(19)}$$

In order to find the constants B, C, D in terms of A, we take account of the conditions at the ends of the rod. When $x = 0$, i.e. at the clamped end of the rod, $y = 0$ and $dy/dx = 0$ for all values of t. Since $y = 0$ at $x = 0$, $A + C = 0$, and since $dy/dx = 0$ at $x = 0$, $B + D = 0$. At the free end, where $x = l$, both the bending moment J and the shearing force F vanish; by (15), $F = dJ/dx$. Hence, by (7), both $d^2y/dx^2 = 0$ and $d^3y/dx^3 = 0$, when $x = l$. If we write $fl = m$, we find from (19), since $C = -A$ and $D = -B$,

$$A(\cosh m + \cos m) + B(\sinh m + \sin m) = 0, \quad \text{......(20)}$$

$$A(\sinh m - \sin m) + B(\cosh m + \cos m) = 0. \quad \text{......(21)}$$

Eliminating the ratio B/A, we find

$$(\cosh m + \cos m)^2 = \sinh^2 m - \sin^2 m,$$

or

$$\cosh m \, . \cos m = -1. \quad \text{..................(22)}$$

Here $m^2 = f^2 l^2 = (2pl^2/r)\sqrt{(\rho/E)}$, and thus we find for the frequency n

$$n = \frac{p}{2\pi} = m^2 \frac{r}{4\pi l^2} \sqrt{\left(\frac{E}{\rho}\right)}. \quad \text{..................(23)}$$

We can therefore calculate the frequencies for the different modes of vibration when we know the values of m which satisfy (22).

When a value of m satisfying (22) has been obtained, we find, by (20), in terms of that root

$$B/A = -(\cosh m + \cos m)/(\sinh m + \sin m) = M. \quad \text{...(24)}$$

Then, since $C = -A$, $D = -B$, and $f = m/l$, (19) becomes

$$y = A\left\{\cosh\frac{mx}{l} - \cos\frac{mx}{l} + M\left(\sinh\frac{mx}{l} - \sin\frac{mx}{l}\right)\right\} \sin pt, \quad \text{......(25)}$$

which gives the form of the rod at any time. The positions of the nodes are given by those values of x, other than $x=0$, which make $y=0$.

34. Roots of cosh m . cos $m=-1$. A careful drawing, such as Fig. 18, will give the values of m with some accuracy. The curve, starting at $m=0$, $z=1$ and representing

$$z=(\cosh m)^{-1}=2/(e^m+e^{-m}),$$

approaches the axis Om very closely as m increases. The wave curve is $z=-\cos m$. Here m is, of course, measured in radians and the first maximum of $z=-\cos m$ is at $m=\pi=3{\cdot}14159$. The roots of (22) are the values of m corresponding to the points of intersection of the curves. It will be seen that the first root is about 1·87 and

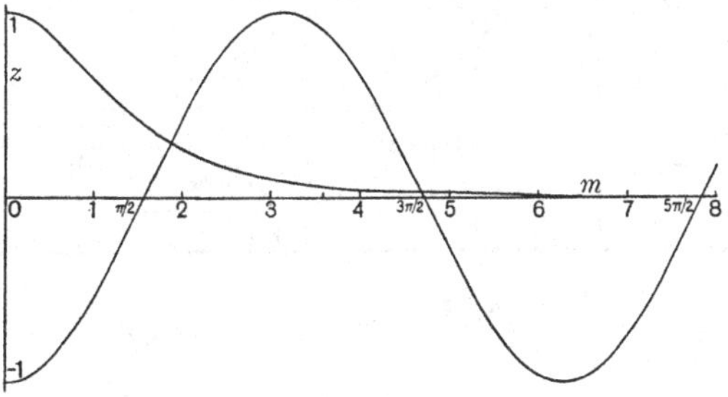

Fig. 18

that the succeeding roots are very nearly $\frac{3}{2}\pi$, $\frac{5}{2}\pi$, ... The roots can be found by appropriate mathematical methods to any desired accuracy.

The successive values of m, viz. m_0, m_1, m_2, ..., correspond to no node, one node, two nodes, ... The Table gives m_0, m_1, ... to four decimal places. The values of m_3, m_4, ... do not, to this accuracy, differ from odd integral multiples of $\frac{1}{2}\pi$.

$m_0=\frac{1}{2}\pi+0{\cdot}3043=1{\cdot}1937\times\pi/2=1{\cdot}8751$	$m_3=\frac{7}{2}\pi$	$m_6=\frac{13}{2}\pi$
$m_1=\frac{3}{2}\pi-0{\cdot}0183=2{\cdot}9884\times\pi/2=4{\cdot}6941$	$m_4=\frac{9}{2}\pi$	$m_7=\frac{15}{2}\pi$
$m_2=\frac{5}{2}\pi+0{\cdot}0008=5{\cdot}0005\times\pi/2=7{\cdot}8548$	$m_5=\frac{11}{2}\pi$	$m_8=\frac{17}{2}\pi$

When the rod, of length l, vibrates with one node, the distance of the node from the free end is $0{\cdot}2165l$, as may be found from (25) by a method of "trial and error"; the calculation is necessarily lengthy. Better mathematical methods of finding the positions of the nodes are available. Seebeck, who first investigated the matter, gave 0·2261*l*, and his erroneous result has been copied by many subsequent writers.

EXPERIMENT 6. **Determination of Young's modulus by vibrations of a rod.**

35. Introduction. A thin steel rod *AB* (Fig. 19), assumed uniform, is held in a firm clamp *C* so that the part *OA* projects. A rod of "silver" steel, about 0·25 cm. in diameter, is suitable for the experiment. The coil *G* and the electromagnet *M* are used in EXPERIMENT 7 and are removed for the present experiment.

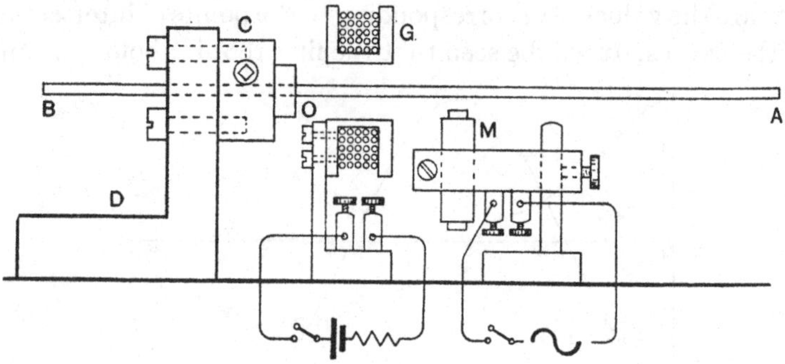

Fig. 19

A self-centering three-jaw lathe chuck forms a convenient clamp and allows the length *OA* to be easily adjusted. The chuck may be fixed to a massive support by bolts screwing into the holes drilled and tapped in the chuck by the makers. A lathe chuck is strongly constructed and will not be damaged if a considerable effort be used in clamping the rod.

If *n* be the frequency of transverse vibrations of a uniform rod of circular section, clamped at one end, as in Fig. 19, we have, by (23), § 33,

$$n = m^2 \frac{r}{4\pi l^2} \sqrt{\left(\frac{E}{\rho}\right)}, \quad \text{....................(1)}$$

where r is the radius of the cross-section, l is the length of OA, E is Young's modulus, ρ is the density and the number m is a root of the equation

$$\cosh m \,.\, \cos m = -1. \quad \text{....................(2)}$$

The values of m, viz. $m_0, m_1, m_2, \ldots$ corresponding to 0, 1, 2, ... nodes, are given in § 34. We have

$$m_0 = 1{\cdot}8751, \qquad m_1 = 4{\cdot}6941.$$

Equation (1) shows that, for a given rod and a given number of nodes, the product nl^2 is constant. Thus, if we adjust l so that n has definite values given by tuning forks of known frequencies, we shall obtain a series of values of nl^2 which, by theory, should be constant. On account of slight imperfections in the clamping, the effective length of the rod is a little greater than the observed free length. The necessary correction is found by plotting l against $1/\sqrt{n}$. When the effective value of l has been found for a known frequency, Young's modulus can be calculated by (1) when r and ρ are known.

36. Calibration by tuning forks. The rod AB (Fig. 19) should be as nearly straight as possible. The end A should be at right angles to the length of the rod and B should be distinguished from A by a file mark at B. The rod is made to vibrate by plucking, or, better, by a light blow with a "hammer" formed by fixing a piece of thick-walled rubber tube about 2 cm. long to one end of a stout aluminium wire about 10 cm. long. The tuning fork and the rod are struck with the rubber-headed hammer and the pitch of the rod is compared with that of the fork. If the pitch of the rod be higher than that of the fork, the free length of the rod is increased. The rod is *firmly* clamped for each test. With a little care, the rod can be tuned to give the octave below the fork and the octave above the fork as well as the fundamental note of the fork. The observations are made with a number of forks. The following Table, covering three octaves, will facilitate the calculations.

Note	n	$1/\sqrt{n}$	n	$1/\sqrt{n}$	n	$1/\sqrt{n}$
C	128	0·08839	256	0·06250	512	0·04420
D	144	0·08333	288	0·05892	576	0·04167
E	160	0·07906	320	0·05590	640	0·03953
F	170·67	0·07655	341·33	0·05413	682·67	0·03827
G	192	0·07217	384	0·05103	768	0·03608
A	213·33	0·06847	426·67	0·04841	853·33	0·03423
B	240	0·06455	480	0·04564	960	0·03228
C	256	0·06250	512	0·04420	1024	0·03125

According to the simple theory, l should be proportional to $1/\sqrt{n}$, for a given number of nodes. But there is some yielding of the rod in the clamp itself, and thus the effective length is somewhat greater than the distance, l, from the end A of the rod to the face of the clamp. When l is plotted against $1/\sqrt{n}$, as in Fig. 20, the points will lie on a straight line PR not passing exactly through the origin but cutting Ol at R on the negative side. The positive constants a, b in the linear relation

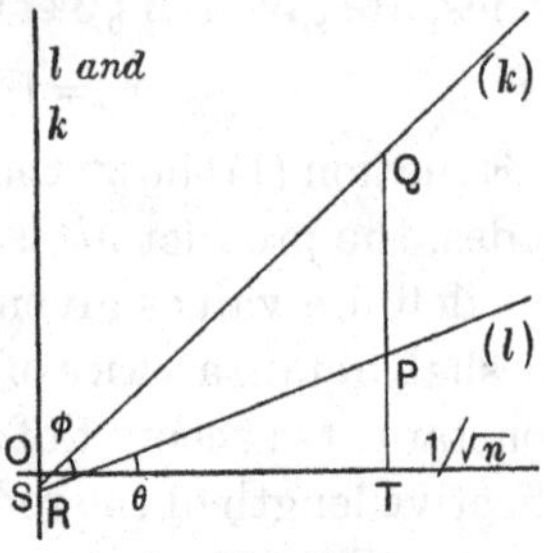

Fig. 20

$$l = -a + b/\sqrt{n}, \quad \text{.......................(3)}$$

between l and $1/\sqrt{n}$, can be obtained from the straight line. If l' be the effective length

$$l' = l + a. \quad \text{..........................(4)}$$

An average value of nl'^2 can be found from the slope of the line, for

$$nl'^2 = b^2. \quad \text{..........................(5)}$$

If the rod be struck with the little hammer, it will give out a note due to vibration with one node, in addition to the lower note due to vibration with no node. By adjusting the length of OA, we can tune the rod so that the higher of the two notes is in unison with the fundamental note of the fork. With care, the rod, vibrating with one node, can be tuned also to the octave above the fork. For forks of higher frequency, a little practice may be needed before the observer can distinguish the note corresponding to one node in the presence of the note corresponding to no node.

If, at the time when the rod is struck, it be lightly touched at a point H, where AH is about $\frac{1}{5}AO$, the amplitude of the vibration with no node will be much diminished. With a little skill, it is possible to leave the rod vibrating with the node clearly visible.

Let the length of OA be now denoted by k. When k is plotted against $1/\sqrt{n}$, as in Fig. 20, the points will lie on a straight line QS not passing exactly through the origin. The (positive) constants c, d in the equation

$$k = -c + d/\sqrt{n} \quad \text{......................(6)}$$

can be obtained from the line. If k' be the effective length,

$$k' = k + c. \quad \text{.........................(7)}$$

An average value of nk'^2 can be found from the slope of the line, for

$$nk'^2 = d^2. \quad \text{.........................(8)}$$

The method depends upon the rod being uniform in section, density and elasticity. If the rod, when tuned to the various forks, be not always clamped with the same firmness, irregular variations in the products nl^2 and nk^2 are, of course, to be expected.

37. Determination of Young's modulus. The diameter $2r$ of the rod is measured in several places and the mean is found. The mass M and the length L of the whole rod AB are found. Then the density $\rho = M/(\pi r^2 L)$.

An average value of the product nl'^2, where l' is the effective length, is found from a series of observations when the rod vibrates with no node. Then, by (1), we have, for Young's modulus,

$$E = \rho\{(4\pi nl'^2)/(m_0{}^2 r)\}^2, \quad \text{.................(9)}$$

where $m_0 = 1{\cdot}8751$.

An average value for the product nk'^2 is found from a series of observations of n and k when the rod vibrates with one node. Then

$$E = \rho\{(4\pi nk'^2)/(m_1{}^2 r)\}^2, \quad \text{.................(10)}$$

where $m_1 = 4{\cdot}6941$.

38. Ratio of m_1 to m_0. For a given frequency, the length OA is, by (1), proportional to m. Thus we can determine experimentally the ratio of m_1 to m_0 by finding the lengths sounding a given note, first with one node (k) and then with no node (l). Accuracy will be gained if we find the ratio of k' to l', where k' and l' are the *effective* lengths in the two cases. By (5) and (8), using effective lengths, we have

$$\frac{m_1}{m_0} = \frac{k'}{l'} = \frac{d}{b} = \frac{\tan\phi}{\tan\theta}, \quad \text{....................(11)}$$

where ϕ and θ are the angles which the straight lines QS, PR (Fig. 20) make with the axis of $1/\sqrt{n}$ when k and l are plotted on the same scale against $1/\sqrt{n}$. The value of k'/l' may be compared with the theoretical value of m_1/m_0, viz. $4{\cdot}6941/1{\cdot}8751$ or $2{\cdot}503$.

39. Practical example.

The diameter ($2r$) of the "silver" steel rod was 0·2538 cm., its whole length 40·0 cm., and its mass 15·87 grm. Hence density $=\rho=7\cdot842$ grm. cm.$^{-3}$.

Results with no node. Four forks C, D, E, F were used; the frequency of C was 256. The rod was tuned to unison with these forks and to the octaves below them.

Fork	n	$1/\sqrt{n}$	l	n	$1/\sqrt{n}$	l
			cm.			cm.
C	128	0·08839	12·20	256	0·06250	8·33
D	144	0·08333	11·26	288	0·05892	7·85
E	160	0·07906	10·61	320	0·05590	7·41
F	170·67	0·07655	10·36	341·33	0·05413	7·15
Sums		0·32733	44·43		0·23145	30·74

When l was plotted against $1/\sqrt{n}$, the points lay approximately on a straight line. Awbery's method (Note I) was used to find the best values for the constants a and b in the equation $y=-a+bx$, where $x=1/\sqrt{n}, y=l$. If X, Y be the centroid of the eight points,

$$X=\tfrac{1}{8}(0\cdot32733+0\cdot23145)=0\cdot06985, \qquad Y=\tfrac{1}{8}(44\cdot43+30\cdot74)=9\cdot3962.$$

This point separates the eight points into the two groups in the Table. If we put

$$p=\tfrac{1}{4}\times0\cdot32733=0\cdot08183, \qquad q=\tfrac{1}{4}\times44\cdot43=11\cdot107,$$

$$u=\tfrac{1}{4}\times0\cdot23145=0\cdot05786, \qquad v=\tfrac{1}{4}\times30\cdot74=7\cdot685,$$

then

$$b=(q-v)/(p-u)=3\cdot422/0\cdot02397=142\cdot8,$$

$$a=bX-Y=142\cdot8\times0\cdot06985-9\cdot3962=0\cdot58.$$

The average value of nl'^2 yielded by these observations is b^2 or $142\cdot8^2$. Hence, by (9), we have for Young's modulus

$$E=\rho\{(4\pi nl'^2)/(m_0{}^2r)\}^2=7\cdot842\{4\pi\times142\cdot8^2/(1\cdot8751^2\times0\cdot1269)\}^2$$

$$=2\cdot59\times10^{12}\,\text{dyne cm.}^{-2}.$$

Results with one node. The rod was tuned to unison with eight forks and also to the octaves above four of the forks.

Fork	n	$1/\sqrt{n}$	k	Fork	n	$1/\sqrt{n}$	k
			cm.				cm.
C	256	0·06250	21·30	B	480	0·04564	15·38
D	288	0·05892	19·78	C	512	0·04420	14·69
E	320	0·05590	18·80	D	576	0·04167	13·99
F	341·33	0·05413	18·21	E	640	0·03953	13·62
G	384	0·05103	17·49	F	682·67	0·03827	13·27
A	426·67	0·04841	16·27	G	768	0·03608	12·07

When k, the length of OA in this case, was plotted against $1/\sqrt{n}$, there were numerous irregularities, but the straight line which cut the axis of k at $k=-0\cdot40$ cm. and had $k=23\cdot80$ for $1/\sqrt{n}=0\cdot07$ lay fairly evenly among the plotted points. For the constants c, d in (6), we have

$$c=0\cdot40, \qquad d=(23\cdot80+0\cdot40)/0\cdot07=345\cdot7.$$

The average value of nk'^2 is d^2 or $345\cdot7^2$. Hence, by (10),

$$E=7\cdot842\,\{4\pi\times345\cdot7^2/(4\cdot6941^2\times0\cdot1269)\}^2=2\cdot26\times10^{12}\,\text{dyne cm.}^{-2}.$$

Ratio of m_1 to m_0. The average values of nl'^2 for no node and of nk'^2 for one node are $142\cdot8^2$ and $345\cdot7^2$. Thus, for a given frequency,

$$k'/l'=345\cdot7/142\cdot8=2\cdot421.$$

The theoretical value is $m_1/m_0=4\cdot6941/1\cdot8751=2\cdot503$.

Some of the discrepancies in the results may be due to imperfect clamping by the chuck, which had been discarded as unfit for further use in the workshop.

EXPERIMENT 7. Determination of frequency of alternating current by vibrations of a rod.*

40. Introduction. One method of measuring frequency depends upon resonance. If a body of unit mass, subject to elastic constraints and experiencing a damping resistance proportional to its speed, be acted on by a periodic driving force $F\cos pt$, the equation of motion may be written

$$\frac{d^2z}{dt^2}+2s\frac{dz}{dt}+q^2z=F\cos pt.$$

The frequency of the undriven ($F=0$) and undamped ($s=0$) vibration is $q/2\pi$. If the motion be damped but undriven ($F=0$), there is, strictly, no frequency. The interval between alternate zero values of the displacement z is, however, constant and we may, for convenience, call its reciprocal the "frequency" of the damped vibration. This frequency is not $q/2\pi$ but $(q^2-s^2)^{\frac{1}{2}}/2\pi$.

If A be the amplitude of the steady vibration ultimately set up by the driving force,

$$A=F\{(q^2-p^2)^2+4s^2p^2\}^{-\frac{1}{2}}.$$

For given values of s and q, the maximum value, A_0, of A occurs when $p^2=q^2-2s^2$, and

$$A_0=\tfrac{1}{2}F\,\{s^2\,(q^2-s^2)\}^{-\frac{1}{2}}.$$

The corresponding frequency is $(q^2-2s^2)^{\frac{1}{2}}/2\pi$.

* "The determination of the frequency of an alternating current supply by the vibrations of rods," *Proc. Camb. Phil. Soc.* Vol. XXII, p. 539 (1925).

If the undriven motion be only slightly damped, s^2 is very small compared with q^2, and little error will be made if the value of p giving maximum amplitude be taken as equal to q. When s/q is small, the resonance is "sharp," i.e. a small discrepancy between p and q, or, more strictly, between p and $(q^2-2s^2)^{\frac{1}{2}}$, causes the amplitude to fall far below its maximum value. Thus, when the damping is small, maximum amplitude, for variations of p, indicates that p is very nearly equal to q.

In the following experiments, resonance is used in measuring the frequency of an alternating current.

41. Method. The vibrating body—a steel rod clamped at one end—is acted on magnetically by the alternating current. As far as magnetic quality is concerned, a soft iron rod would suffice, but its vibrations would die away quickly unless maintained. The vibrations of a rod of tool steel, with its smaller elastic hysteresis, are more persistent. A rod of "silver" steel, about 0·25 cm. in diameter, may be used.

The apparatus is described in §35. The alternating current, whose frequency is to be measured, traverses the windings of the small electromagnet M, which is adjustable in position. The current from the mains passes through the primary of a transformer; the secondary, giving about 8 volts, is connected to the windings on M. The two currents have, of course, the same frequency. If the core of M be formed of a bundle of iron wires or of strips of transformer sheet, eddy current heating will be diminished. The coil of the electromagnet should be wound on a non-conducting bobbin; if a conducting bobbin be used, the phase of the powerful currents, which will be set up in it, will be nearly opposite to that of the current in the wire, and thus the efficiency of the electromagnet will be diminished, and, further, much heat will be generated. The electromagnet may be made in horseshoe form.

A coil G, shown in section, is threaded over the rod when necessary and can be joined to a battery, a resistance and a key. The coil, which does not touch the rod, is used to magnetise it when desired.

When the length of OA is suitably adjusted, the rod will be thrown into strong vibration by the action of the electromagnet.

The resonance is sharply marked and may be missed unless the length OA be adjusted by small steps. If the rod be just not clamped, so that it can slide through the chuck, it exhibits marked "uneasiness" when OA is approximately of the proper length for resonance. If the engines at the power station do not run at an absolutely constant speed, it may happen that a length which gives strong resonance at one moment fails to do so shortly afterwards. Generally, however, the changes of frequency will be small in the case of a modern power station.

When the magnetising coil G is not used, the steel rod is magnetised by induction by the electromagnet M and, consequently, will be strongly attracted to M whenever the strength of M, irrespective of sign, has a maximum value. The action on the rod can thus be represented by a constant attraction on which is superposed an alternating force making twice as many cycles per second as the alternating current. If the current have frequency N, so that it makes N cycles per second, the alternating force will have frequency $2N$.

If a continuous current be sent through G, the rod will be subjected to two magnetising forces, one constant and the other alternating. Hence the mechanical action on the rod will be the resultant of three forces—one constant, one of frequency N and one of frequency $2N$. If the rod be suitably adjusted, it will respond to the force of frequency N, the response increasing with the current in G. If the magnet be of horseshoe form, it is so placed that only one pole has effective action on the rod.

Sometimes the rod is sufficiently magnetised to vibrate with frequency N under the action of the electromagnet, though no current flows through G. This state does not, as a rule, continue long, since the alternating magnetic force shakes out the "permanent" magnetism. A current sent through G makes the response certain.

The rod can be made to vibrate steadily with frequency N by the action of the electromagnet without the help of the coil G. If a horseshoe magnet be used, it is placed with its limbs vertical and the line joining the poles parallel to and vertically below the rod near its free end. If the end of the rod be displaced horizontally from its zero position, the force which the electromagnet exerts on the rod may be large or small, but it always has a horizontal component tending to restore the rod to its zero position. If this restoring force be large when the horizontal displacement is greatest, the vibration will be maintained. If the frequency of the horizontal vibrations be N, the displacement will have a maximum value $2N$ times per second. The poles of the electromagnet have maximum strength $2N$ times per second, and thus it is possible for the electromagnet, when supplied with current of frequency N, to maintain a horizontal vibration of the rod of the same frequency. The free length OA of the rod for this mode of vibration is somewhat greater than that found when the rod, magnetised by the coil G, vibrates *vertically* with frequency N, because a restoring effect, due to the electromagnet, additional to that arising from the elasticity of the rod, has been introduced. A little care may be needed in the adjustment of the rod. Unless the adjustment of the free length of the rod be made in small steps, this horizontal vibration may elude detection. Since we cannot estimate the restoring effect due to the electromagnet, we cannot make any quantitative use of this mode of vibration.

The vertical component of the force exerted on the rod by the magnet may be analysed into a steady vertical force and an alternating force of frequency $2N$. The latter has no appreciable effect on the rod, for the length is such that the frequency of free vibrations of the rod is nearly N.

It is shown in § 32 that, if n be the frequency of transverse (undamped) vibrations of a thin uniform rod of circular section, clamped at one end, as in Fig. 19,

$$n = \Omega \frac{r}{l^2} \sqrt{\left(\frac{E}{\rho}\right)}, \qquad \text{(1)}$$

where r is the radius of the section, l the length OA, E Young's modulus, ρ the density and Ω a number depending upon the number of nodes. The full theory is given in § 33. For our present purpose it is sufficient to know that, for a given rod and a given number of nodes, nl^2 is constant. Hence, if the length l_0 for a known frequency n_0 be determined, we can at once find the value of n corresponding to any other observed length l, for

$$n = n_0 l_0^2 / l^2. \qquad \text{(2)}$$

Thus, if the length be l_0 when the rod is in unison with a tuning fork of frequency n_0, the frequency of the current can be calculated as soon as the length giving resonance has been found.

42. Determination of frequency of current. The rod is first calibrated by aid of a series of tuning forks for no node and for one node, just as in EXPERIMENT 6, § 36. The calibration curves PR and QS are drawn as in § 36, or, alternatively, the constants a, b and c, d in (3) and (6) of § 36 are found.

The electromagnet M and the coil G are then put into place, and a constant current is sent through G. The length of OA is adjusted so that the rod, vibrating with no node, responds to the magnet, and the length (l) of OA is measured. Then the frequency, N, of the rod is calculated by (5), § 36, or from the value of $1/\sqrt{N}$, which is read off from the l-line. The frequency of the magnetic driving force is now known, for it equals N.

While the current flows in G, the length of OA is re-adjusted so that the rod, vibrating with one node, responds to the electromagnet, and the length (k) of OA is measured. Then N is found by (8), § 36, or from the k-line.

The current in G is now stopped, or the coil is removed, and the rod, vibrating first with no node and then with one node, is adjusted to respond, with frequency $2N$, to the electromagnet. The length of OA is measured in each case. The value of $2N$ for the cases of no node and of one node are found from (5) and (8), § 36, or from the l-line and the k-line respectively.

When the rod has been adjusted so as to vibrate with one node, the distances AO, AH, where H is the node, are measured and the value of h, where $h = AH/AO$, is compared with the theoretical value $h = 0{\cdot}2165$. It is perhaps best to take the *effective* length of AO. This comparison is made when the frequency of the rod is N and again when it is $2N$.

43. Practical example.

The rod was calibrated against tuning forks, as in EXPERIMENT 6, with the results recorded in § 39. These give, for no node and one node respectively,

$$l = -0{\cdot}58 + 142{\cdot}8/\sqrt{n}, \quad \text{or} \quad n = 142{\cdot}8^2/(l + 0{\cdot}58)^2,$$
$$k = -0{\cdot}40 + 345{\cdot}7/\sqrt{n}, \quad \text{or} \quad n = 345{\cdot}7^2/(k + 0{\cdot}40)^2.$$

Results for no node. Rod responded to current with $l = 14{\cdot}02$ cm. Hence

$$N = 142{\cdot}8^2/(14{\cdot}02 + 0{\cdot}58)^2 = 95{\cdot}67 \text{ cycles per sec.}$$

Rod responded to octave above current with $l = 9{\cdot}88$ cm. Hence

$$2N = 142{\cdot}8^2/(9{\cdot}88 + 0{\cdot}58)^2 = 186{\cdot}38, \qquad N = 93{\cdot}19 \text{ cycles per sec.}$$

Results for one node. Rod responded to current with $k = 35{\cdot}31$ cm. Hence

$$N = 345{\cdot}7^2/(35{\cdot}31 + 0{\cdot}40)^2 = 93{\cdot}72 \text{ cycles per sec.}$$

Rod responded to octave above current with $k = 24{\cdot}86$ cm. Hence

$$2N = 345{\cdot}7^2/(24{\cdot}86 + 0{\cdot}40)^2 = 187{\cdot}3, \qquad N = 93{\cdot}65 \text{ cycles per sec.}$$

Position of node. When the rod vibrated with one node in response to the current, AH was 7·6 cm. and the *effective* length was 35·71 cm. Hence $h = 7{\cdot}6/35{\cdot}71 = 0{\cdot}213$. When the rod responded to the octave above the current, AH was 5·3 cm. and the effective length was 25·26 cm. Hence $h = 5{\cdot}3/25{\cdot}26 = 0{\cdot}210$. The theoretical value is 0·2165.

The frequency of the Cambridge supply at the date of the experiment was liable to serious variations.

44. Experiment with several nodes. A long rod can be adjusted to vibrate with 0, 1, 2, ... nodes under the action of the alternating current, the number of nodes being limited by the finite length of the rod. Equation (23), § 33, shows that, when n is given, l/m is constant. When there are many nodes, the rod will be too long to remain approximately horizontal when clamped as in Fig. 19, § 35. The clamp is therefore placed so that the part OA of the rod is vertically below the clamp. The electromagnet may be placed opposite the middle of the "loop" which is next the clamp. Gravity now comes into play, and, in effect, aids the stiffness of the rod, the increase in effective stiffness increasing with l, and causing l/m, for a given frequency, to increase slightly as l is increased. The effect only becomes appreciable when the tension, $\dot{T}$ dynes cm.$^{-2}$, is appreciable compared with Young's modulus. In the experiment of § 45, T/E was always less than 5×10^{-7}.

Measurements, similar to those described in § 36, may be made with a short length of rod, using a set of tuning forks, to determine the correction due to imperfect clamping. The quotient $\pi l'/2m$ may then be calculated for the cases of 0, 1, 2, ... nodes. We use $2m/\pi$ in preference to m, because, as § 34 shows, $2m/\pi$ may be taken as an integer for m_3, m_4, ...

As an alternative, we may plot the actual length $l = OA$ against

$2m/\pi$ for 0, 1, 2, ... nodes. The points will lie on a straight line passing nearly through the origin.

Any given rod is not necessarily uniform along its length, and thus some irregularities may occur in the results.

45. Practical example.

A steel rod, 0·204 cm. in diameter, was used. An allowance of 0·3 cm. was made for imperfect clamping. If l cm. be the observed length of OA, and l' cm. be the effective length, $l'=l+0{\cdot}3$.

Nodes	l'	$2m/\pi$	$\pi l'/2m$	Nodes	l'	$2m/\pi$	$\pi l'/2m$
0	9·05	1·194	7·58	5	82·0	11	7·45
1	22·10	2·988	7·40	6	96·8	13	7·45
2	37·51	5	7·50	7	112·6	15	7·51
3	52·4	7	7·49	8	128·4	17	7·55
4	67·1	9	7·46				

EXPERIMENT 8. **A bifilar method of measuring the rigidity of wires.***

46. Introduction. In this method, the couple due to the torsion of two similar wires is balanced against the couple due to the load carried by the wires and arising from bifilar action. The wires hang from two torsion heads and are twisted by them. The wires are twisted beyond the elastic limit and consequently hysteresis effects appear. There are other difficulties, as will appear later. The method is therefore unsuitable for accurate measurements of rigidity, although it forms a useful exercise in the use of a bifilar suspension.

47. Bifilar couple. We first consider two light flexible strings. Let the strings AB, CD, each l cm. in length, hang from two fixed points A, C, which are at a distance $2a_1$ cm. apart in a horizontal plane. The lower ends B, D of the strings are attached to a rigid body of mass M grm., the points B, D being $2a_2$ cm. apart. The centre of gravity of the body is symmetrical with regard to B and D and thus the tensions of the strings are equal. The line BD will then be horizontal. If, now, a couple, whose axis is vertical, be applied to the body, the body will be in equilibrium when the

* *Camb. Phil. Soc. Proc.* Vol. xx, p. 61 (1920).

couple due to the obliquity of the strings balances the applied couple.*

In Fig. 21, A', B', C', D' are the projections of A, B, C, D on a horizontal plane. In our symmetrical case, $A'C'$, $B'D'$ bisect each other in O. When the body has turned through θ radians from the zero position, in which the strings are in the vertical plane through $A'C'$, the line $B'D'$ will make an angle θ with $A'C'$. Let ON be the perpendicular from O on $A'B'$. Let the tension in each string be T dynes.

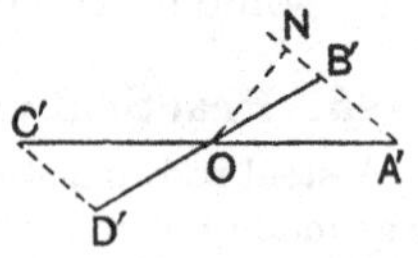

Fig. 21

If the vertical distance of BD below AC be h cm., the vertical component of the tension is Th/l. Since $A'B'$ is the projection of the string AB, the horizontal component is $T \, . \, A'B'/l$. The weight of the body equals the sum of the vertical components, and thus

$$Mg = 2Th/l. \qquad \text{..........................(1)}$$

The horizontal component of the tension at B acts along a line whose projection is $A'B'$, and hence its moment about the vertical axis through O is $T \, . \, A'B' \, . \, ON/l$. Since the moment due to the two tensions equals that of the applied couple, G dyne cm.,

$$G = 2T \, . \, A'B' \, . \, ON/l. \qquad \text{.......................(2)}$$

But $A'B' \, . \, ON$ is twice the area $OA'B'$ and thus is $a_1 a_2 \sin\theta$. Hence, by (1) and (2),

$$\frac{G}{Mg} = \frac{A'B' \, . \, ON}{h} = \frac{a_1 a_2 \sin\theta}{h}. \qquad \text{...............(3)}$$

The vertical distance h depends upon θ. We have

$$h^2 = l^2 - A'B'^2 = l^2 - a_1{}^2 - a_2{}^2 + 2a_1 a_2 \cos\theta.$$

When, however, $A'B'$ is small compared with l, $A'B'^2$ is negligible compared with l^2. We may then put $h = l$ in (3) and so obtain

$$G = \sin\theta \, . \, Mga_1 a_2/l. \qquad \text{......................(4)}$$

In the examples of § 52, h never differed from l by as much as 1 in 4000.

* For the general theory of the bifilar suspension, see Maxwell, *Electricity and Magnetism*, Vol. II, § 459; A. Gray, *Absolute Measurements in Electricity and Magnetism*, Vol. I, p. 242; Kohlrausch, *Physical Measurements*, p. 226 (1894).

48. Apparatus. The upper ends of the wires under test are soldered into torsion heads *S*, *T* (Fig. 22) which pass through a board *XY* held in a firm support.

The lower ends of the wires are soldered into screws which pass through "clearing" holes in the bar *EF*, and are secured with nuts.

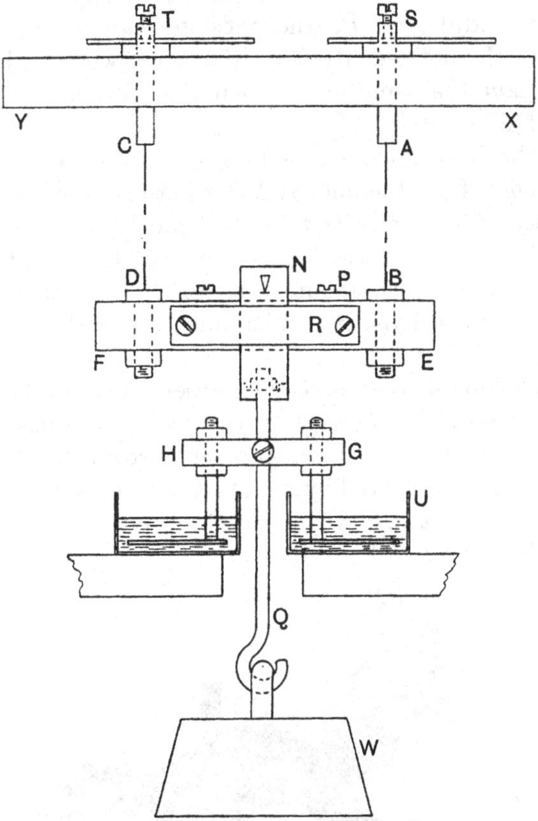

Fig. 22

The heads of the screws are made with "flats" to fit a spanner. Before the screws are secured to *EF*, the torsion heads are set to their zeros. The head of each screw is held with a spanner while its nut is tightened, the purpose being that the wires should be free from torsion when the bar is not subject to any actions except those of the wires and of gravity. This will be attained if, after the nuts have been tightened, the flats on *both* screws have the same directions as when the wires hung freely. The distance BD is, as nearly as may be, equal to AC.

The load is carried by a knife edge forming part of the link N, Figs. 22, 23. The knife edge rests in a V-groove in a plate, P, fixed to EF by screws passing through slots. By adjusting P, the tensions can be equalised; the notes emitted by the wires when plucked have the same pitch when the tensions are equal.

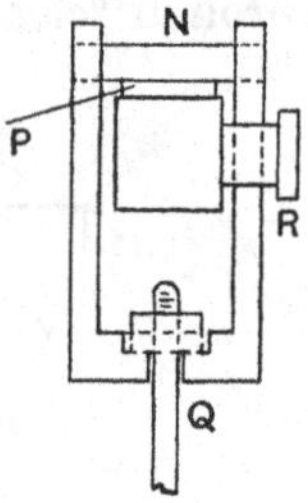

Fig. 23

A weight W (a few kilogrammes) is suspended by the rod Q from the link N. A slot in the lower cross-piece of N allows Q to be put into place; the nut drops into a recess. The weight should be so attached to Q that it cannot turn about a vertical axis relative to Q with any freedom; otherwise it will be difficult to reduce the system to rest.

The deflexion of the bar is best observed optically. A metal strip R is screwed to EF, packing pieces being interposed to allow the link N free movement, and a plane mirror is fixed to R. The deflexion can be observed by aid of a telescope and scale, or of a lamp and scale. It is, however, simpler to employ a goniometer* such as that shown in Fig. 24.

Fig. 24

The base has a spherical pivot at one end and a cross-bar carrying a scale at the other. Angles are measured by a movable

* For a fuller account of the goniometer, see *Experimental Optics*, p. 62.

arm which turns at one end about the pivot while the other end moves over the scale on the cross-bar. The optical system consists of an achromatic lens fixed to the arm near the pivot and of a fine vertical wire held in a frame at the other end of the arm and adjusted to be in the focal plane of the lens. The scale is divided into millimetres and the pivot is adjusted so that its centre is 40 cm. from the edge of the scale. The readings are taken along the *edge* of the scale by means of a fine wire passing across an opening in the arm and stretched by a spring. This wire is at right angles to the scale when the arm is in its central, or zero, position.

When the goniometer is used in mechanical experiments to determine the angle turned through by a body about a vertical axis, a plane mirror is attached to the body and the image of the goniometer wire is made to coincide with the wire itself. If this coincidence be restored after the body has turned, the angle turned through by the body is equal to that turned through by the goniometer arm. In this use of the goniometer, all that is necessary is that the mirror should be nearly vertical and that the rays from the wire, after passing through the lens, should fall upon the mirror. No other adjustments are required, and the centre of the spherical pivot need not lie on the vertical axis about which the body turns. The goniometer is placed so that its lens is three or four centimetres from the mirror carried by the suspended system.

If the reading on the scale of the goniometer for any position of the arm differ from the reading when the arm is in the central position by x cm., and if the distance from the centre of the spherical pivot to the *edge* of the scale be p cm., then the angular displacement of the arm is θ radians, where $\tan\theta = x/p$. Thus the goniometer measures tangents of angles and not angles themselves. A conversion table is given in § 51.

The motion of the suspended system, as so far described, would be only slightly damped, and it would not be easy to reduce the system to rest; the vibrations of the building would add to the difficulty. A simple damping device is therefore used. An annulus of thin sheet metal is carried by the bar GH, which is clamped to the rod Q. The annulus is immersed in motor lubricating oil, or other highly viscous liquid, contained in the annular trough U, which rests on the table. The rod Q passes through a hole in the table. By adjusting the height of GH, the annulus can be brought close to the bottom of the trough, and then the motion is so highly damped that the system is little affected by vibrations of the floor or the table.

If the wires be overstrained by turning the torsion heads through too large angles, they will no longer be vertical when the heads read zero, and it will be necessary to readjust the screws in the bar EF. To prevent overstrain, and at the same time to allow the heads to be turned through π in either direction from their zeros, a *movable* safety device is used. A horizontal disk of metal, about 1 cm. in diameter, with a vertical pin fixed eccentrically to it, turns about its centre on a vertical pivot screwed into the board XY (Fig. 22). The pivot is fixed at such a distance from the torsion head that the greatest distance from the pin to the axis of the head is small enough to prevent the longer arm of the steel wire, which forms the index of the head, from passing the pin; the shorter arm cannot touch the pin. The torsion head can thus be turned through more (but only a little more) than π in either direction from zero.

Care must be taken not to bend the wires near the soldered joints. A bend at B or D will alter the effective value of a_2. If the wire AB be bent near A, the effect, when the torsion head is turned, will be the same as if the point A describes a small horizontal circle. This effective displacement of A has two effects. It (1) changes a_1 as the head is turned, and (2) causes the bar EF to turn through an angle in addition to that directly due to the torsion of the wires. Annealed wires are, therefore, more suitable for the experiment than hard drawn wires, as they are more easily straightened. When the bend is small compared with $2a_1$, the first effect is small and may be neglected. But the maximum angular displacement of the bar is only small and the second effect is not negligible; it is considered in § 50.

The torsion heads are read on circles divided at intervals of 45°, the dividing lines being scribed on the board XY.

49. Theory of method. If each torsion head be turned from its zero through ϕ radians in the same direction, the bar EF will turn in that direction until the bifilar and torsional couples are equal. If EF turn through θ, the whole twist of each wire is $\phi-\theta$.

Let the radius and the length of each wire be r cm. and l cm., and let the rigidity of the metal be n dyne cm.$^{-2}$. Since the wires are *nearly* vertical, the couple due to torsion, is, to a close approximation, the same as if the wires were vertical. Hence, if G be the couple exerted by the *pair* of wires upon the bar,

$$G=\pi n r^4 (\phi-\theta)/l.^* \quad \text{.....................(5)}$$

The small couple due to the *bending* of the wires assists the bifilar couple; Kohlrausch† takes account of this small couple by writing, in place of (4),

$$G=\sin\theta\,.\,Mga_1a_2/l', \quad \text{.....................(6)}$$

where

$$l'=l-r^2\{2\pi E/Mg\}^{\frac{1}{2}}, \quad \text{.....................(7)}$$

and E is Young's modulus.

Equating the torsional to the (corrected) bifilar couple, we have

$$n=\frac{ga_1a_2l}{\pi r^4 l'}\,.\,MJ, \quad \text{.......................(8)}$$

where

$$J=\sin\theta/(\phi-\theta). \quad \text{.......................(9)}$$

50. Experimental details. The distances $AC=2a_1$, $BD=2a_2$ are measured. The diameters of the wires are taken at a number of points and the mean radius is found.

The total mass, M grm., of the system carried by the wires is found. The masses of the screws are found before they are soldered to the wires.

The torsion heads are first set to zero, and the goniometer is adjusted so that its arm is in the central position when the goniometer wire coincides with its own image.

The readings must be taken over the range $-\pi$ to π for ϕ, in order to eliminate errors due to slight bends in the wires; the theory assumes absence of hysteresis. But in experimental work in elasticity we must realise that, when the strains are more than infinitesimal, hysteresis effects are unavoidable. To ensure that the effects of hysteresis shall be orderly and not irregular, the

* G. F. C. Searle, *Experimental Elasticity*, § 39.

† Kohlrausch, *Wied. Ann.* Bd. XVII, p. 737 (1882).

torsion heads are taken through a *complete cycle* from π to $-\pi$ and back to π; to make the two readings for $\phi=\pi$ agree as closely as possible, a preliminary half cycle from $-\pi$ to π is done. In order that the conditions may be as nearly uniform as possible throughout the cycle and a half, the readings for the preliminary settings are taken and recorded; this will secure approximately constant time intervals between successive readings. Thus the heads are set in succession at the following multiples of $\pi/4$:

$$-4, -3, -2, -1, 0, 1, 2, 3,$$

$$4, 3, 2, 1, 0, -1. -2, -3, -4, -3, -2, -1, 0, 1, 2, 3, 4.$$

The first 8 settings give the preliminary readings; only the readings for the last 17 settings are used in the calculation of n.

If ϕ go through a complete cycle, and θ be plotted against ϕ, a narrow hysteresis loop will be obtained. When readings are taken

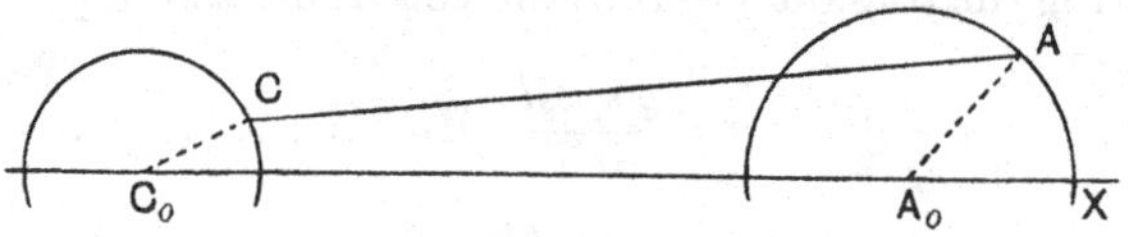

Fig. 25

as above, there will be *two* values of θ for each value of ϕ except $\phi=-\pi$. With careful work, the two values of θ for $\phi=\pi$ will be exactly or very nearly identical; for the wires used in § 52, the difference between these two values was seldom as great as one minute. As a rough method of eliminating the effects of hysteresis, the mean of the two values of θ for each value of ϕ is taken as *the* value of θ for that value of ϕ.

The effect of bends in the wires near their upper ends, A, C (Fig. 22), will be the same as if these points described small horizontal circles about the centres A_0, C_0, as in Fig. 25. Let the axes of the torsion heads cut the plane of Fig. 25 in A_0, C_0, and let ϕ, the angle through which each torsion head is turned, be measured from C_0A_0X. The effective points of suspension are not A_0 and C_0 but A and C, which move round with the heads. Thus $AA_0X=\phi+\alpha$, $CC_0X=\phi+\gamma$, where α, γ depend upon the directions of the bends relative to the torsion heads. Let $A_0A=s$, $C_0C=t$, $A_0C_0=2a_1$. When the reading of the torsion head s is ϕ

the line AC through the points of suspension makes a small angle ϵ with the line A_0C_0 through the axes of the heads. We have

$$\tan\epsilon = \frac{s\sin(\phi+\alpha) - t\sin(\phi+\gamma)}{2a_1 + s\cos(\phi+\alpha) - t\cos(\phi+\gamma)}.$$

When s and t are small compared with $2a_1$, $\tan\epsilon$ may be replaced by ϵ, and the variable terms in the denominator may be neglected. Then, putting

$$(s\sin\alpha - t\sin\gamma)/2a_1 = P, \qquad (s\cos\alpha - t\cos\gamma)/2a_1 = Q,$$

we have

$$\epsilon = P\cos\phi + Q\sin\phi. \quad \text{.................(10)}$$

Here P and Q are the values of ϵ when $\phi = 0$ and when $\phi = \frac{1}{2}\pi$.

If the line BD make an angle θ with A_0C_0 when the heads read ϕ, the angle between BD and AC is $\theta - \epsilon$. The wires will not be quite free from torsion when the heads read zero; let η be the mean twist of the wires when $\phi = 0$. We must thus write $\sin(\theta - \epsilon)$ for $\sin\theta$ and $\phi + \eta - \theta$ for $\phi - \theta$ in equation (9), which thus becomes

$$\sin(\theta - \epsilon) = J(\phi + \eta - \theta). \quad \text{...............(11)}$$

To evade difficulties, θ is kept small. Then, since ϵ is also small, we may replace the sine by the angle in (11), and thus obtain

$$\theta = H(\phi + \eta) + F\epsilon, \quad \text{.....................(12)}$$

where

$$H = J/(1+J), \qquad F = 1/(1+J).$$

If θ_0, ϵ_0 correspond to $\phi = 0$, we have, since $\epsilon_0 = P$,

$$\theta_0 = H\eta + F\epsilon_0 = H\eta + FP. \quad \text{...............(13)}$$

Thus

$$\theta - \theta_0 = H\phi + F(\epsilon - P)$$
$$= H\phi + F(P\cos\phi + Q\sin\phi - P). \quad \text{.........(14)}$$

Since this equation is *linear* in θ, we may take θ_0 as corresponding to *any* initial position of the bar which is *near* its ideal zero position.

Thus, if β be the angle at any time between the bar and some nearly ideal zero position,

$$\beta = \theta - \theta_0. \quad \text{.......................(15)}$$

Since β, though small—say less than 0·2 radian—is not infinitesimal, some correction should be made. An exact solution cannot be given, but accuracy is gained by writing $\sin\beta$ for β, and then the final formula becomes

$$\sin\beta = H\phi + F(P\cos\phi + Q\sin\phi - P). \quad \text{.........(16)}$$

To eliminate P and Q, we combine the observations. Let β_m correspond to $\phi = m\pi/4$. Then, putting $\phi = \pi$ and $\phi = -\pi$, so that $m = 4$ and $m = -4$, we have

$$\pi H = \tfrac{1}{2}(\sin\beta_4 - \sin\beta_{-4}). \quad \text{...............(17)}$$

A second value for πH is found by giving m the values 3, −3, 1, −1. Then

$$\pi H = \sin\beta_3 - \sin\beta_{-3} - (\sin\beta_1 - \sin\beta_{-1}). \quad \text{.....(18)}$$

The two values of H are usually in good agreement, although, when β is plotted against ϕ, the curve differs considerably from a straight line. The mean value of πH is used to find J. Thus

$$J = \pi H/(\pi - \pi H). \quad \text{.....................(19)}$$

Then n is found by (8).

The values of FP, FQ are easily found. Thus

$$FP = -\tfrac{1}{4}(\sin\beta_4 + \sin\beta_{-4}), \quad \text{...............(20)}$$

$$FQ = \tfrac{1}{2}(\sin\beta_2 - \sin\beta_{-2} - \pi H). \quad \text{...........(21)}$$

51. Conversion table. A goniometer, such as that described in § 48, gives the *tangent* of the angle ψ through which the arm is turned from its zero. To find $\sin\psi$ we subtract from $\tan\psi$ the small quantity s given in the Table, and to find the angle ψ in radians we subtract from $\tan\psi$ the small quantity c.

$\tan\psi$	s	c	$\tan\psi$	s	c	$\tan\psi$	s	c
0·02	0·00000	0·00000	0·10	0·00050	0·00033	0·18	0·00285	0·00191
0·03	0·00001	0·00001	0·11	0·00066	0·00044	0·19	0·00334	0·00224
0·04	0·00003	0·00002	0·12	0·00085	0·00057	0·20	0·00388	0·00260
0·05	0·00006	0·00004	0·13	0·00108	0·00072	0·21	0·00448	0·00301
0·06	0·00011	0·00007	0·14	0·00135	0·00090	0·22	0·00514	0·00345
0·07	0·00017	0·00011	0·15	0·00166	0·00111	0·23	0·00585	0·00393
0·08	0·00025	0·00017	0·16	0·00201	0·00134	0·24	0·00663	0·00446
0·09	0·00036	0·00024	0·17	0·00240	0·00161	0·25	0·00746	0·00502

Simple interpolation, by "proportional parts," will give s and c with an error not exceeding unity in the fifth decimal place. Thus, when $\tan\psi = 0{\cdot}205$, $s = 0{\cdot}00418$ and $c = 0{\cdot}00280$. Hence

$$\sin\psi = 0{\cdot}205 - 0{\cdot}00418 = 0{\cdot}20082,$$

$$\psi = 0{\cdot}205 - 0{\cdot}00280 = 0{\cdot}20220.$$

52. Practical example.

Soft brass wires were used.

The distances AC, BD were each 6·0 cm. Hence $a_1 = a_2 = 3\cdot0$ cm.

Mean radius of wires $= r = 0\cdot0352$ cm. Length of wires $= l = 47\cdot30$ cm.

Mass of suspended system, *excluding* weight W (Fig. 22) $= 417\cdot6$ grm.

The small correction for buoyancy of damper was neglected.

Deflexions were observed by a goniometer. Distance from centre of pivot to scale was 40·0 cm. Central, or zero, reading is 10·0 cm. The following goniometer readings were obtained for the last 17 of the values of ϕ specified in § 50.

ϕ	Readings		Mean	x	$\tan\beta$ $= x/40$	$\sin\beta$ obsd.	$\sin\beta$ calcd.
	cm.	cm.		cm.			
π	14·10 ↓	14·10	14·100	4·115	0·1029	0·1023	0·1033
$\frac{3}{4}\pi$	12·96	13·04	13·000	3·015	0·0754	0·0752	0·0749
$\frac{1}{2}\pi$	11·85	11·98	11·915	1·930	0·0482	0·0482	0·0471
$\frac{1}{4}\pi$	10·80	10·96	10·880	0·895	0·0224	0·0224	0·0220
0	9·90	10·07	9·985	0·000	0·0000	0·0000	0·0000
$-\frac{1}{4}\pi$	9·16	9·33	9·245	−0·740	−0·0185	−0·0185	−0·0204
$-\frac{1}{2}\pi$	8·30	↑ 8·43	8·365	−1·620	−0·0405	−0·0405	−0·0415
$-\frac{3}{4}\pi$	7·37	7·44	7·405	−2·580	−0·0645	−0·0644	−0·0653
$-\pi$	6·33		6·330	−3·655	−0·0914	−0·0910	−0·0921

The value of x was found, with sufficient accuracy, by subtracting from the mean reading for a given ϕ, as shown in column 4, the mean reading 9·985 cm. corresponding to $\phi = 0$. The differences between the readings in columns 2 and 3 are due to hysteresis. The seventh column shows that $\sin\beta$ is not proportional to ϕ.

By (17), $\pi H = \frac{1}{2}(0\cdot1023 + 0\cdot0910) = 0\cdot09665$,

and by (18), $\pi H = 0\cdot0752 + 0\cdot0644 - (0\cdot0224 + 0\cdot0185) = 0\cdot09870$.

Mean value of $\pi H = 0\cdot0977$.

Then, by (19), $$J = \frac{\pi H}{\pi - \pi H} = 0\cdot03210.$$

By (20), $FP = -\frac{1}{4}(0\cdot1023 - 0\cdot0910) = -0\cdot0028$,

and by (21), $FQ = \frac{1}{2}(0\cdot0482 + 0\cdot0405 - 0\cdot0977) = -0\cdot0045$.

In Table, column "$\sin\beta$ calcd." gives $\sin\beta$ as calculated by (16), using values of πH, FP and FQ just found; there is fair agreement between calculated and observed values of $\sin\beta$.

Total load M was $417\cdot6 + 4999 = 5416\cdot6$ grm.

Taking $E = 10^{12}$ dyne cm.$^{-2}$, we have $r^2(2\pi E/Mg)^{\frac{1}{2}} = 1\cdot35$ cm., and hence, by (7), $l' = 47\cdot30 - 1\cdot35 = 45\cdot95$ cm.

By (8), $$n = \frac{g a_1 a_2 l}{\pi r^4 l'} \,.\, MJ = \frac{981 \times 3^2 \times 47\cdot30}{\pi \times 0\cdot0352^4 \times 45\cdot95} \times 5416\cdot6 \times 0\cdot03210$$
$$= 3\cdot277 \times 10^{11} \text{ dyne cm.}^{-2}.$$

Observations, with $M = 3417\cdot1$ grm., gave $n = 3\cdot326 \times 10^{11}$ dyne cm.$^{-2}$.

An independent determination of n was made by attaching a bar, of moment of inertia $K = 4\cdot766 \times 10^4$ grm. cm.^{2}, to each wire in turn; mean periodic time of torsional vibrations was $T = 10\cdot55$ secs. Hence

$$n = \frac{8\pi K l}{T^2 r^4} = \frac{8\pi \times 4\cdot766 \times 10^4 \times 47\cdot30}{10\cdot55^2 \times 0\cdot0352^4} = 3\cdot316 \times 10^{11} \text{ dyne cm.}^{-2}.$$

CHAPTER IV

OPTICAL METHODS OF DETERMINING ELASTIC CONSTANTS

53. Introduction. When the deformation of an elastic body is fully known, so that the strain at each point is known, we can, at least in theory, find for each element of the surface of the body the magnitude and direction of the force per unit area which must be applied to that element.

An alternative method, which we will now describe, of finding the connexion between the deformation and the forces required to produce it may be used with advantage in some cases.

We suppose the body divided into elements of small volume. Some are "interior" elements and are entirely surrounded by other elements and some are "exposed" elements; each of the latter elements and the body have part of the surface of the body in common. When the body is deformed, the forces applied to the surface of the body on an "exposed" element do work upon that element. The "exposed" element, acting by the remainder of its surface, does work on the elements which touch it. The work which either an "exposed" or an "interior" element does on contiguous elements is exactly the negative of that which they do on it. Hence the total of all the amounts of work which are done on all the elements, either by other elements or by forces applied to the surface of the body, is precisely equal to the work done by the latter forces alone. Each element can be taken so small that the strain may be taken as uniform throughout it, and then the mechanical work spent upon it per unit volume is a quadratic function of the strain. Hence, if we know the strain at each point of the body and calculate the mechanical work U required per unit volume to produce this strain, we have, on integration throughout the whole body,

$$W = \int U dv, \qquad \text{.........................(1)}$$

where W is the mechanical work done by the forces applied to the *surface* of the body during the strain and dv is an element of volume. When we keep well within the elastic limit, the deformation is so small that no appreciable error is caused by taking dv as an element of the *unstrained* body.

When the strain can be specified in terms of a single variable, differentiation of (1) with regard to that variable will determine the relation between the deformation and the system of forces required to produce it.

The discussions throughout this chapter are, for the sake of simplicity, limited to the case of isotropic and homogeneous material.

54. Thermal effects due to strain. If no heat enter or leave any element of the body, the temperature of that element will, in general, change when the strain is produced. The size of the element will change with the change of temperature, and thus the forces causing the adiabatic strain will differ from those required for the same strain when the temperature is kept constant by imparting or abstracting heat. Thus k_ϕ and E_ϕ, the adiabatic values of the bulk modulus and of Young's modulus, differ from the isothermal values k_t and E_t.

The case of the rigidity, n, is exceptional. A rise of temperature does not alter the angles of an unstrained cube of isotropic substance, and consequently the shearing stress required to produce a small shear is the same for an adiabatic as for an isothermal strain. Hence $n_\phi = n_t$. In the case of a simple shear it also follows that no change of temperature occurs when the strain is adiabatic and that no supply of heat is necessary to ensure that the strain is isothermal. The case of rigidity is considered at greater length in *Experimental Elasticity*, §§ 21, 22.

The rise or fall of temperature in adiabatic strain and the heat supplied or abstracted in isothermal strain can be calculated by the principles of thermodynamics, but the results are not required for our present purpose.

It is important to notice that it is only in the adiabatic case that U equals the increase of internal energy of the element. In the isothermal case heat is imparted to or abstracted from the element, and hence the increase of energy of the element is due partly to the mechanical work spent upon it and partly to the heat imparted to it. It is only the *mechanical* work spent upon the element which concerns us in (1). To avoid confusion, we shall call U the mechanical work spent per unit volume.

55. Mechanical work spent in stretching. Let a straight uniform rod of unstrained length l and cross-section α be stretched to length $l+\lambda$ by equal and opposite forces of amount F per unit area applied to the two ends. Then $F=E\lambda/l$, where E is Young's modulus. The work done by the force during the stretching is half the product of the final force and the extension and thus is $\frac{1}{2}F\alpha\lambda$ or $\frac{1}{2}E\alpha\lambda^2/l$. The volume of the rod is αl, and thus, if U be the mechanical work spent per unit volume, $U=\frac{1}{2}E\lambda^2/l^2$.

56. Application to bending of bar. The bar, which is assumed to have a plane of symmetry parallel to its length, is bent by equal and opposite couples with their axes perpendicular to that plane, which we may call the plane of bending. The longitudinal fibres will be bent into circular arcs in planes parallel to the plane of bending. A transverse section intersects the plane of bending in a straight line which we take as Oy, the axis of y. Of the longitudinal fibres which intersect Oy only one is unstretched and O, its point of intersection with Oy, we take as origin. The axis Ox is in the transverse section and, of course, perpendicular to the plane of bending. Let the radius of curvature of the unstretched fibre through O be ρ. On this fibre we take another point O' such that OO' is of unit length. The line of intersection of the two transverse planes through O and O' (Fig. 26) meets the plane of bending in R and thus $OR=\rho$. If the positive direction of y be from R to O, the length of the fibre through x, y which is included between the two transverse planes is $(\rho+y)/\rho$ or $1+y/\rho$; the projection of this fibre on the plane of bending is PP'. The increase of length per unit length is y/ρ. Hence, by § 55, the mechanical work spent per unit volume at the point x, y is $\frac{1}{2}Ey^2/\rho^2$. If W be the expenditure of mechanical work corresponding to the part of the bar contained between the two transverse planes,

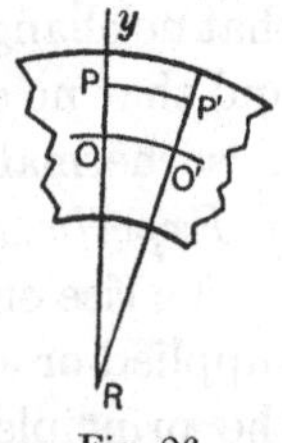

Fig. 26

$$W=\tfrac{1}{2}(E/\rho^2)\iint y^2\,dx\,dy,$$

where the integration extends over the area of the transverse section. The value of the integral* is I, the "moment of inertia" of the section about Ox. Hence

$$W = \tfrac{1}{2}EI/\rho^2. \quad \text{.......................(2)}$$

If either couple be G, the work done on unit length of the bar during the bending is half the product of the final couple and the angle $1/\rho$ through which unit length has been bent. Hence

$$W = \tfrac{1}{2}G/\rho.$$

Thus, by (2), $$G = EI/\rho. \quad \text{...........................(3)}$$

For a bar of circular section, $I = \frac{1}{4}\pi r^4$ and then

$$G = \tfrac{1}{4}\pi E r^4/\rho. \quad \text{.......................(3a)}$$

Since the increase of length per unit length of the fibre through x, y is y/ρ, the tension in it, i.e. the pull per unit area, is Ey/ρ, and thus the resultant pull across the section is $(E/\rho)\iint y\,dx\,dy$. Since the bar is bent by couples, this resultant pull is zero, and hence $\iint y\,dx\,dy = 0$. Hence the "centre of gravity" of the section lies on Ox and, since the section is symmetrical about Oy, the "centre of gravity" is actually at O. In other words, the fibre which passes through the "centre of gravity" of the section is not stretched when the bar is bent by the couples.†

57. Mechanical work spent in shearing. Let a plane, the plane of the paper, parallel to two faces of a cube of elastic material of side h, cut the other four faces in AB, BC, CD, DA (Fig. 27). Let tangential forces of amount F per unit area be applied, in the direction indicated by the arrows, to the four faces corresponding to AB, BC, CD, DA; the forces are parallel to the plane of the paper. Let the cube be sheared by the action of the forces

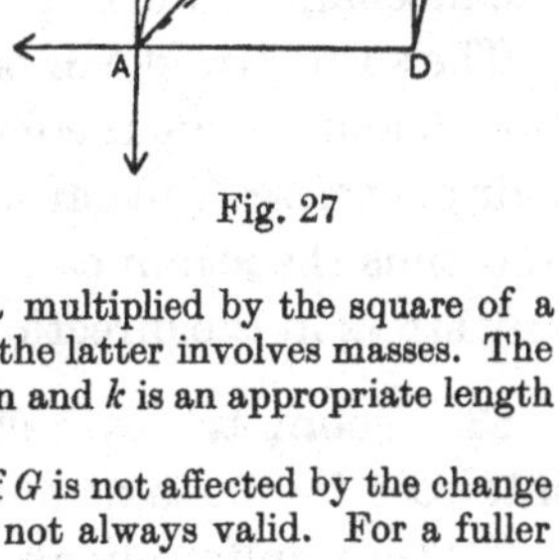

Fig. 27

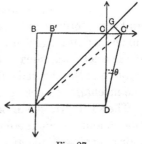

* The integral is of the "dimensions" of an area multiplied by the square of a length and is not, strictly, a "moment of inertia"; the latter involves masses. The integral equals Ak^2, where A is the area of the section and k is an appropriate length called the radius of gyration of the area about Ox.

† It has been assumed, as is usual, that the value of G is not affected by the change of section due to bending—an assumption which is not always valid. For a fuller discussion, see *Experimental Elasticity*, §§ 27–38.

through the *small* angle θ, so that the square $ABCD$ becomes the parallelogram $AB'C'D$, with $BAB' = CDC' = \theta$. Then, if n be the rigidity of the material, $F = n\theta$. Since it is only the motion of the parts of the cube relative to one another which concern us, we may suppose the position of the face AD to be unchanged. Then the forces applied to AB, DA, CD do no work. The work done by the force applied to the face BC is half the product of the final force, Fh^2, and the distance BB', or $h\theta$, through which the face BC has moved. Hence the work done during the shearing is $\frac{1}{2}Fh^2 . h\theta$ or $\frac{1}{2}h^3 n\theta^2$. Since the volume of the cube is h^3, the mechanical work spent per unit volume in shearing the substance is $\frac{1}{2}n\theta^2$.

The shear converts the diagonal AC, of length $h\sqrt{2}$, into the diagonal AC'. If $C'G$ be the perpendicular from C' on AC, we have, to the first order of small quantities,

$$AC' = AG = h\sqrt{2} + CC'/\sqrt{2}.$$

Since $CC' = h\theta$, $AC' - h\sqrt{2} = h\theta/\sqrt{2}$. If the elongation, i.e. the increase of length per unit length, along the diagonal AC be e, we have $e = (h\theta/\sqrt{2})/(h\sqrt{2})$ or $e = \frac{1}{2}\theta$. The contraction, i.e. the diminution of length per unit length, along the perpendicular diagonal BD has the same value, viz. $\frac{1}{2}\theta$. The elongation is a maximum along lines parallel to AC and the contraction is a maximum along lines parallel to BD.

If a portion of the substance have elongation e in a direction L and contraction e in a direction M perpendicular to L and suffer no change of length in a direction N perpendicular to L and M, it experiences a shear $2e$. The mechanical work spent per unit volume of the sheared substance is $\frac{1}{2}n\,(\text{shear})^2$. Hence in our case the mechanical work spent per unit volume is $\frac{1}{2}n\,(2e)^2$ or $2ne^2$.

The shear strain can be caused by a pulling stress in the direction L and an equal compressive stress in the direction M. This pair of stresses will cause no change of length in the direction N and thus the strain can be produced without the application of any stress in the direction N.

58. Geometry of helicoid. If a straight line EFG (Fig. 28) intersect Oz, the axis of z, at right angles in F, and if the plane EFO turn uniformly about Oz through τ radians while F moves uniformly along Oz through unit distance, the moving line generates a helicoid surface which has a twist of τ radians per unit length. We take the axis Ox so that Ox is one position of the moving line, and Oy so that $Oxyz$ forms a right-handed system of axes.

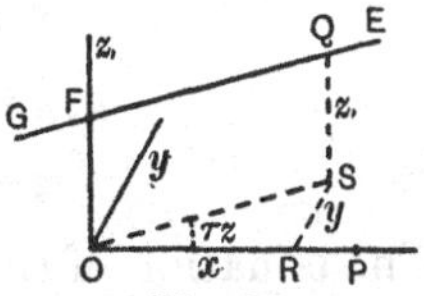

Fig. 28

Let Q be a point on EFG with coordinates x, y, z. Then OS, the projection of FQ on the plane Oxy, makes angle $SOx = \tau z$ with Ox, and hence the equation to the helicoid is

$$y = x \tan(\tau z). \quad \text{.......................(4)}$$

The section of the helicoid by the plane $x = p$, which cuts Ox in P, is the plane curve

$$y = p \tan(\tau z),$$

which cuts OPx at right angles at P.

Along this curve

$$dy/dz = p\tau \sec^2(\tau z) = p\tau(1 + y^2/p^2),$$

$$d^2y/dz^2 = 2(\tau/p)\, y\, dy/dz = 2\tau^2 y(1 + y^2/p^2).$$

At P, $y = 0$ and $d^2y/dz^2 = 0$. Hence the plane section has a point of inflexion at P, and the curvature of the arc at P is zero. Since the straight line OPx lies in the surface, there are two plane curves which lie on the surface and intersect at right angles at P and have zero curvature at P.

At the point $y = 0$, $z = 0$ in the plane section $x = p$, we have $dy/dz = p\tau$. This fixes the direction of the normal at P which is perpendicular both to Ox and to the tangent at P to the section of the helicoid by the plane $x = p$.

Now take new axes Px' (along OPx), Py', Pz' (Fig. 29), where Py', Pz' have been turned round Ox, in the plane $x = p$, through such an angle ϕ, in the negative direction, that Py' is normal to the helicoid at P. Then $\tan\phi = p\tau$ and

$$\cos\phi = (1 + p^2\tau^2)^{-\frac{1}{2}} = q.$$

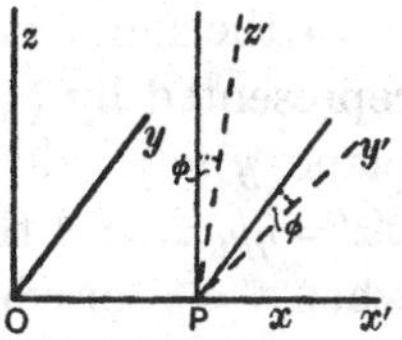

Fig. 29

The figure corresponds to positive values of p and τ. With the right-handed axes employed, positive rotation about Oz through $\frac{1}{2}\pi$ turns Ox to Oy, and positive rotation about

Oy through $\frac{1}{2}\pi$ turns Oz to Ox. The old coordinates of a point are given in terms of the new by

$$x = x' + p,$$
$$y = y'\cos\phi + z'\sin\phi = q(y' + p\tau z'),$$
$$z = z'\cos\phi - y'\sin\phi = q(z' - p\tau y').$$

The equation of the helicoid referred to the new axes is

$$q(y' + p\tau z') = (x' + p)\tan\{q\tau(z' - p\tau y')\}. \quad \text{......(4a)}$$

At points very near P we may replace $\tan\{q\tau(z' - p\tau y')\}$ by $q\tau(z' - p\tau y')$. Then the equation of the surface becomes

$$y'(1 + p^2\tau^2 + p\tau^2 x') = \tau x' z'. \quad \text{..............(4b)}$$

In finding the curvature at P of the section by any plane containing the normal at P we consider infinitesimal values of x', y', z' and accordingly neglect $p\tau^2 x'$ in comparison with unity. Then

$$y'(1 + p^2\tau^2) = y'/q^2 = \tau x' z'. \quad \text{..............(4c)}$$

If we write $x' = r\cos\theta$, $z' = r\sin\theta$, we have $x'^2 + z'^2 = r^2$, and then

$$y'(1 + p^2\tau^2) = \tfrac{1}{2}\tau r^2 \sin 2\theta.$$

If ρ be the radius of curvature of the section by a plane containing the normal Py' and making angle θ with Ox,

$$\frac{1}{\rho} = \frac{2y'}{r^2} = \frac{\tau \sin 2\theta}{1 + p^2\tau^2} = q^2\tau \sin 2\theta.$$

The curvature, $1/\rho$, is therefore zero when $\theta = 0$ and when $\theta = \frac{1}{2}\pi$, as we have already found. When $\theta = \frac{1}{4}\pi$, $1/\rho = q^2\tau$, a maximum positive value, and when $\theta = -\frac{1}{4}\pi$, $1/\rho = -q^2\tau$, a maximum negative value. The centres of curvature of the sections corresponding to $\frac{1}{4}\pi$ and to $-\frac{1}{4}\pi$ lie respectively on the positive and negative parts of the axis Py'.

In the immediate neighbourhood of P, the helicoid may be represented by (4c) and then the section of the surface by the plane $y' = \eta$, where η is small, is the rectangular hyperbola $x'z' = \eta/q^2\tau$, and thus, when τ is positive, r has a minimum value when $x' = z'$, or when $\theta = \pi/4$ or $\theta = 5\pi/4$. Thus, if K be the point on the hyperbola for which $\theta = \pi/4$, the tangent KT to the hyperbola at K is at right angles to KN, where N is the point in which the plane $y' = \eta$ cuts the axis Py'. Since KT lies in the plane $y' = \eta$

and is perpendicular to KN, KT is *perpendicular* to the plane $PKNy'$. Hence the normal to the surface at K, being perpendicular to KT, must lie *in* the plane $PKNy'$ and therefore intersects the axis PNy' which is the normal to the surface at P. Hence the arc PK is an element of one of the two lines of curvature* through P. The other line of curvature corresponds to $\theta = -\frac{1}{4}\pi$.

Each line of curvature on the helicoid throughout its length cuts at $\frac{1}{4}\pi$ the straight lines on the surface which are perpendicular to the axis Oz.

If we limit the helicoid to a twisted strip of width $2p$, where p is very small compared with $1/\tau$, the length along the central line Oz corresponding to a twist of one radian, we can neglect $p^2\tau^2$ in comparison with unity and put $q^2 = 1$. Then at all points on the twisted strip the principal curvatures are given by $1/\rho = \pm\tau$.

59. Application to torsion of blade. The blade is a strip of metal or glass of considerable length $2c$, of width $2a$ (small compared with $2c$) and of thickness $2b$ (small compared with $2a$). The two large faces of length $2c$ and width $2a$ will, for brevity, be called "the faces" of the blade. If the blade be twisted by couples suitably applied to its ends in planes perpendicular to the long ($2c$) axis of the blade, the particles which in the unstrained blade lay on the plane midway between "the faces" will after the twisting lie on a median surface which is a helicoid. The twist is assumed so small that $a\tau$ is small and $a^2\tau^2$ is negligible in comparison with unity.

Let P be a point (Fig. 30) on the median helicoid, at shortest distance $PO = p$ from the axis Oz. The principal section* of the median helicoid which contains the normal at P and the tangent PR (making $\frac{1}{4}\pi$ with OP produced) has radius of curvature $\rho = 1/\tau$ at P; the centre of curvature lies on the positive part of Py' (Fig. 29).

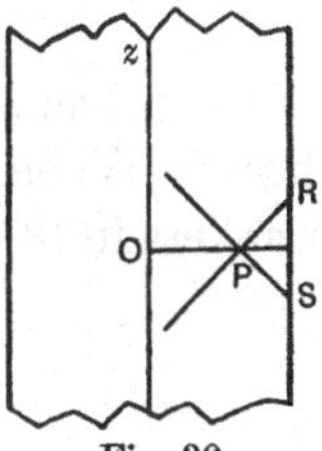

Fig. 30

The particles which, in the unstrained blade, lay on a plane parallel to the median plane will, in the strained blade, lie on a surface "parallel" to the median

* See *Experimental Optics*, Chapter XI, §§ 268, 269.

helicoid, the normal distance between the two surfaces being constant. The filament which lies (1) in a surface "parallel" to the median helicoid and (2) in the plane containing the normal Py' and PR, and passes through Q on Py' at distance h from P, measured towards the centre of curvature of the filament, has radius $\rho - h$. The investigation of § 56 applies, and thus parallel to PR there is a contraction h/ρ or $h\tau$. In the other principal section there is an elongation $h\tau$. The faces of the blade are free from pressure and thus, since the blade is very thin, we may assume that at Q there is no stress parallel to the normal. Thus the conditions of § 57 are fulfilled and the strain is a simple shear. The mechanical work spent per unit volume at Q is $2nh^2\tau^2$. If the mechanical work spent per unit length of the blade be W, we have

$$W = 2n\tau^2 \, . \, 2a \int_{-b}^{b} h^2 dh = \frac{8n\tau^2 ab^3}{3}. \quad \text{...........(5)}$$

If J be the couple corresponding to the twist τ per unit length, the mechanical work spent per unit length during the twisting is half the product of the final couple and the twist τ. Hence

$$W = \tfrac{1}{2} J\tau. \quad \text{..............................(6)}$$

Thus, by (5), $$J = 2W/\tau = \tfrac{16}{3} n\tau ab^3. \quad \text{..................(7)}$$

This result agrees with that obtained in *Experimental Elasticity*, § 45, by a different method.

Barré de Saint-Venant obtained the complete solution for the torsion of a uniform bar whose cross-section is a rectangle of sides $2a$, $2b$. The ratio of the couple J to $n\tau$ is given by

$$\frac{J}{n\tau} = \frac{16ab^3}{3} - b^4 \left(\frac{4}{\pi}\right)^5 \left\{ \sum_{m=0}^{m=\infty} \frac{1}{(2m+1)^5} \tanh \frac{(2m+1)\pi a}{2b} \right\}, \quad \text{...(8)}$$

where m has the values 0, 1, 2, ..., and a is equal to or greater than b. When $a > 3b$, the sum of the infinite series within the brackets in (8) differs by less than two parts in 10,000 from

$$\frac{1}{1^5} + \frac{1}{3^5} + \frac{1}{5^5} + \ldots = 1{\cdot}00452,$$

its value when b/a is zero. Thus, when $a > 3b$, we may put

$$\frac{J}{n\tau} = ab^3 \left\{ \frac{16}{3} - \frac{b}{a} \left(\frac{4}{\pi}\right)^5 \times 1{\cdot}00452 \right\} = ab^3 \left(\frac{16}{3} - 3{\cdot}361 \frac{b}{a} \right). \quad \text{...(9)}$$

EXPERIMENT 9. **Determination of elastic constants of glass by Cornu's* method.**

60. Introduction. When a plane "test-plate" of glass is nearly in contact with the surface of a specimen in the form of a bar of glass, interference fringes will be seen under suitable illumination. These fringes will be easily visible when the surface of the bar differs only slightly from a plane. The fringes form contour lines, each fringe corresponding to a definite distance between the surface of the test-plate and that of the bar. By observations on the fringes, the principal radii of curvature of the surface of the bar can be determined. If the surface of the unstrained bar be plane and if the bar be deformed by some simple system of forces, the principal curvatures can be calculated in terms of the dimensions of the bar and the elastic constants. By equating the calculated curvatures to those found from observations on the fringes, the elastic constants can be found.

A bar of plate glass of rectangular section is used as the specimen. The width is $2a$ and the thickness $2b$; the length $2c$ is great compared with $2a$, and $2a$ is considerable compared with $2b$. The two faces of length $2c$ and width $2a$ will be called, for brevity, "the faces" of the bar. The ratio b/a, though much less than unity, must be sufficiently great to ensure that, when the bar is bent, the relation between the two principal curvatures of the anticlastically curved faces is that given by the ordinary theory. The necessary condition† is that ρ, the radius of curvature of the longitudinal axis of the bent bar, should be great compared with $a^2/2b$. For the bar used in § 67, $a^2/2b = 6{\cdot}38$ cm. The least value of ρ used in the experiment was 2953 cm.

When the bar is unstrained, one of the faces should be so nearly flat that when it is tested against a small test-plate the fringes seen under suitable illumination are very wide. The expense of an optically worked bar would be prohibitive, but if a number of bars be cut by a glazier from the plate glass used for shop windows,

* Cornu, *Comptes Rendus*, Vol. LXIX, p. 333.

† *Experimental Elasticity*, § 33. The condition found here is that ρ/σ should be large compared with $a^2/2b$, where σ is Poisson's ratio. Since σ is about $0{\cdot}25$ for glass, this condition will certainly be satisfied if, as in the text, ρ be great compared with $a^2/2b$.

some bars will be found with fairly good faces. The test-plate may be of the same quality of plate glass and should be tested against a prism of good quality.

The experiment is much simplified if a scale divided to millimetres be etched on the best surface of the test-plate. This could be done by an instrument maker when a sufficiently good test-plate has been found.

61. Optical arrangements. In Fig. 31 the cross-section of the bar is marked *A*, the test-plate is marked *B*. Light from the sodium flame *F*, after passing through the lens *M*, is reflected by the sloping glass plate *C* on to the combination of bar and test-

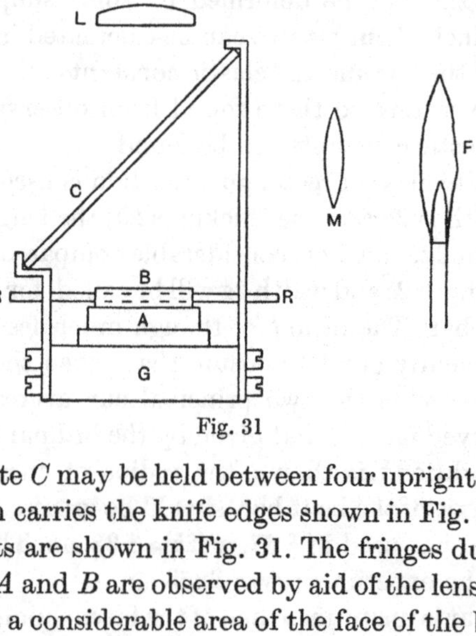

Fig. 31

plate. The plate *C* may be held between four uprights fixed to the block *G* which carries the knife edges shown in Fig. 32; only two of the uprights are shown in Fig. 31. The fringes due to the air-film between *A* and *B* are observed by aid of the lens *L*, or by the naked eye. As a considerable area of the face of the bar has to be illuminated, the flame is placed approximately in the focal plane of *M*. The lower face of the bar *A* may be coated with black varnish to diminish the reflexion at that face.

The divided face of the test-plate *B* rests upon the bar. As the opposed faces of test-plate and bar are not absolutely parallel, the fringes are formed approximately in the air space, and hence there will be little, if any, parallax between the dividing lines of the scale and the fringes.

62. Method for measurement of Young's modulus and Poisson's ratio. The bar rests upon two knife edges H, K (Fig. 32) which are fixed to a block G attached to a base-board J. The base J is bolted to a heavy tripod stand. The knife edges are adjusted so that the bar rests evenly on both edges. The bar will rattle when tapped if this adjustment be poor. The "rattling" test should be made when the bar is not loaded. Knife edges U, V rest in transverse grooves cut near the ends of the bar by a triangular file moistened with turpentine. The test-plate rests upon the bar at its centre.

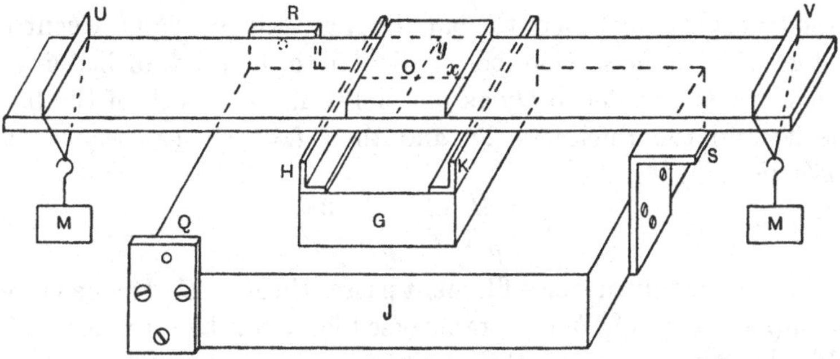

Fig. 32

To protect the bar, two safety rods are provided. They pass through holes in the uprights (Fig. 31) which support the sloping plate; one rod, marked RR, is seen. The rods are parallel to the knife edges and just so far above them as to allow the bar to pass freely between them and the knife edges. The rods prevent the bar from being flung down, and perhaps broken, when a load is placed at one end only; the rods should be in position before the loads are put on.

The heavy tripod should rest upon a board standing upon four "sorbo" rubber balls partially let into holes near its corners on its under-side. This arrangement will, to a great extent, prevent vibrations of the table from affecting the bar and will make for steadiness of the fringes.

The distances UH, VK are adjusted to equality and then $UH = VK = l$. Equal masses M, M are hung from U, V. Convenient values of M are 1·5, 2·0, 2·5, 3·0 kilos. With a bar of the dimensions given in § 67, it is hardly safe to use 4 kilos.

For the sake of the geometry of the fringe pattern, we take as axes, on the upper face of the bar, a line Ox parallel to the length of the bar and Oy across the bar. The axis Oz points vertically upwards.

Between the knife edges, the bending moment experienced by the bar has the constant value Mgl. Hence, by § 56, if the neutral longitudinal filament which passes through C, the "centre of gravity" of the transverse section, have radius of curvature ρ, then

$$EI/\rho = Mgl, \qquad\qquad (1)$$

where E is Young's modulus and I is the "moment of inertia" of the transverse section of the bar about an axis through the centre of gravity of the section, perpendicular to the plane of bending; this axis is parallel to Oy as just defined. The width of the bar is $2a$ and the thickness $2b$, and thus $I = \frac{4}{3}ab^3$. (*Experimental Elasticity*, § 66.)

Hence

$$E = \frac{M}{\rho^{-1}} \cdot \frac{gl}{I} = \frac{M}{\rho^{-1}} \cdot \frac{3gl}{4ab^3}. \qquad\qquad (2)$$

The central transverse filament passes through C. If ρ be large compared with $a^2/2b$, this transverse filament is bent to radius ρ'. If σ be Poisson's ratio, this radius is given by*

$$\rho' = \rho/\sigma. \qquad\qquad (3)$$

The centre of curvature of this filament and that of the neutral longitudinal filament lie on opposite sides of C. The radii of curvature of the corresponding lines on the upper face are $\rho + b$ and $(\rho/\sigma) - b$. But b is so small compared with ρ and ρ/σ that we may treat ρ and ρ/σ as equal to the radii of the two lines on the face. A suitable correction is easily applied if desired.

The bar is slightly bent by its own weight. The curvature due to this cause is superposed on that due to the loading; it is eliminated by the method of reducing the observations when a series of loads is used.

When the bar is bent, the fringe pattern will be as in Fig. 33 and will consist of two sets of hyperbolas having common axes

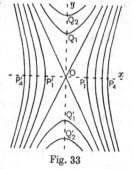

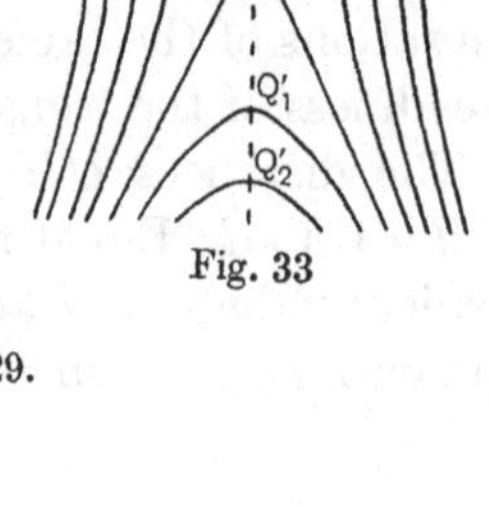

Fig. 33

* *Experimental Elasticity*, § 29.

and common asymptotes. The transverse axes P_1P_1', P_2P_2', ... of one set lie along Ox and the transverse axes Q_1Q_1', ... of the other set lie along Oy. The foci of a hyperbola lie on its transverse axis.*

63. Theory of fringes for case of bending. If O be a point on an "ordinary" surface (*Experimental Optics*, § 268) and O be taken as origin, the equation of the surface near O will be of the form

$$\tfrac{1}{2}Fx^2 + Hxy + \tfrac{1}{2}Gy^2 = z, \quad \text{.................}(4)$$

where the axes of x and y are in the tangent plane at O and z is the distance of a point on the surface from that plane. The section of the surface by the plane $z = h$ is the conic

$$\tfrac{1}{2}Fx^2 + Hxy + \tfrac{1}{2}Gy^2 = h. \quad \text{.................}(5)$$

By turning the axes of x and y round in the tangent plane through an angle ϕ, given by $\tan 2\phi = 2H/(F - G)$, equation (4) can be deprived of the term in xy. When, as in our case, the conic is a hyperbola, the surface (4) and the conic (5) take the forms

$$\tfrac{1}{2}Ax^2 - \tfrac{1}{2}By^2 = z, \text{......................}(6)$$

$$\tfrac{1}{2}Ax^2 - \tfrac{1}{2}By^2 = h, \text{......................}(7)$$

where A and B are taken to be positive quantities. When h is positive, the transverse axis of the hyperbola lies along Ox; when h is negative, the transverse axis lies along Oy.

If the asymptotes make angle θ with Ox, $\tan\theta = \sqrt{(A/B)}$. The plane $y = 0$ cuts the surface in the curve $x^2 = 2z/A$. If ρ_1 be the radius of curvature of this curve at O, we have $\rho_1 = \tfrac{1}{2}x^2/z = 1/A$. Similarly, for ρ_2, the radius of curvature of the section of the surface by the plane $x = 0$, we have $\rho_2 = 1/B$.

If the upper "face" of the bar be represented by (6), the principal curvatures of the "face" are $1/\rho_1$, or A, and $1/\rho_2$, or B. It is these curvatures which are measured by aid of the fringe pattern.

* Since the bar has anticlastic curvature and the test-plate is plane, the plate cannot touch the bar at the centre of the fringe system, where the tangent plane to the bar is parallel to the plate. The position of the fringe centre will therefore depend on how the plate rests on the bar, but the form and orientation of the fringes will be independent of the position of their centre.

The principal radii of curvature of the "face", viz. ρ_1 and ρ_2, are very great compared with $2b$, the thickness of the bar, and we may write

$$\rho_1 = \rho, \quad \rho_2 = \rho',$$

where ρ and ρ' are the radii of curvature of the neutral longitudinal filament and of the central transverse filament.

If the directions of the axes of x and y be the same as in §62, we have $\rho = \rho_1 = 1/A$ and $\rho' = \rho/\sigma = \rho_2 = 1/B$. Hence

$$\sigma = \rho_1/\rho_2 = B/A = \cot^2\theta. \quad \text{.................(8)}$$

Thus, in theory, Poisson's ratio σ can be determined at once when θ is known, without the measurement of any distances or any loads. In practice, the small curvatures of the test-plate and the small initial curvatures of the face of the bar defeat this plan. We must therefore use a method in which a *series* of loads is employed.

To measure ρ, we make measurements on those fringes which cut, at right angles, the longitudinal line Ox. The test-plate is easily adjusted so that the length of its scale is parallel to Ox, but it may happen that the dividing lines do not actually intersect Ox.* It is generally possible to obtain the intersection by putting a load of a few grammes on the test-plate in a suitable position. This, of course, alters the form of the bar, as do also the weights of the test-plate and of the bar itself, but the effect may be neglected since the experiment is not capable of great accuracy. The scale readings of the dark fringes are then observed, and the distances $P_1P_1' = 2p_1$, $P_2P_2' = 2p_2$, ... (Fig. 33) are found. If the distance between the bar and the test-plate at O be k cm., at P_1, P_1' be k_1 and at P_n, P_n' be k_n, we have $k_1 = k + \frac{1}{2}p_1^2/\rho$, $k_n = k + \frac{1}{2}p_n^2/\rho$. But, by optical theory, $k_n - k_1 = (n-1)\lambda/2$, where λ is the wavelength of the monochromatic light employed. Thus

$$\tfrac{1}{2}(p_n^2 - p_1^2)/\rho = \tfrac{1}{2}(n-1)\lambda,$$

or

$$\rho = \frac{1}{\lambda} \cdot \frac{p_n^2 - p_1^2}{n-1}. \quad \text{......................(9)}$$

Hence, if we plot p_1^2, p_2^2, ... against the numbers 1, 2, ..., as in

* If the dividing lines be longer than is usual on a scale—say 2 cm. long—it is unlikely that they will not intersect Ox.

Fig. 34, we should find a straight line if the surfaces were perfect and the observations free from error. The plotted points will lie nearly on a straight line. If the distance k be such that the inmost hyperbolas of Fig. 33 reduce to the asymptotes, OP_1 has its maximum value $\sqrt{(\lambda\rho)}$. For any other value of k, $OP_1 < \sqrt{(\lambda\rho)}$.

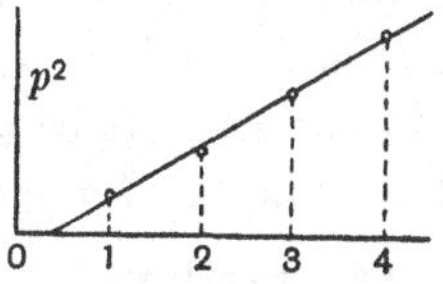

Fig. 34

Hence the straight line (Fig. 34) cuts the axis of numbers either at the origin or between 0 and 1. By aid of the straight line drawn to pass as evenly as possible among the plotted points, we can find an average value of

$$(p_n^2 - p_1^2)/(n-1).$$

We may also use an arithmetical method, as in § 67. Then, by (9), ρ is found. Young's modulus is then given by (2).

Similarly, we measure the dark fringes which cut Oy at right angles and find $2q_1 = Q_1Q_1'$, $2q_2 = Q_2Q_2'$, ... (Fig. 33). From the corresponding line on the graph, or by arithmetical work, we find

$$(q_n^2 - q_1^2)/(n-1)$$

and deduce the radius of curvature, ρ', of the transverse section of the face of the bar. Thus

$$\rho' = \frac{1}{\lambda} \cdot \frac{q_n^2 - q_1^2}{n-1}. \quad \text{.....................(10)}$$

By (9) and (10) we obtain (in theory) Poisson's ratio. Thus

$$\sigma = \rho/\rho'. \quad \text{.........................(11)}$$

The mean wave-length of sodium light is $5{\cdot}893 \times 10^{-5}$ cm.

If time allow, a series of masses 1·5, 2·0, 2·5, 3·0 kilos may be used for M. Then ρ^{-1} is plotted against M and the best value of M/ρ^{-1} is found from the slope of the straight line lying most evenly among the plotted points. This "best value" is used in (2).

Reliable values of ρ and ρ' will not be obtained unless the curvatures $1/\rho$ and $1/\rho'$ due to the bending be large compared with those due to imperfections of the surfaces of the bar and of the test-plate. The results of § 67 show that, unless the principal radii of curvature of the face of the unstrained bar and of the face of the test-plate be great compared with 100 metres, the corresponding curvatures will not be negligible compared with the curvatures produced in the face of the bar when the bar is strained. For this reason, M should not be less than 1·5 kilos. For safety M should not be greater than 3 kilos with specimens similar to that of § 67. The loads should not be left suspended from the bar longer than is necessary.

64. Method for measurement of rigidity. In order to twist the bar about its longitudinal axis, a couple is applied by means of two wooden clamps fixed to the bar by bolts. One clamp is provided with two perforated plates; a steel rod AB (Figs. 35, 36) passes through the holes in these plates and through holes in the plates Q, R (Figs. 32, 36) rising from the base J. The bar can turn freely about the horizontal axle AB. Pins D, E (Fig. 35) pass across slots cut in the upper member of the other clamp; a mass M can be hung from either pin. The lower member of this clamp carries a stud C which rests upon a projecting bracket S (Figs. 32, 37) fixed to the base J. Figs. 36 and 37 show transverse sections of the bar

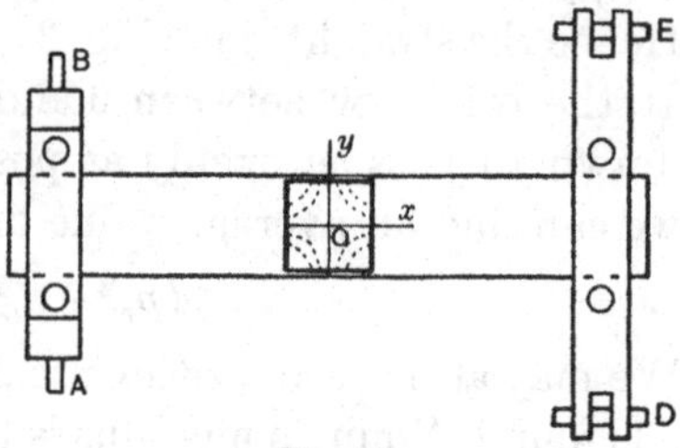

Fig. 35

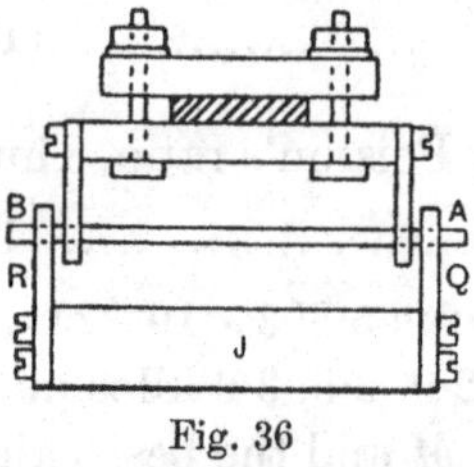

Fig. 36

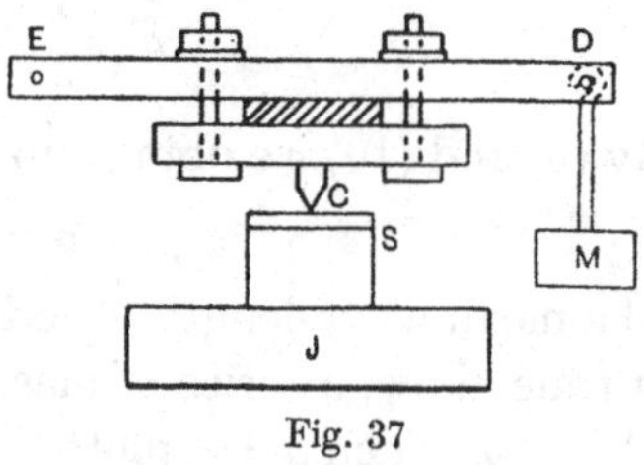

Fig. 37

through the clamps, as seen by an observer situated to the left of AB (Fig. 35). The plates Q, R and the bracket S are so arranged that, when the bar is in position, its lower surface clears the knife edges H, K (Fig. 32) by a few millimetres. The safety rods used in the bending experiment are not needed now and are removed. The test-plate rests on the centre of the bar.

Since the bar can turn freely about the horizontal axle AB, the axle cannot exert any moment upon the bar tending to *bend* the longitudinal axis of the bar.

The clamp ED, being symmetrical about the axis of the bar, does not by its own weight exert any twisting couple upon the bar. To twist the bar, a load M is hung from D. If we suppose that equal and opposite vertical forces of Mg dynes are applied to the point of the stud C, we see that the load at D may be replaced by a downward force Mg at C and a couple $Mg \times CD$, where CD denotes the *horizontal* distance between C and D. The downward force is balanced by the reaction of the bracket. If the plane DCE be perpendicular to the (horizontal) axis Ox of the bar, the couple exerts no *bending* moment upon the bar.

65. Theory of fringes for case of twisting. Under the action of the couple, the central plane of the bar is twisted through τ radians per unit length and takes the form of a helicoid surface. If we measure z vertically upwards and take Ox, Oy parallel and perpendicular to the length of the bar and in the tangent plane, assumed horizontal, to the helicoid at O, the equation of the helicoid is, by § 58,

$$z = y \tan \tau x.$$

The twist τ is positive when the load hangs from the pin D (Fig. 35). For small values of τx, we have

$$z = \tau xy.$$

When either $x=0$ or $y=0$, we have $z=0$. Thus the plane $z=0$ intersects the helicoid along Ox, and along Oy. To find the section of the helicoid by a plane containing the normal Oz and making an angle θ with Ox, we put $x = r\cos\theta$, $y = r\sin\theta$. Then

$$z = \tfrac{1}{2}\tau r^2 \sin 2\theta.$$

If R be the radius of curvature of this plane section at O, and if R be counted positive when the corresponding centre of curvature is *above* O, we have $2Rz = r^2$, or $R = \tfrac{1}{2}r^2/z$. Thus the curvature, i.e. $1/R$, is given by

$$1/R = 2z/r^2 = \tau \sin 2\theta.$$

Thus, when $\theta = \tfrac{1}{4}\pi$, $1/R = \tau$, and when $\theta = -\tfrac{1}{4}\pi$, $1/R = -\tau$. When $\theta = 0$, or $\theta = \tfrac{1}{2}\pi$, $1/R = 0$. The last two cases correspond to the axes Ox and Oy, which, as we have already seen, lie in the surface.

The section of the helicoid by the plane $z = h$ is the rectangular hyperbola $xy = h/\tau$. When h is positive, the hyperbola, which, of course, has two branches, lies in the quadrants marked U, U' in Fig. 38. When h is negative, the hyperbola lies in the quadrants V, V'. If the twist be reversed, so that τ changes sign, the two hyperbolas change places.

The section of the helicoid by the plane $y=x$, which contains UU', has curvature τ and the section by the plane $y=-x$, which contains VV', has curvature $-\tau$. When the load is hung from D (Fig. 37), τ is positive.

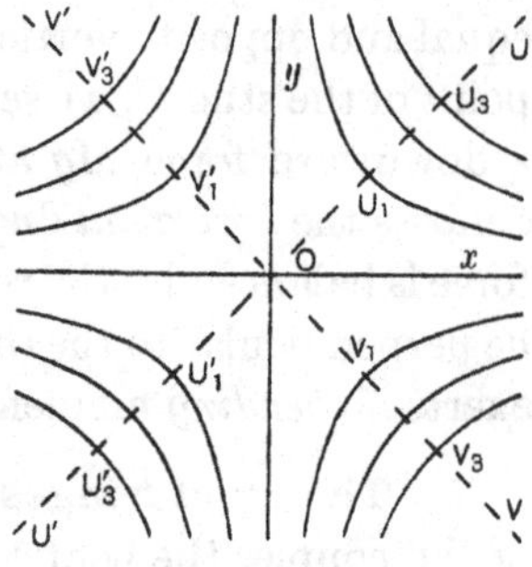

Fig. 38

The actual face of the bar will not be strictly a helicoid, but, when the radius $1/\tau$ is very large compared with the thickness of the bar, we may treat the face as a helicoid of pitch $2\pi/\tau$.

To determine the curvature τ, we make measurements on those dark fringes which cut at right angles (1) UU' and (2) VV' (Fig. 38). The test-plate is first adjusted so that the length of the scale is parallel to UU'. The distances $U_1U_1'=2u_1$, $U_2U_2'=2u_2$, ... are measured. Since the radius of curvature is $1/\tau$, we have, as in (9),

$$\tau_{DU}=\lambda(n-1)/(u_n^2-u_1^2), \quad \ldots\ldots\ldots\ldots(12)$$

where τ_{DU} denotes the curvature of the section of the surface by the plane $y=x$ (which contains UU'), when the load hangs from D.

Observations are next made along VV', and $V_1V_1'=2v_1$, $V_2V_2'=2v_2$, ... are measured. Then

$$\tau_{DV}=\lambda(n-1)/(v_n^2-v_1^2). \quad \ldots\ldots\ldots\ldots(13)$$

The load is now transferred to E and observations along UU' and VV' give τ_{EU} and τ_{EV}. The change of curvature along UU', due to the transference of the load from D to E, is $\tau_{EU}+\tau_{DU}$, and this will be independent of any curvature of the unstrained bar along that line. Hence τ_U, the mean of τ_{DU} and τ_{EU}, will fairly represent the value of τ which we seek. Similarly τ_V, the mean of τ_{DV} and τ_{EV}, will nearly equal τ. Hence we may write

$$\tau=\tfrac{1}{2}(\tau_U+\tau_V). \quad \ldots\ldots\ldots\ldots\ldots\ldots(14)$$

It is here assumed that $DC=CE$; any small difference between DC and CE will not affect the value of τ given by (14).

By (9), § 59, we have, in terms of the rigidity n,

$$\text{Couple}/(n\tau)=\tfrac{1}{2}Mg\,.\,DE/(n\tau)=ab^3\left(\tfrac{16}{3}-3{\cdot}361\,b/a\right),$$

and thus, if $DE=2h$,

$$n=\frac{Mgh}{\tau ab^3(\frac{16}{3}-3\cdot361\,b/a)}. \qquad \text{..............(15)}$$

If time allow, a series of masses 1·5, 2·0, 2·5, 3·0 kilos should be used. We may plot τ against M and so obtain the best value of M/τ.

66. Comparison of results. Poisson's ratio has been found from the bending of the bar by equation (11). It can also be found in terms of E and n. Thus, *Experimental Elasticity*, p. 20,

$$\sigma=\tfrac{1}{2}E/n-1. \qquad \text{......................(16)}$$

This value of σ may be compared with that given by (11).

67. Practical example.

Measurements by Mr J. G. Wilson.

Dimensions of bar. Width $=2a=4\cdot175$ cm. Hence $a=2\cdot087$ cm. Measured thickness of bar varied from 0·6804 to 0·6847 cm. Mean of ten measurements gave $2b=0\cdot6822$ cm., and $b=0\cdot3411$ cm.

Young's modulus. When bar was symmetrically supported on knife edges, distance from each knife edge to point of support of nearer load was $l=11\cdot19$ cm.

The transverse axes P_1P_1', P_2P_2', ... (Fig. 33) of the hyperbolas whose foci lay on the longitudinal axis of the bar were measured for loads of 1·5, 2·0, 2·5 kilos. The results for 1·5 kilos are given in the Table. Sodium light with $\lambda=5\cdot893\times10^{-5}$ cm. was used.

Load $=M=1500$ grm.

Fringe n	Reading right	Reading left	$2p_n$	p_n	p_n^2	$\frac{p_n^2-p_1^2}{n-1}$
	cm.	cm.	cm.	cm.	cm.^{2}	cm.^{2}
1	7·36	8·41	1·05	0·525	0·276	
2	7·12	8·62	1·50	0·750	0·562	0·2860
3	6·96	8·80	1·84	0·920	0·846	0·2850
4	6·80	8·94	2·14	1·070	1·145	0·2897
5	6·65	9·06	2·41	1·205	1·452	0·2940
6	6·52	9·18	2·66	1·330	1·769	0·2986
7	6·42	9·29	2·87	1·435	2·059	0·2972
8	6·32	9·39	3·07	1·535	2·356	0·2971
9	6·21	9·48	3·27	1·635	2·673	0·2996
10	6·12	9·57	3·45	1·725	2·976	0·3000

Mean value of $(p_n^2-p_1^2)/(n-1)=0\cdot2941$ cm.^{2} Radius of curvature, ρ, of longitudinal axis of bar is, by (9), given by

$$\rho=\frac{1}{\lambda}\cdot\frac{p_n^2-p_1^2}{n-1}=\frac{0\cdot2941}{5\cdot893\times10^{-5}}=4990\,\text{cm}.$$

These observations, with those for two other loads, give

M	$(p_n^2-p_1^2)/(n-1)$	ρ	ρ^{-1}	$M\rho$
grm.	cm.^{2}	cm.	cm.$^{-1}$	grm. cm.
1500	0·2941	4990	$2{\cdot}004\times10^{-4}$	$7{\cdot}485\times10^6$
2000	0·2241	3803	$2{\cdot}630\times10^{-4}$	$7{\cdot}606\times10^6$
2500	0·1740	2953	$2{\cdot}386\times10^{-4}$	$7{\cdot}382\times10^6$

Mean value $M\rho=7{\cdot}49\times10^6$ grm. cm.

The Table shows that the curvature $1/\rho$ is roughly proportional to the load M. When $1/\rho$ is plotted against M, the straight line which lies most evenly (as judged by eye) among the points will furnish a linear relation nearly agreeing with

$$1/\rho=1{\cdot}382\times10^{-7}\times M-0{\cdot}0907\times10^{-4},$$

which is given by the method of least squares. Here ρ is in cm. and M in grm. The relative curvature of the test-plate and the face of the bar, along the longitudinal axis of the bar, when the bar rests on the knife edges but carries no loads at its ends, is $-0{\cdot}0907\times10^{-4}$ cm.$^{-1}$. The corresponding radius is $1{\cdot}10\times10^5$ cm., or 1·10 kilometres. The negative sign shows that the face of the unloaded bar is concave relative to the test-plate. A load of 1 grm. changes the curvature by $1{\cdot}382\times10^{-7}$ cm.$^{-1}$ and hence, in (2), § 62, we put

$$(1/\rho)\,M^{-1}=1{\cdot}382\times10^{-7} \quad\text{or}\quad M/\rho^{-1}=7{\cdot}231\times10^6 \text{ grm. cm.}$$

This differs appreciably from $7{\cdot}49\times10^6$, the mean of the values of $M\rho$ in the Table. We then have for Young's modulus

$$E=\frac{M}{\rho^{-1}}\cdot\frac{3gl}{4ab^3}=\frac{7{\cdot}231\times10^6\times3\times981\times11{\cdot}19}{4\times2{\cdot}087\times0{\cdot}3411^3}=7{\cdot}19\times10^{11}\text{ dyne cm.}^{-2}.$$

Poisson's ratio. The transverse axes Q_1Q_1', Q_2Q_2', ... (Fig. 33) of the hyperbolas whose foci lay on Oy, which is at right angles to the longitudinal axis of the bar, were measured for loads of 2·0 and 2·5 kilos. On account of the finite size of the test-plate, only a few fringes could be observed and they were broad and difficult to measure accurately. The radius of transverse curvature is ρ'. The results were:

M	$(q_n^2-q_1^2)/(n-1)$	ρ'	$(\rho')^{-1}$	$M\rho'$
grm.	cm.^{2}	cm.	cm.$^{-1}$	grm. cm.
2000	0·4528	7684	$1{\cdot}3014\times10^{-4}$	$1{\cdot}537\times10^7$
2500	0·4001	6789	$1{\cdot}4730\times10^{-4}$	$1{\cdot}697\times10^7$

The values of $M\rho'$ are discrepant—a warning that the relative curvature of the test-plate and the unloaded bar along the transverse axis of the bar is serious. The linear relation corresponding to the two observations is

$$1/\rho'=3{\cdot}432\times10^{-8}\,M+0{\cdot}615\times10^{-4}.$$

The relative curvature of the unloaded bar along the transverse axis is $0{\cdot}615\times10^{-4}$ cm.$^{-1}$. The corresponding radius is $1{\cdot}63\times10^4$ cm. or 163 metres. The face of the unloaded bar is concave towards the test-plate. The reciprocal of $3{\cdot}432\times10^{-8}$ is $2{\cdot}914\times10^7$, and we must use $2{\cdot}914\times10^7$ for $M/(\rho')^{-1}$ in place of $1{\cdot}617\times10^7$, the mean of the two values of $M\rho'$.

With $M/(\rho')^{-1} = 2{\cdot}914 \times 10^7$ and $M/\rho^{-1} = 7{\cdot}231 \times 10^6$ grm. cm., we have

$$\sigma = \rho/\rho' = 7{\cdot}231/29{\cdot}17 = 0{\cdot}248.$$

Rigidity. Load $M = 2500$ grm. was used. Length of arm

$$DE = 2h = 20{\cdot}07 \text{ cm.}, \quad \text{and} \quad h = 10{\cdot}04 \text{ cm.}$$

Measurements of transverse axes $2u_1 = U_1 U_1'$, ... (Fig. 38) of hyperbolas with foci on line UU', i.e. on $y = x$, were made when load was on D, giving τ_{DU}, and when it was on E, giving τ_{EU}. Similar measurements for transverse axes of hyperbolas with foci on VV', with load first on D and then on E, gave τ_{DV} and τ_{EV}. The values of τ_{DU} and τ_{DV} are given by (12) and (13), § 65. Results were:

[DU]	$(u_n^2 - u_1^2)/(n-1) = 0{\cdot}3340$ cm.^{2}	$\tau_{DU} = 1{\cdot}764 \times 10^{-4}$ cm.$^{-1}$
[EU]	$(u_n^2 - u_1^2)/(n-1) = 0{\cdot}2513$	$\tau_{EU} = 2{\cdot}345 \times 10^{-4}$
[DV]	$(v_n^2 - v_1^2)/(n-1) = 0{\cdot}2214$	$\tau_{DV} = 2{\cdot}662 \times 10^{-4}$
[EV]	$(v_n^2 - v_1^2)/(n-1) = 0{\cdot}4052$	$\tau_{EV} = 1{\cdot}454 \times 10^{-4}$

Hence

$$\tau_U = \tfrac{1}{2}(\tau_{DU} + \tau_{EU}) = 2{\cdot}054 \times 10^{-4}, \quad \tau_V = \tfrac{1}{2}(\tau_{DV} + \tau_{EV}) = 2{\cdot}058 \times 10^{-4} \text{ cm.}^{-1}.$$

Mean $\tau = 2{\cdot}056 \times 10^{-4}$ cm.$^{-1}$.

From dimensions of bar, $\frac{16}{3} - 3{\cdot}361 b/a = 4{\cdot}784$. Then, by (15), § 65,

$$n = \frac{Mgh}{\tau ab^3 (\frac{16}{3} - 3{\cdot}361 b/a)} = \frac{2500 \times 981 \times 10{\cdot}04}{2{\cdot}056 \times 10^{-4} \times 2{\cdot}087 \times 0{\cdot}3411^3 \times 4{\cdot}784}$$

$$= 3{\cdot}02 \times 10^{11} \text{ dyne cm.}^{-2}.$$

By § 66, $\qquad \sigma = \tfrac{1}{2}E/n - 1 = 0{\cdot}5 \times 7{\cdot}19/3{\cdot}02 - 1 = 0{\cdot}190.$

Experiment 10. **Determination of elastic constants of glass by focal line method.***

68. Introduction. The specimen is a bar of glass, but the method is not restricted to that material. A bar of metal could be employed if the face used for the optical reflexion were (1) sufficiently well polished and (2) sufficiently nearly plane when the bar is unstrained; the second requirement is not easily satisfied. It is therefore convenient to use a strip of plate glass which is well polished. If a number of strips be examined, one may be found which, when unstrained, has a face sufficiently nearly plane.

We suppose that the glass bar is rectangular in form, having a considerable length $2c$, width $2a$ and comparatively small thickness $2b$ cm.; in § 73, $c = 8$, $a = 1{\cdot}25$, $b = 0{\cdot}0902$ cm. Those surfaces which contain the edges $2c$, $2a$ will be called the "faces" of the

* Searle, "A focal line method of determining the elastic constants of glass," *Proc. Camb. Phil. Soc.* Vol. XXI, p. 772.

bar. The curvatures of one of these faces are determined by optical observations on focal lines.

Let the bar be *bent* by equal and opposite couples applied at the ends of the bar, the magnitude of each couple being G dyne cm.; the axis of each couple is parallel to the edges $2a$. Let the unstretched longitudinal filament, which, in the un-bent bar, passes through the centres of gravity of the transverse sections, be bent by these couples into an arc of a circle of radius ρ cm. Then, by § 56, if E be Young's modulus,

$$EI/\rho = \text{bending moment} = G, \quad \text{..............(1)}$$

where I is the "moment of inertia" of the transverse section of the bar about an axis through the centre of the section parallel to the edges $2a$; thus $I = \frac{4}{3}ab^3$. Hence

$$E = 3G\rho/(4ab^3). \quad \text{........................(2)}$$

A transverse filament which intersects at right angles the longitudinal filament just mentioned and, when the bar is unstrained, is parallel to the edges $2a$, will be bent into an arc of a circle of radius ρ', where

$$\rho' = \rho/\sigma, \quad \text{.........................(3)}$$

and σ is Poisson's ratio.* The point of intersection of the two filaments lies *between* the two centres of curvature.

When $2b/\rho$ is small, it will be sufficiently accurate to consider that the principal curvatures† of either face of the bar are $1/\rho$ and $1/\rho'$. The plane of one principal section of a face bisects the edges $2a$ at right angles and the plane of the other principal section of that face cuts the longitudinal filaments at right angles. Since the two curvatures have opposite signs, the face is anticlastic.

Now let the bar be *twisted* by equal and opposite couples of J dyne cm. applied at the ends of the bar, the axis of each being parallel to the edges $2c$. The particles, which in the unstrained bar lay in the median plane bisecting the edges $2b$, now lie on a helicoid. Let the helicoid have a twist of τ radians per unit length. By § 58, the lines of curvature‡ of the distorted median surface cut at $\frac{1}{4}\pi$ that median line of this surface to which, before the strain,

* For the conditions under which (1) and (3) are good approximations, see *Experimental Elasticity*, §§ 27–34.

† *Experimental Optics*, § 268. ‡ *Experimental Optics*, § 269.

the edges $2c$ were parallel. The principal radii of curvature are equal in length and opposite in sign. Let these radii be h and $-h$. Then, by § 58, $|h| = 1/|\tau|$. Hence, by § 59, if n be the rigidity, we have, to a sufficient approximation, when b/a is small,

$$n = \frac{|Jh|}{ab^3(\frac{16}{3} - 3{\cdot}361b/a)} = \frac{3|Jh|}{16ab^3(1 - 0{\cdot}630b/a)}. \quad \ldots(4)$$

When $2b/h$ is small, either face may be considered as having the same curvatures as the median surface.

Unless the faces of the unstrained bar be, to a reasonable approximation, optically plane, the focal lines will be ill-defined. Bars cut from thin "patent plate" glass are generally satisfactory.

69. Focal lines due to reflexion at a curved surface. In place of the general investigation given in books on optics, we use a "stationary time" method, adapting it to the special circumstances of the problem for (1) the bending, (2) the twisting of the bar.

Take "right-handed" rectangular axes Ox, Oy, Oz (Fig. 39), the origin O being on the surface, and let Ox coincide with the normal to the surface at O.

In the first case, we suppose that the lines of curvature at O lie in the planes Oxy, Oxz respectively. Then, for points near the origin, the equation to the surface can be written

$$x = \tfrac{1}{2}y^2/s + \tfrac{1}{2}z^2/t. \quad \ldots\ldots\ldots\ldots(5)$$

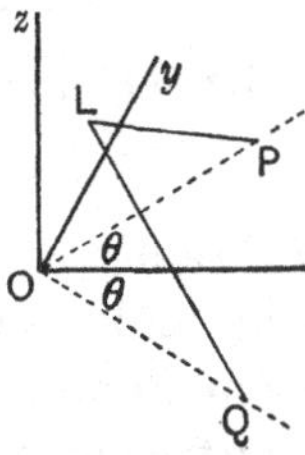

Fig. 39

Here s, t are the radii of curvature of the sections by the planes Oxy, Oxz, and are counted *positive* when the centres of curvature lie on the *positive* part of the axis of x.

Let P be a luminous point in the plane Oxy, the line OP lying in the quadrant yOx; let $OP = p$ and let the acute angle POx be θ. Let OQ be a straight line in the plane Oxy, such that Ox bisects the angle between OP and OQ. Then the chief incident ray PO becomes, after reflexion, the chief reflected ray OQ. Let $OQ = q$. Let p, q be positive when x is positive for P, Q.

Take any point L on the reflecting surface near O. Then, if the second and third coordinates of L be y, z, the first is $\frac{1}{2}y^2/s + \frac{1}{2}z^2/t$. Let $PL = u$, $QL = v$. Then

$$u^2 = (p\cos\theta - \tfrac{1}{2}y^2/s - \tfrac{1}{2}z^2/t)^2 + (p\sin\theta - y)^2 + z^2.$$

Hence, as far as squares of y and z,

$$u^2 = p^2\{1 - 2y\sin\theta/p - \cos\theta\,(y^2/s + z^2/t)/p + (y^2+z^2)/p^2\}.$$

Taking the square root, and expanding as far as the squares of y, z, we have

$$u = p\{1 - y\sin\theta/p - \tfrac{1}{2}\cos\theta\,(y^2/s + z^2/t)/p + \tfrac{1}{2}(y^2+z^2)/p^2 - \tfrac{1}{2}y^2\sin^2\theta/p^2\}$$

$$= p - y\sin\theta - \tfrac{1}{2}\cos\theta\,(y^2/s + z^2/t) + \tfrac{1}{2}(y^2\cos^2\theta + z^2)/p.$$

For v we change the sign of θ and write q for p. Thus

$$v = q + y\sin\theta - \tfrac{1}{2}\cos\theta\,(y^2/s + z^2/t) + \tfrac{1}{2}(y^2\cos^2\theta + z^2)/q.$$

For the total distance $PL + LQ = u + v$, we have

$$u + v = p + q - \cos\theta\,(y^2/s + z^2/t) + \tfrac{1}{2}(y^2\cos^2\theta + z^2)(1/p + 1/q). \quad \text{......(6)}$$

If $u+v$ have a stationary value, i.e. if $d(u+v)/dy = 0$, and $d(u+v)/dz = 0$, simultaneously, the ray reflected at L will meet the chief ray OQ in Q. We have $d(u+v)/dy = 0$, if either

$$y = 0, \quad \text{or} \quad 1/p + 1/q - 2/(s\cos\theta) \equiv \beta = 0,$$

and $d(u+v)/dz = 0$, if either

$$z = 0, \quad \text{or} \quad 1/p + 1/q - 2\cos\theta/t \equiv \gamma = 0.$$

If $y = 0$ and $z = 0$, the reflexion is restricted to a single point on the surface and the positions of P and Q along the lines defined by θ are immaterial. If $\beta = 0$ and $\gamma = 0$, *all* the rays from P which meet the surface near O pass, after reflexion, through the single point Q on the chief reflected ray. This case, however, requires the special conditions $\cos^2\theta = t/s$, and $1/p + 1/q = 2/\sqrt{(st)}$. The points P, Q are then the foci of a quadric surface of revolution, with PQ as axis, the reflecting surface coinciding with the quadric near O. When, as in the case of the bent bar, the surface is anticlastic, t/s is negative and this special case cannot arise.

We are left with the two remaining pairs of conditions, viz. (1) $z = 0$ and $\beta = 0$, and (2) $y = 0$ and $\gamma = 0$. When $z = 0$ and $\beta = 0$, all the rays from P, which meet the surface near O in the plane $z = 0$, pass after reflexion through that point Q on the chief reflected ray which is defined by

$$1/p + 1/q = 2/(s\cos\theta). \quad \text{.....................(7)}$$

Similarly, all the rays from P, which meet the surface near O in the plane $y=0$, pass after reflexion through the point Q' defined by

$$1/p+1/q'=2\cos\theta/t. \qquad\qquad (8)$$

In the experiment, P is made (by a lens) to be, in effect, at infinity and hence $1/p=0$. We then have for Q,

$$s=2q/\cos\theta, \qquad\qquad (9)$$

and for Q',

$$t=2q'\cos\theta. \qquad\qquad (10)$$

The planes containing (1) Q and Oz, and (2) Q' and Oy are the principal planes of the reflected wave front and contain the focal lines* through Q, Q' respectively.

In the second case, we suppose that the surface is a helicoid having Oz for axis and that the principal radii of curvature at O are h and $-h$. Then the lines of curvature at O are inclined at $\frac{1}{4}\pi$ to Oy and Oz. For points near O, the equation to the surface may be written

$$x=yz/h. \qquad\qquad (11)$$

The line of curvature which, at O, bisects the angle between the positive directions of Oy and Oz has its centre of curvature on the positive part of Ox, if h be positive.

Writing $2yz/h$ in place of $y^2/s+z^2/t$ in (6), we have

$$u+v=p+q-2\cos\theta\,.\,yz/h+\tfrac{1}{2}(y^2\cos^2\theta+z^2)(1/p+1/q). \qquad\qquad (12)$$

If $d(u+v)/dy=0$, and $d(u+v)/dz=0$, simultaneously, the ray reflected at L will meet the chief reflected ray in Q. These conditions give

$$2z/h=y(1/p+1/q)\cos\theta, \quad 2y/h=z(1/p+1/q)/\cos\theta. \qquad\qquad (13)$$

The values of z/y satisfying these equations are given by

$$z/y=\mp\cos\theta.$$

Thus, if Q and Q' correspond to $z=-y\cos\theta$ and to $z=y\cos\theta$ respectively, we have

$$1/p+1/q=-2/h, \quad 1/p+1/q'=2/h. \qquad\qquad (14)$$

When p is given, the values of q, q' depend only on h and not on θ.

* *Experimental Optics*, § 270.

In the experiment, P is, in effect, at infinity. Thus $1/p=0$, and then

$$h=-2q, \quad \text{.........................(15)}$$

$$h=2q'. \quad \text{.........................(16)}$$

We will now find the planes which contain the chief reflected ray OQ and the rays reflected at points on the curves in which the surface is cut by the planes $z=-y\cos\theta$, $z=y\cos\theta$ respectively.

On yO produced take a point M (Fig. 40) and draw KMK' parallel to Oz to meet the plane $z=-y\cos\theta$ at K, and the plane $z=y\cos\theta$ at K'. Then

$$KM=K'M=OM\cos\theta.$$

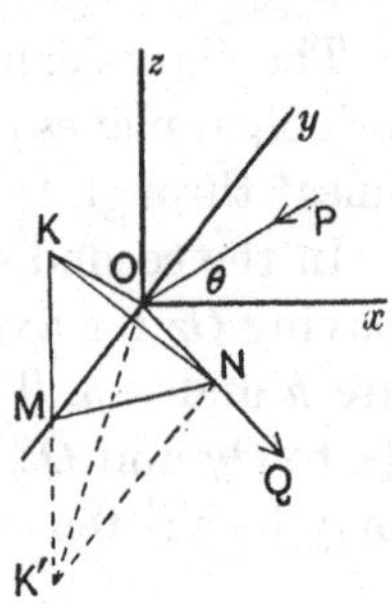

Fig. 40

Through KM draw a plane KMN to meet OQ at right angles at N. Then KN, MN, $K'N$ are perpendicular to ON, and KNM, $K'NM$ are the angles which the planes OKN, $OK'N$ make with the plane Oxy. Since $NOx=POx=\theta$, and $ONM=\frac{1}{2}\pi$, we have $OMN=\theta$. Hence

$$MN=OM\cos\theta=KM=K'M.$$

Since $KMN=\frac{1}{2}\pi$, we have

$$KNM=K'NM=\tfrac{1}{4}\pi.$$

The planes OKN, $OK'N$ contain the rays which meet in Q, Q' respectively and are thus the principal planes of the reflected wave front. The focal lines through Q, Q' lie in the planes $OK'N$, OKN respectively; their distances from O depend upon h, for $q=-\frac{1}{2}h$, $q'=\frac{1}{2}h$.

70. Optical method of measuring curvatures of reflecting surface. The surface S (Fig. 41) is placed so that the tangent plane at its central point O is vertical. The normal Ox and the axis Oy are horizontal and Oz is vertical. The surface is placed so that its lines of curvature at O are either vertical and horizontal or are inclined at $\frac{1}{4}\pi$ to Oz. A collimator, consisting of a lens L and cross-wires, which intersect at K in the

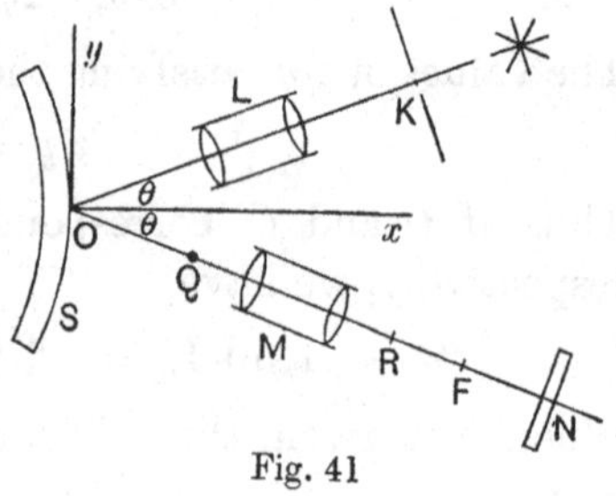

Fig. 41

focal plane of L, is directed to O, the line of collimation KO being horizontal. The wires are stretched across an opening in a plate and are illuminated by a lamp. The rays from K fall on S as a parallel beam and thus $1/p=0$ in (7), (8) and (14). The rays reflected at S are received by a converging lens M of focal length f, whose axis passes through O and is horizontal, and are brought to a point-focus or to a focal line on the ground-glass screen N, fixed to a carriage sliding on a short optical bench. If the reflecting surface S be one surface of a glass plate, the other surface is smeared with soap or vaseline to prevent regular reflexion there.

The positions of the focal lines are found by Newton's method.* For this we must know the distance of the object from the appropriate focus and the distance of the image from the other focus. It is, therefore, convenient that one focus of M should coincide with O. A plane mirror is substituted for N, and then M is set so that a speck of paper placed on S at O coincides with its own image formed by rays which pass (twice) through M and are reflected normally by the plane mirror.

If S be plane, both wires will be sharply focused on N when N is in the focal plane through F, the other focus of M.

If S be curved, and θ, the angle of incidence, be finite, it will not be possible to obtain an "image" of either cross-wire, unless the directions of the wires be properly adjusted. It will still be impossible to obtain images of both wires simultaneously, but each can be focused in turn by adjusting N.

The cross-wires are soldered to the plate. To ensure that the angle between the wires does not differ appreciably from $\frac{1}{2}\pi$, two lines at right angles are scribed on the plate and the wires are laid along these lines when they are soldered to the plate. If the plate can be turned in its own plane, the directions of the wires can be adjusted. Careful setting of the cross-wires is needed if good "images" are to be obtained.

If, instead of cross-wires, there were a fine hole at K, a parallel beam would fall on S and the reflected rays would pass through two focal lines meeting the chief reflected ray in Q, Q'. If a luminous point were placed at Q, it would give rise to two focal

* *Experimental Optics*, § 152.

lines, one passing through K and one elsewhere. A luminous point at Q' would give rise to two focal lines, one passing through K and one elsewhere. If the cross-wires at K be placed in the directions of those focal lines due to Q and Q' which pass through K, it will be possible to focus each wire in turn by adjusting the screen N. Hence, when the lines of curvature at O are horizontal and vertical, the cross-wires must also be horizontal and vertical. When the lines of curvature at O are inclined at $\frac{1}{4}\pi$ to the vertical axis Oz, the cross-wires must also be inclined at $\frac{1}{4}\pi$ to Oz.

Let R be the image of Q formed by M, let $FR = \xi$, and let ξ be counted positive when R lies on the same side of F as the focus O. Then, as in § 69, counting OQ $(=q)$ positive when Q is "real," we have, by Newton's formula,* $\xi q = f^2$, or

$$q = f^2/\xi. \quad \text{.........................(17)}$$

The distance ξ is easily measured, as it is merely the displacement of the screen N from the position in which both wires are in focus on it, when a good plane surface, such as that of a prism, is used as the reflecting surface.

In the case of (15) and (16), we can, without knowing θ, find h from (17) when f and ξ are known. But in the case of (9) and (10), in addition to f and ξ, the value of $\cos\theta$ is required. A *convex* spherical surface, e.g. the surface of a lens, of radius r is substituted for S, and is adjusted so that (1) the chief reflected ray has the same direction as before and (2) the surface of the lens passes through the same point O as did the surface S. Then in (9) and (10) $s = -r$, $t = -r$, and thus, if η be the value of ξ for the vertical focal line and η' the value for the horizontal focal line, we have

$$f^2 = \eta\,(-\tfrac{1}{2}r\cos\theta), \qquad f^2 = \eta'\,(-\tfrac{1}{2}r/\cos\theta).$$

Hence $$\cos^2\theta = \eta'/\eta. \quad \text{.......................(18)}$$

A convex is used in preference to a concave spherical surface, for, unless the curvature of the latter be slight, the images formed by M of the focal lines will not be "real."

In practice f is more easily determined than r, but, if desired, f can be found in terms of r, since $f^4 = \frac{1}{4}r^2\eta\eta'$.

* *Experimental Optics*, § 190.

71. Determination of Young's modulus and Poisson's ratio. The glass bar rests with one face against two vertical pillars, whose sections are C, D (Fig. 42), rising from a base H (see Fig. 44); the base is fixed to the table in any convenient manner. The bar is bent by two strings which pass over ball-bearing pulleys U, V and are attached at A', B' (Fig. 43) to a light yoke Y. A loop at the end of the U-string passes round the bar at A, and similarly for the other string. From the centre of the yoke hangs a light pan in which loads are placed. For each value of the load, the screen N is adjusted so that (1) the vertical wire, (2) the horizontal wire is in focus.

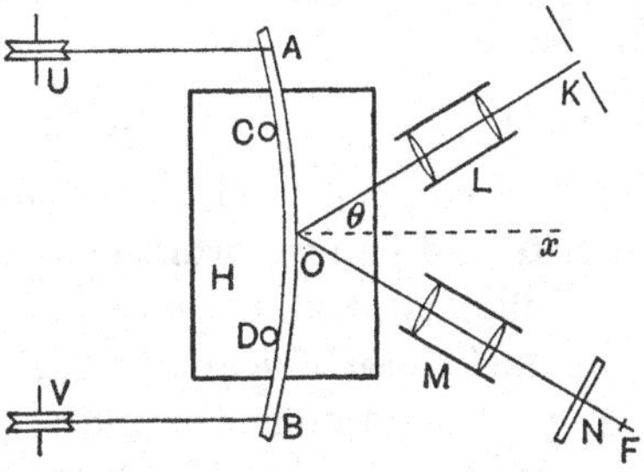

Fig. 42

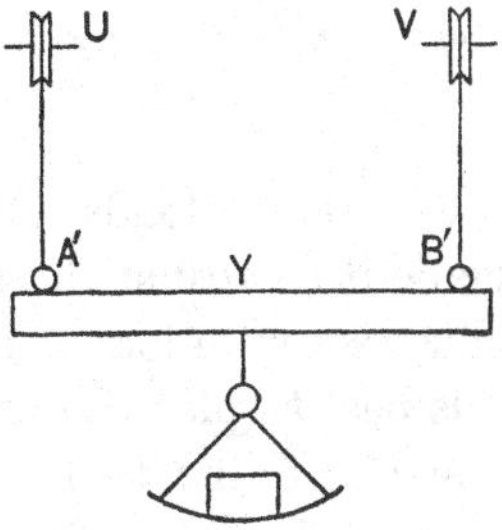

Fig. 43

In Fig. 42 the axes of y and z are in the tangent plane to the reflecting surface of the bar at O, and the axis of z is vertical. The radius of curvature of the section of the surface by the horizontal plane Oxy is ρ and the centre of curvature lies on the negative part of Ox. The radius of curvature of the section by the vertical plane Oxz is ρ' and the centre of curvature lies on the positive part of Ox. Comparing these results with the notation of (5) and using (3), we find

$$s = -\rho, \quad \text{.........................(19)}$$

$$t = \rho' = \rho/\sigma. \quad \text{......................(20)}$$

Thus s is negative and t is positive.

Let ξ, ξ' be the values of FN for the vertical and horizontal lines respectively, when the *total* load carried by the two strings is m grm. When the bar is bent, as in Fig. 42, s is negative and thus, by (9), q is negative, and, by (17), ξ is negative; hence ON is greater than OF for the vertical focal line. Let

$$\tfrac{1}{2}(AB - CD) = d \text{ cm.},$$

where CD is measured between the lines of contact of the bar with the pillars. If the strings be symmetrically arranged so that $AC = BD = d$, the bending moment, G, at any part of the bar between C and D is constant and $G = \frac{1}{2}mgd$.

By (2), $E = 3G\rho/(4ab^3)$, by (19) and (9), $\rho = -s = -2q/\cos\theta$; by (17), $q = f^2/\xi$. Hence, we have for Young's modulus

$$E = \frac{3gdf^2}{4ab^3\cos\theta} \cdot \left(\frac{m}{-\xi}\right). \qquad \text{.................(21)}$$

The value of $\cos\theta$ is given by (18).

By (3), (19) and (20), we have $\sigma = \rho/\rho' = -s/t$. By (9) and (10), $s/t = q/(q'\cos^2\theta)$. Since, by (17), $q/q' = \xi'/\xi$, and, by (18), $\cos^2\theta = \eta'/\eta$, we have for Poisson's ratio

$$\sigma = -\xi'\eta/(\xi\eta'). \qquad \text{....................(22)}$$

A series of loads, increasing by equal steps, is used. It is convenient to adjust the mass of the yoke and pan to form the first of the series of loads. In view of the difficulty of the observations, it is best to plot the values of $-\xi$ against those of m and to draw a straight line to lie as evenly as possible among the plotted points. The ratio of the rate of increase of $-\xi$ to that of m is then found from the straight line, and this ratio is used in place of $-\xi/m$ in (21). The ratio $-\xi'/\xi$ is found by plotting ξ' against $-\xi$.

72. Determination of rigidity. The base and pillars used in § 71 are again employed. On the pillars C, D slides a bridge EF (Fig. 44) which can be clamped to C, D. The upper end A of the glass bar is placed between the bridge EF and a metal plate RS and is clamped by screws passing through the ends of RS. The lower end B of the bar rests in a groove cut in the drum W, which turns freely about a vertical pivot rising from a plate T, which is attached to the base H in such a position that the edges of the bar are vertical. A fine string attached to W passes over a ball-bearing pulley and supports a mass m grm. Let the radius of the drum on which the string is wound be k cm., and let J be the couple applied to the glass bar; then $J = gkm$ dyne cm. To distinguish between positive and negative couples, we count the loads positive or negative according as they strain the bar into a right-handed or a left-handed helicoid.

The small table L (Fig. 45), having an angle-piece K, which can be clamped between EF and RS in place of the bar, is used for supporting a prism or a lens; the height is adjusted by sliding EF on the pillars.

We suppose that the positive direction of the axis of x is from O towards the reader, that the axes of y and z are as shown in Fig. 44, and that, in the standard case, the load m turns the drum W in a clockwise direction as seen from the end A of the bar. The bar is thus twisted into a right-handed helicoid with Oz as axis,

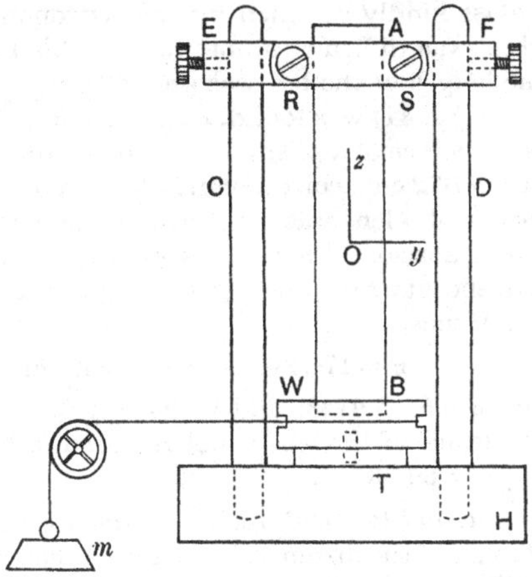

Fig. 44

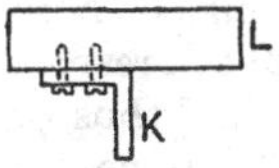

Fig. 45

and hence the quantity h in (11) is negative. By § 69, $q = -\frac{1}{2}h$ for that focal line which has a north-east direction,* as seen on the ground-glass screen N, when N lies between the lens M and the observer, and, by (17), ξq is positive. In the standard case, h is negative and hence ξ is positive for the north-east focal line and ξ' is negative for the north-west line.

If the direction of the couple be reversed, ξ is negative for the north-east line and ξ' is positive for the north-west line.

A series of loads is applied for each direction of the couple; the loads which produce a left-handed helicoid are counted as negative. From the straight lines lying most evenly among the points, when ξ and $-\xi'$ are plotted against m, the ratios of the rates of

* "North" is upwards and "east" is towards the observer's right.

increase of ξ and of $-\xi'$ to that of m are found; these ratios are used for ξ/m and for $-\xi'/m$. The mean of the positive quantities ξ/m and $-\xi'/m$ is denoted by ξ_0/m_0; this mean is used in (23) below. By (15), (16) and (17), $h=-2f^2/\xi=2f^2/\xi'$. For the couple we have $J=gkm$, and thus, by (4), we find for the rigidity

$$n=\frac{3gkf^2}{8ab^3(1-0{\cdot}630b/a)}\cdot\frac{m_0}{\xi_0}. \quad \text{.................(23)}$$

73. Practical example.

Mr C. F. Sharman, of King's College, used a bar cut from a photographic plate of thin "patent plate" glass kindly supplied by the Astronomer Royal at a time when such glass was difficult to obtain. For this bar, $a=1{\cdot}25$ cm., $b=0{\cdot}0902$ cm. The length of the bar was about 16 cm.

The focal length of the lens M (Fig. 41) was found, by the goniometer method,* to be $f=16{\cdot}13$ cm. When a face of a prism was the reflector, the bench reading of the screen N was 17·72 cm., corresponding to the point F (Fig. 41). When the convex surface of a lens was the reflector, the bench readings of N for the vertical and horizontal focal lines were 29·12 and 27·21 cm. respectively. In each of the latter readings, NO was greater than FO, and thus η, η' are negative. Hence

$$\eta=17{\cdot}72-29{\cdot}12=-11{\cdot}40\text{ cm.},\qquad \eta'=17{\cdot}72-27{\cdot}21=-9{\cdot}49\text{ cm.}$$

Then $\cos^2\theta=\eta'/\eta=0{\cdot}83246$; thus $\cos\theta=0{\cdot}91239$ and $\theta=24°\,9'\,44''$.

The radius, r, of the convex surface of the lens is not required in this experiment. By § 70, $r=2f^2/\sqrt{(\eta\eta')}=50{\cdot}03$ cm.

Determination of Young's modulus and Poisson's ratio. The distance AB between the strings was 15·20 cm., and the distance CD between the lines of contact of the bar with the pillars was 6·98 cm. Hence

$$d=\tfrac{1}{2}(AB-CD)=4{\cdot}11\text{ cm.}$$

To allow for any slight initial curvature of the bar, the zero readings of the screen N for the vertical (ξ) and the horizontal (ξ') focal lines were obtained, not by aid of the prism, but from the bar itself when the load was zero.

The values found for ξ, ξ' were as follows:

Load $=m$	ξ	ξ'	Load $=m$	ξ	ξ'
grm.	cm.	cm.	grm.	cm.	cm.
0	0	0	800	−1·06	0·26
200	−0·23	0·02	1000	−1·33	0·31
400	−0·53	0·15	1200	−1·56	0·38
600	−0·80	0·21	1400	−1·84	0·45

When $-\xi$ was plotted against m, and ξ' against $-\xi$, the slopes of the best straight lines (§ 71) gave the values

$$-\xi/m=1{\cdot}321\times10^{-3}\text{ cm. grm.}^{-1},\qquad \xi'/\xi=-0{\cdot}230.$$

* *Experimental Optics*, § 234.

Then, by (21), we have for Young's modulus,

$$E=\frac{3gdf^2}{4ab^3\cos\theta\,(-\xi/m)}=\frac{3\times 981\times 4{\cdot}11\times 16{\cdot}13^2}{4\times 1{\cdot}25\times 0{\cdot}0902^3\times 0{\cdot}91239\times 1{\cdot}321\times 10^{-3}}$$

$$=7{\cdot}12\times 10^{11}\,\text{dyne cm.}^{-2}.$$

Also, by (22), we have for Poisson's ratio,

$$\sigma=-\frac{\xi'\eta}{\xi\eta'}=\frac{0{\cdot}230}{0{\cdot}83246}=0{\cdot}276.$$

Poisson's ratio can also be found from the formula $\sigma=\frac{1}{2}E/n-1$. With $E=7{\cdot}12\times 10^{11}$ and $n=2{\cdot}93\times 10^{11}$, as found below, we have

$$\sigma=0{\cdot}215.$$

Determination of rigidity. The radius of the drum on which the string was wound was (with allowance for thickness of string) $k=2{\cdot}535$ cm. In the Table, couples straining the bar into a left-handed helicoid are indicated by the negative sign prefixed to the loads.

Positive couples			Negative couples		
Load m	N.E. image ξ	N.W. image ξ'	Load m	N.E. image ξ	N.W. image ξ'
grm.	cm.	cm.	grm.	cm.	cm.
200	0·15	−0·17	−200	−0·17	0·22
400	0·29	−0·35	−400	−0·40	0·42
600	0·53	−0·59	−600	−0·58	0·57
800	0·72	−0·72	−800	−0·77	0·81

When ξ and $-\xi'$ were plotted against m, the slopes of the best straight lines (§ 72) gave the values $\xi/m=9{\cdot}30\times 10^{-4}$, and $-\xi'/m=9{\cdot}60\times 10^{-4}$. The mean is $\xi_0/m_0=9{\cdot}45\times 10^{-4}$ cm. grm.$^{-1}$. Then, by (23), we have for the rigidity,

$$n=\frac{3gkf^2}{8ab^3(1-0{\cdot}630b/a)}\cdot\frac{m_0}{\xi_0}=\frac{3\times 981\times 2{\cdot}535\times 16{\cdot}13^2}{8\times 1{\cdot}25\times 0{\cdot}0902^3\times 0{\cdot}9545}\cdot\frac{1}{9{\cdot}45\times 10^{-4}}$$

$$=2{\cdot}93\times 10^{11}\,\text{dyne cm.}^{-2}.$$

Experiment 11. Measurement of Young's modulus by method of optical interference.*

74. Introduction. When a small distance is measured by aid of a screw with a micrometer head, it is given by the number of complete turns and a fraction of a turn of the screw. Each complete turn of a perfect screw steps out a definite distance—the pitch of the screw. Screws of high quality can be obtained, but it is, of course, impossible to construct a truly perfect screw. Of recent years the demands of precise mechanical work have led increas-

* Searle, "An optical interference method of measuring Young's modulus for rods," *Proc. Camb. Phil. Soc.* Vol. xxii, p. 475.

ingly to the use of waves of light as measures of distance, and now some of the most accurate comparisons of standards of length and tests of engineers' gauges are made by means of waves of light. The wave-length takes the place of the pitch of the screw. The fractions of wave-lengths are obtained from measurements of the diameters of interference rings, or by an equivalent method.

When the measurements are made not in a vacuum but in air, a very small correction, depending upon the wave-length, is needed to allow for the refractive index of the air corresponding to the density of the air at the time. Some spectrum lines, such as the red cadmium line of wave-length $6{\cdot}4384696 \times 10^{-5}$ cm., in dry air at 15° C. and 760 mm. of mercury, are very narrow and furnish the most reliable standards available.

Two marks on a bar of metal have long been used to define a length, but, with increasing accuracy of measurement, it has been found that the lengths of metal bars are by no means invariable over long periods of time. The conclusion has been reached that it is better to depend upon a very definite radiation emitted by practically free atoms than to rely upon the distance between the centres of two marks, which are grooves, each covering hundreds of atoms in its width, ploughed on a bar composed of vast numbers of atoms (mixed atoms in the case of alloys) and subject to strains, perhaps progressively changing, due to manufacture or to the mode of support.

The following experiment may serve as an introduction to work of this class. The method of detecting and measuring small movements by optical interference is easily applied, and has been used by makers of microscopes and other instruments in testing the stiffness of their apparatus.

75. Method. Young's modulus can be found by observations on the bending of a rod. In the present experiment, two plates are fixed to the rod at right angles to it, a few centimetres apart, and the change in the distance between two points, one on each plate, due to the bending of the rod is found by an optical interference method. The curvature is easily deduced from the change of distance.

The general arrangement of the apparatus is shown in Fig. 46.

A vertical steel rod *AA* of circular cross-section is fixed to a heavy cast-iron base *Z* shown diagrammatically in the figure. The rod *AA* is bent by a load applied to the end of the horizontal cross-bar *BC* carried by the upper end of *AA*. The point of application of the load is defined by a transverse groove cut in the bar. The rod

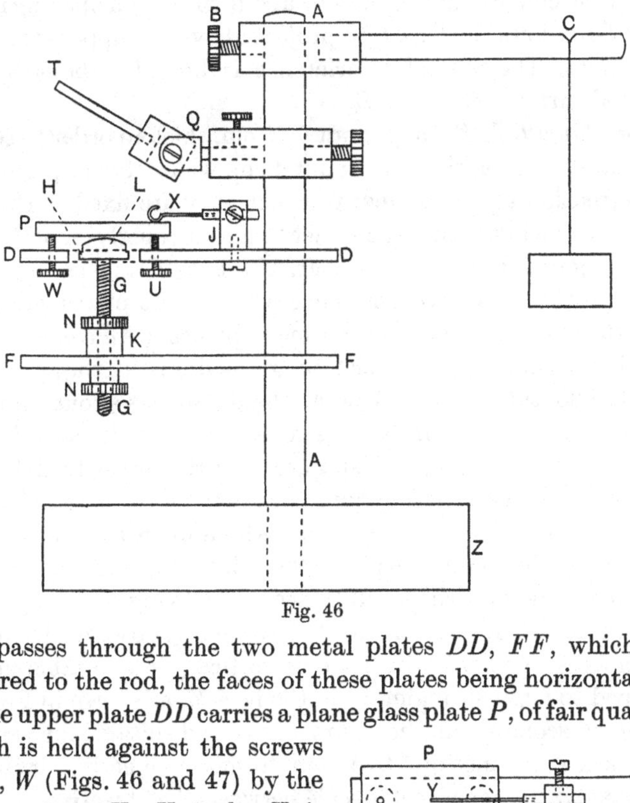

Fig. 46

AA passes through the two metal plates *DD*, *FF*, which are soldered to the rod, the faces of these plates being horizontal.

The upper plate *DD* carries a plane glass plate *P*, of fair quality, which is held against the screws *U*, *V*, *W* (Figs. 46 and 47) by the wire springs *X*, *Y* (only *X* is shown in Fig. 46). These springs are soldered into short brass cylinders held by set-screws in the fittings *J*, *J*. The spring *X* presses *P* against the screw *U*; the spring *Y* presses on *P* between the screws *V*, *W*. Thus the screw *U* turns the plate about an axis through the tips of *V* and *W*, while the screw *W* turns it about an axis through the tips of *U*, *V*.

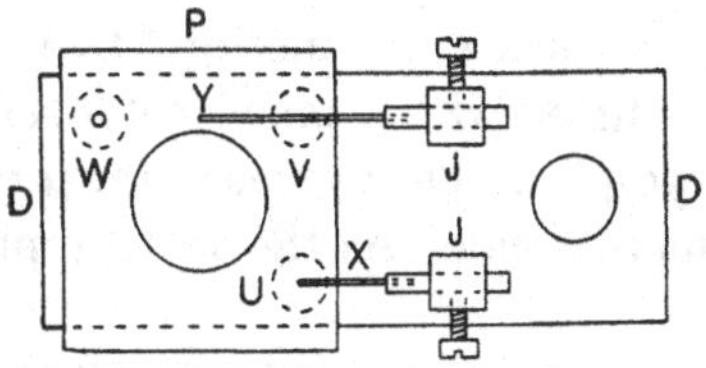

Fig. 47

The lower plate *FF* carries a pillar *K* through which passes the screwed rod *GG*. This rod is locked in position by the nuts *N*, *N*, which have knurled edges.* The rod carries a small metal table *H*. A plano-convex lens *L*, with convex surface upwards, is fixed by black sealing-wax in a recess turned in *H*; the radius of curvature of the convex surface may be about 50 cm. An opening in the plate *DD* allows the lens *L* to project above the upper surface of that plate. The optical interference is brought about by the opposed surfaces of *P* and *L*.

The plates *DD*, *FF* are about 0·25 cm. thick in order to secure adequate stiffness; they are about 4 cm. wide. It is important to pay attention to the manner in which they are fixed to the rod. Care must be taken that the solder "runs" properly on both sides of each joint. It is obvious that, if the soldering be not symmetrical, the "effective" distance between the plates will differ from the distance between their mean planes, with a consequent error in the value obtained for Young's modulus. In the apparatus used in § 80, the distance between the plates was about 2·7 cm.

Light from a sodium flame is reflected on to the lens *L* by a plane glass reflecting plate *T* supported by the vertical rod *AA* and suitably adjusted. A converging lens may be so placed between the flame and *T* as to concentrate the light upon *L* after reflexion at *T*. When the lens *L*, the plate *P* and the reflector *T* are properly adjusted, Newton's rings will be seen on looking vertically down upon the lens. The rings are most conveniently observed by means of a microscope held, with its axis vertical, in a suitable holder clamped to a rod springing from the base *Z*, as shown in Fig. 49. The microscope should be of low power and suitable for viewing objects at a distance of 5 to 10 cm. from the objective; it should be fitted with a micrometer scale between the objective and the eye-piece. The eye-piece must be adjustable relative to the micrometer scale. As the load applied to the cross-bar is increased,

* Since the first instruments were made at the Cavendish Laboratory, it has been found unnecessary to provide any vertical adjustment for the lens. In later models, the pillar supporting the lens is a solid rod secured to the plate *FF* by a nut, as suggested in Fig. 50. If the pillar be of steel, the gap between the lens and the plate will be little affected by a rise of temperature which is the same for the pillar as for the vertical rod *AA* (Fig. 50). Unequal heating, such as may be caused by touching one side of *AA* with the finger, will change the gap and should be avoided.

the ring system will contract and rings will disappear one by one at the centre of the pattern. As the load is diminished, the ring system will expand and new rings will start out from the centre.

If the fitting holding the reflector T be provided with a hinge Q (Fig. 46), or if it be fixed to the rod supporting the microscope, the line joining the flame to T may be either at right angles to BC or parallel to BC. The latter arrangement is convenient when the width of the bench is limited.

It is necessary that the load should be increased or diminished in a gradual manner. If the load be increased by placing a mass in a pan, it is impossible to avoid setting up vibrations of the system which cause such rapid motions of the Newton's rings that it is impossible to count how many rings have disappeared. The requisite gradual variation of the load suspended from the cross-bar BC from zero to some determinate value, or the reverse process, is easily accomplished by the device shown in Fig. 48. A light frame is formed by riveting and soldering the ends of a bent stout wire W into a yoke L. Into a block attached to L is soldered the lower end of a helical spring such that a load of a kilogramme extends it about 2 cm. The spring is stiff enough to remain upright when the frame is upright, although the upper end of the spring is free. A loop R is formed at the upper end of the spring and the cross-bar BC (Fig. 46), seen in section in Fig. 48, passes through R with plenty of clearance. The frame is suspended by its apex T from a hook H. This hook can be raised or lowered by a screw arrangement supported by a bridge which is carried by two vertical rods springing from the massive base of the instrument, as shown in Fig. 49. The hook is fixed to the lower end of the screw. The screw, which has a groove cut in it parallel to its axis, is prevented from turning in the bridge by a pin projecting into the groove and is raised or lowered by a nut which bears on the bridge and is furnished with a large head.

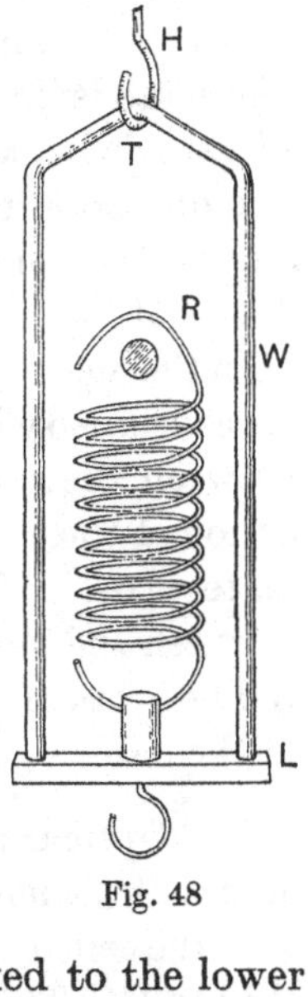

Fig. 48

A load, conveniently formed of slotted weights resting on a

suitable hanger, is suspended from a hook carried by the yoke L (Fig. 48). When the hook H is sufficiently lowered, the whole load of weight, frame and spring is borne by the cross-bar. When H is sufficiently raised, the loop R will rise clear of the cross-bar, which will then be relieved of all load.

The effects of the vibrations of the room upon the rings are much diminished when the instrument stands on a stone slab resting on a thick pad of felt.

The complete apparatus, as made by W. G. Pye and Co., is shown in Fig. 49. In Pye's model the spring arrangement has been modified, and the loading weight hangs from a metal plate having a hole through which the cross-bar passes with ample clearance. The plate is connected to the adjustable hook by the spring. The operation of this arrangement is the same as that of the device of Fig. 48.

Fig. 49

In the mechanical theory we are concerned with the distance between the (vertical) line of action of the load and the axis of the part of the rod between the plates DD, FF (Fig. 50). This distance will not equal that between the notch C and the axis of the rod near A unless the rod be vertical. It may be made vertical by the levelling screws in the base.

76. Optical adjustments. For good workmanship, it is desirable that the lens L (Fig. 46) should be so mounted on the table H that the centre of curvature of its upper surface lies on the axis of the rod GG. This adjustment may be tested by holding the rod in a self-centering chuck in a lathe and observing the image of a flame formed by reflexion at the convex surface. If the image do not remain stationary as the lathe mandrel revolves, the centre of curvature lies off the axis of GG. Moderate accuracy in this adjustment suffices.

The height of the pillar *G* (Fig. 46) should be such that the vertex of the lens is one or two millimetres above the upper surface of the plate *DD*. The screws *U*, *V*, *W* are set slightly higher than the vertex so that the glass plate *P* can be slipped into place without rubbing and perhaps damaging the lens. The screws are then lowered until the gap between the lens and the plate is less than one-tenth of a millimetre. The gap is easily observed by aid of a magnifying glass if it be illuminated by a roughly horizontal beam of light. The plate *P* is then adjusted by the screws *U* and *W* so that the plate appears parallel to the tangent plane of the lens at its highest point. If this preliminary setting be carefully made, the rings, if not actually in view, may be brought into view by slight adjustments of the screws *U* and *W*. Once the rings have been seen, it is easy to bring their centre to any desired point by turning the screws *U* and *W*. At this stage, it is more convenient to observe the rings by eye or with a magnifying lens than to use the microscope.

The screws *U*, *V*, *W* may now be slightly readjusted so as to reduce the gap to the order of a hundredth of a millimetre, care being taken to keep the centre of the rings approximately on the axis of the lens pillar *GG*. This adjustment is made when the cross-bar carries no load. The ring system is most vigorous when the gap is very small. If there be not actual contact between lens and plate, a slight *upward* pressure of the finger upon the cross-bar will cause the rings to expand in a lively manner. If this slight pressure fail to affect the rings, the lens is *touching* the plate, and the screws *U*, *V*, *W* must be turned in the direction tending to raise the plate, until the rings respond to the pressure.

The wave-length λ_1, of the D_1 line of sodium, exceeds λ_2, the wave-length of D_2, by about 1/10 per cent. The ring system for D_1 will reinforce that for D_2, near the centre of the system, when the width t of the gap and the number n are given by

$$2t = n\lambda_1 = (n+h)\lambda_2,$$

where h is a positive integer or zero. Then

$$n = h\lambda_2/(\lambda_1 - \lambda_2), \quad t = \tfrac{1}{2}h\lambda_1\lambda_2/(\lambda_1 - \lambda_2).$$

Now $\lambda_1 = 5{\cdot}89593 \times 10^{-5}$ cm., $\lambda_2 = 5{\cdot}88996 \times 10^{-5}$ cm., and hence, for $h=0, 1, 2, \ldots$ we have $n=0, 986{\cdot}6, 1973{\cdot}2, \ldots$, and t has the values

$$0, \quad 0{\cdot}02908, \quad 0{\cdot}05817, \ldots \text{ centimetres.}$$

The bright rings for D_1 will fall midway between those for D_2 near the centre of the pattern, and thus, near the centre, the pattern will be inconspicuous, when

$$2t = n\lambda_1 = (n + \tfrac{1}{2}k)\,\lambda_2,$$

where k is an odd positive integer. For $k=1, 3, 5, \ldots$ we have $n=493{\cdot}3, 1479{\cdot}9, 2466{\cdot}5, \ldots$, and t has the values

$$0{\cdot}01454, \quad 0{\cdot}04363, \quad 0{\cdot}07271, \ldots \text{ centimetres.}$$

The rings will not *quite* disappear, for the D_2 is more intense than the D_1 line.

The maxima of contrast at $t=0$, $t=0{\cdot}029$, $t=0{\cdot}058$ cm. and the minima of contrast at $t=0{\cdot}015$, $t=0{\cdot}044$ cm. may be observed, but the apparatus is not designed for the measurement of the distances.

A sodium flame provides the most *convenient* source of light. If, however, a mercury lamp be used and the green line ($\lambda = 5{\cdot}460742 \times 10^{-5}$ cm. in air at 15° C. and 760 mm. pressure) be isolated by a suitable filter, the ring system will be more intense. Since the line is single, the rings will not disappear and reappear periodically as the distance between the plate and the lens is increased. The rings may still be seen when the distance is as great as 0·3 cm.

77. Experimental details. When the optical adjustments have been made, the observations are taken. A load is hung from the frame, and the hook H (Fig. 48) is lowered so far that the whole weight of load and frame is supported by the cross-bar BC. If, now, H be gradually raised by its screw, rings will begin to move outwards from the centre as soon as the hook H begins to take any part of the load. The formation of each new ring is noted—it starts as a dark central spot—and the counting is continued until further raising of H ceases to produce fresh rings. It will then be found that the loop R (Fig. 48) has been lifted clear of the cross-bar. The process can be reversed, and the load can be *imposed* by lowering H, but the formation of a new ring is somewhat easier to observe than the disappearance of an old ring. With a little practice, rough estimates of "fractions" of rings can be made. Thus, if initially there be a dark spot at the centre and finally there be a light spot at the centre, the number of rings is an integer plus one half. The method of making true estimates of these fractions, with the necessary theory, is given in §79.

The distance of the centre of the ring system from the nearest point of the rod is measured by a divided scale. The distance a cm. (Fig. 50) from the centre of the rings to the axis of the rod is found by adding to the distance just measured the radius of the rod. It is assumed that the centre of the rings lies approximately in the vertical plane CAA.

If a series of observations with different loads be made, it will be found that the number of rings which appear or disappear is proportional to the load. In each case, the "load" includes the device of Fig. 48.

78. Mechanical theory. A load of M grammes suspended from BC at C (Fig. 50) is equivalent to a load of Mg dynes acting along the axis of the vertical rod AA together with a couple of Mgc dyne cm., where c cm. is the distance of C, the point of support, from the axis of AA.* The couple bends the rod into an arc of large radius ρ cm., where

$$EI/\rho = Mgc. \quad ......(1)$$

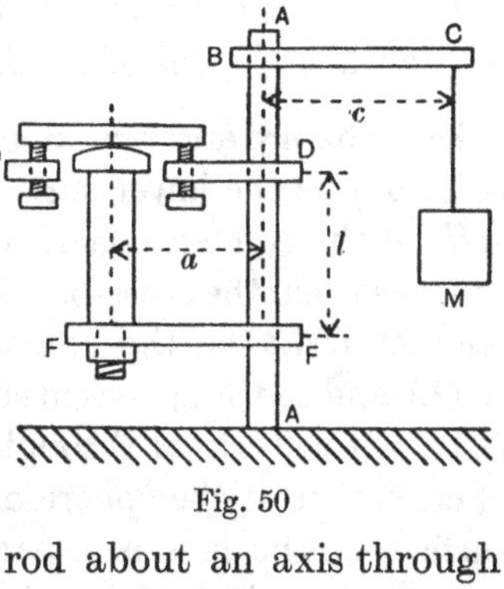

Fig. 50

Here E is Young's modulus for the metal and I is the "moment of inertia" of the cross-section of the rod about an axis through the centre of gravity of the section perpendicular to the plane of bending, and thus $I = \frac{1}{4}\pi r^4$, where r cm. is the radius of the rod. Hence

$$1/\rho = 4Mgc/(\pi E r^4). \quad ....(2)$$

* The rod AA may not be accurately vertical and, in any case, it is bent by the loading. The distance c, on which the couple depends, is the horizontal distance of the vertical line through C from the mid-point of the axis of the rod between the plates D and F.

If the distance between the points in which the mean planes of the plates *DD*, *FF* cut the axis of the rod be l cm., the angle between the plates when the rod is bent will be l/ρ radians. If the distance of the central spot of the Newton's ring system from the axis of *AA* be a cm., the *increase* of distance between the lens and the plate due to the bending will be al/ρ cm. or

$$4Mgacl/(\pi Er^4)\text{ cm.}$$

The load Mg causes the rod to *shorten* by $Mg/(E\pi r^2)$ cm. per cm. and thus reduces the distance between the lens and the plate by

$$Mgl/(\pi Er^2)\text{ cm.}$$

Hence, if x cm. be the resultant increase of distance between the lens and the plate due to the load Mg applied at C,

$$x = \frac{Mgl}{\pi Er^4}(4ac - r^2). \quad \text{......................(3)}$$

If n new rings be formed when the load Mg is removed, we have $x = \frac{1}{2}n\lambda$. Hence

$$E = \frac{2Mgl(4ac - r^2)}{n\pi\lambda r^4}. \quad \text{......................(4)}$$

For sodium light we may take $\lambda = 5{\cdot}893 \times 10^{-5}$ cm. as the mean of the wave-lengths of the D_1 and D_2 lines.

79. Theory and practice of optical measurements. Let *OX* (Fig. 51) be the lower surface of the horizontal glass plate. Let *AH* be the convex spherical surface of the lens when the cross-bar is not loaded and let *OAZ* be the common normal to *OX* and to the spherical surface; then *OAZ* is the vertical through the centre of curvature of the sphere of which the surface of the lens is a part. Let the innermost dark ring* in this initial condition of zero load be of radius *OU* and let the vertical through *U* meet the

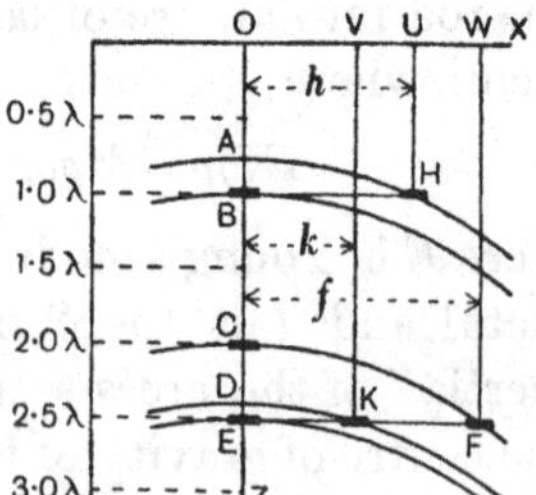

Fig. 51

* Many students suppose that in a system of Newton's rings there are *sharp* boundaries, which they call "edges," between brightness and darkness. Since, in actual fact, the transition from brightness to darkness is *gradual*, it is impossible to take a reading on the "edge" of a dark ring. We can, however, estimate with considerable accuracy the position of the line of maximum darkness in any dark ring.

sphere in H. Let HB be perpendicular to OA. For brevity, we may say that the innermost dark ring is *at* H. Let $BH = h$ cm. Here, and throughout this discussion, "dark ring" is used as an abbreviation for "circle of maximum darkness."

If the load be now applied gradually, the lens will move away from the plate, the rings will contract, and, when the vertex of the lens has sunk to B, where $BO = HU$, the innermost ring will have shrunk to a dark central spot at B. The short heavy lines at H and B indicate the dark ring and the dark central spot respectively.

Let the radius of the spherical surface be R cm. Then, since AB is very small compared with R, we have $2R\,.\,AB = h^2$, and

$$AB = \tfrac{1}{2}h^2/R.$$

Let the central dark spot, when the vertex is at B, be called the "first dark centre," corresponding to $N = 1$, and suppose that the last dark centre which forms before the load is fully applied is the Nth and that C is the corresponding position of the vertex. Let the load be then $M - m$ grm. When the remainder, m, of the load is applied, the rings contract, and for the complete load, M, the innermost dark ring is at K with radius $KE = OV = k$ cm. During the application of the "remainder" m, the vertex descends from C to D. Since the distance between the lens and the plate changes by equal steps of $\tfrac{1}{2}\lambda$ for successive dark centres, $BC = \tfrac{1}{2}(N-1)\lambda$. (In Fig. 51, $N = 3$.)

We have now to find CD. If the load were increased beyond M grm. so that the vertex descended below D to E to produce the $(N+1)$th dark centre at E, at a depth $EO = KV$ below the plate, we should have $CE = \tfrac{1}{2}\lambda$. Since $KE = k$, we have $DE = \tfrac{1}{2}k^2/R$, and hence

$$CD = CE - DE = \tfrac{1}{2}\lambda - \tfrac{1}{2}k^2/R.$$

In Fig. 51, the normal OZ is graduated in half wave-lengths to emphasise the fact that, whether the radius r of a dark ring be null or finite, the vertical distance, such as UH, from the plate to the lens at radius r from OZ is an integral multiple of $\tfrac{1}{2}\lambda$.

The actual distance travelled by the vertex during the increase of the load from zero to M grm. is AD. But

$$AD = AB + BC + CD = \tfrac{1}{2}h^2/R + \tfrac{1}{2}(N-1)\lambda + \tfrac{1}{2}\lambda - \tfrac{1}{2}k^2/R,$$

or

$$AD = \tfrac{1}{2}\{N\lambda + (h^2 - k^2)/R\}. \qquad \text{.................. (5)}$$

If the innermost dark ring be at F when there is a dark centre at E, and if $FE = OW = f$ cm., we have $CE = \frac{1}{2}\lambda = \frac{1}{2}f^2/R$, and $1/R = \lambda/f^2$. Hence,

$$AD = \tfrac{1}{2}\{N + (h^2 - k^2)/f^2\}\lambda.$$

If $AD = \frac{1}{2}n\lambda$, we have

$$n = N + (h^2 - k^2)/f^2, \quad \text{..................(6)}$$

which gives n in terms of the *integer* N and the positive or negative fraction $(h^2 - k^2)/f^2$.

We can use (6) to find n, if we measure the actual diameters $2h$, $2k$, $2f$ of the rings H, K, F with a travelling microscope. Since, however, it is only the ratios h/f, k/f that are needed, we may observe the rings with a microscope fitted with a micrometer scale in front of the eye-piece. If the *diameters* $2h$, $2k$, $2f$, as seen through the microscope, cover s, t, q divisions of the micrometer scale, $s/h = t/k = q/f$, and thus

$$n = N + (s^2 - t^2)/q^2. \quad \text{..................(7)}$$

When there is actually a dark centre for zero load, this is counted as the first dark centre and consequently $s = 0$. If the Nth dark centre form when the load is complete, $t = 0$. Care is needed in these cases to avoid an error of unity in the estimation of n.

A graphical method may be used for finding q^2. The diameters of 4 or 5 rings, from the innermost dark ring outwards, are measured when the cross-bar is free from load. If the number of micrometer divisions corresponding to successive diameters be $d_1, d_2, \ldots$, then $d_2{}^2 - d_1{}^2$, $d_3{}^2 - d_2{}^2, \ldots$ are theoretically equal to q^2. If $d_1{}^2, d_2{}^2, \ldots$ be set off as ordinates against abscissae 1, 2, ..., as in Fig. 52, q^2 is found from the slope of the straight line which lies most evenly among the plotted points. Thus, if $[d_5]^2$ and $[d_1]^2$ be the values of d^2 for the integers 5 and 1, *as given by the straight line*,*

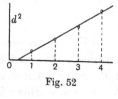

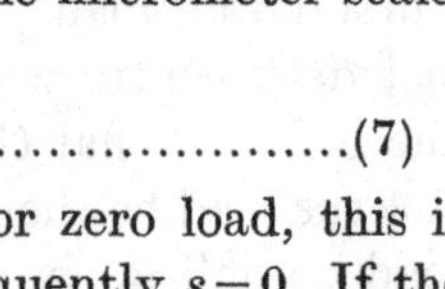

Fig. 52

$$q^2 = \tfrac{1}{4}\{[d_5]^2 - [d_1]^2\}.$$

* Students sometimes find the mean of $d_2{}^2 - d_1{}^2$, $d_3{}^2 - d_2{}^2$ and $d_4{}^2 - d_3{}^2$ under the impression that they are gaining accuracy. But the mean of these quantities, viz. $\frac{1}{3}(d_4{}^2 - d_1{}^2)$, depends only upon d_1 and d_4 and not upon the intermediate values d_2, d_3, whose influence is therefore lost. The straight line, when drawn with judgment, takes account of these values.

If there be a dark *centre* for zero load, the diameter of the *first* dark *ring* has the maximum value $2\sqrt{(\lambda R)}$. Hence the straight line (Fig. 52) cuts the axis of numbers between 0 and 1.

An arithmetical calculation is quicker than the graphical method. When 5 rings are observed, the mean of $\frac{1}{3}(d_4^2-d_1^2)$ and of $\frac{1}{3}(d_5^2-d_2^2)$, viz.

$$q^2=\tfrac{1}{6}(d_5^2+d_4^2-d_2^2-d_1^2),$$

is taken as the value of q^2, and a knowledge of d_3 is not required.

We can also find q^2 from the mean of

$$\tfrac{1}{2}(d_3^2-d_1^2), \quad \tfrac{1}{2}(d_4^2-d_2^2) \text{ and } \tfrac{1}{2}(d_5^2-d_3^2).$$

This requires a knowledge of d_3, is more laborious, and has no advantage, since it leads to the same result as above.

In an experiment, the diameters, in micrometer divisions, and their squares were

$$d_1=13{\cdot}5 \quad d_2=27{\cdot}5 \quad d_3=37{\cdot}2 \quad d_4=44{\cdot}4 \quad d_5=50{\cdot}9$$
$$d_1^2=182{\cdot}2 \quad d_2^2=756{\cdot}2 \quad d_3^2=1383{\cdot}8 \quad d_4^2=1971{\cdot}4 \quad d_5^2=2590{\cdot}8$$

Then $\frac{1}{3}(d_4^2-d_1^2)=596{\cdot}4$, $\frac{1}{3}(d_5^2-d_2^2)=611{\cdot}5$. The mean gives $q^2=604{\cdot}0$.

The value of n obtained from the innermost dark rings for zero and for full load, whose diameters we shall now denote by s_1, t_1, may be checked by measurements on the second and third dark rings of diameters s_2, t_2 and s_3, t_3 for zero and full loads respectively. The second immediately surrounds the first or innermost dark ring. The method is illustrated by the following example, in which n was determined for a load of 805·4 grm. The integer N was 26. The diameters are in micrometer divisions.

Load = 0	$s_1=22{\cdot}9$	$s_2=33{\cdot}6$	$s_3=41{\cdot}5$
Load = 805·4 grm.	$t_1=23{\cdot}7$	$t_2=33{\cdot}9$	$t_3=41{\cdot}8$
	$s_1^2-t_1^2=-37{\cdot}3$	$s_2^2-t_2^2=-20{\cdot}2$	$s_3^2-t_3^2=-25{\cdot}0$

The mean value of s^2-t^2 is $-27{\cdot}5$. Using the value $q^2=604$, we have

$$n=N+(s^2-t^2)/q^2=26-27{\cdot}5/604=26-0{\cdot}046=25{\cdot}954.$$

The greatest deviation of s^2-t^2 from the mean is 9·8. The corresponding error in n is 9·8/604 or 0·016.

If, instead of loading the cross-bar, we gradually reduce the load from its full value to zero, the rings will expand, new rings spreading out from the centre. For full load, the vertex is at D (Fig. 51) and the innermost dark ring is at K with radius $EK=k$. When the load is a little reduced, this ring expands and a dark centre appears when the ring has spread out to F. This dark centre is counted as the first. The vertex is now at C. If the load be sufficiently reduced, the vertex moves to B through $\frac{1}{2}(N-1)\lambda$ from C and the Nth dark centre appears. The removal of the remainder of the load brings the vertex to A and causes the dark centre at B to spread out into a ring at H, of radius $BH=h$. Hence

$$AD=AB+\tfrac{1}{2}(N-1)\lambda+CD,$$

and, consequently, AD is again given by (5).

It must be remembered that, whether the method of loading or the method of unloading be used, the radius of the innermost ring, when there is zero load, is h, and that k is the radius when the cross-bar bears the full load.

80. Practical example.

Mr J. G. Wilson tested a bar of mild steel. Sodium light of mean wavelength $\lambda=5{\cdot}893\times10^{-5}$ cm. was used.

Dimensions of apparatus. Diameter of rod $=2r=1{\cdot}1100$ cm. Hence $r=0{\cdot}555$ cm. Distance between central sections of plates DD, FF (Fig. 50) $=l=2{\cdot}685$ cm. Horizontal distance between line of action of load and axis of vertical rod $AA=c=12{\cdot}67$ cm. Distance between axis of rod and centre of system of Newton's rings $=a=5{\cdot}19$ cm. Hence

$$4ac-r^2=262{\cdot}72\text{ cm.}^2.$$

Determination of q^2. The readings are those observed on the micrometer scale of the microscope. Two sets of observations were made, the cross-bar carried no load.

Exp.	Ring	Reading left	Reading right	Diam. d	d^2	$d^2_{n+3}-d_n{}^2=3q^2$	Mean q^2
1	1	59·0	34·2	24·8	615		
	2	64·0	28·6	35·4	1253		
	3	67·8	24·9	42·9	1840		
	4	71·0	21·8	49·2	2421	1806	
	5	73·9	18·9	55·0	3025	1772	596·3
2	1	58·9	33·2	25·7	660		
	2	63·9	28·6	35·3	1246		
	3	67·7	24·7	43·0	1849		
	4	70·9	21·4	49·5	2450	1790	
	5	73·8	18·4	55·4	3069	1823	602·2

Mean value of $q^2=599{\cdot}2$.

Determination of n. Three loads were used. The number N of new dark centres is noted. In each case the Left and Right readings of five rings were observed.

Load $= M = 981{\cdot}22$ grm. $N = 31$.

Ring	Readings for s^2 before loading				Readings for t^2 after loading				s^2-t^2
	Left	Right	s	s^2	Left	Right	t	t^2	
1	53·9	38·8	15·1	228	57·0	42·0	15·0	225	3
2	61·7	31·9	29·8	888	63·3	34·2	29·1	847	41
3	65·1	27·3	37·8	1429	67·4	29·9	37·5	1406	23
4	68·9	23·9	45·0	2025	71·1	26·3	44·8	2007	18
5	71·8	20·6	51·2	2621	74·3	23·3	51·0	2601	20

Mean value of $s^2-t^2 = 21{\cdot}0$. Hence, by (7),

$$n = N + (s^2-t^2)/q^2 = 31 + 21{\cdot}0/599{\cdot}2 = 31{\cdot}035,$$

and $$M/n = 981{\cdot}22/31{\cdot}035 = 31{\cdot}62 \text{ grm.}$$

Load $= M = 1203{\cdot}7$ grm. $N = 39$.

Ring	Readings for s^2 before loading				Readings for t^2 after loading				s^2-t^2
	Left	Right	s	s^2	Left	Right	t	t^2	
1	53·0	41·0	12·0	144	54·2	31·5	22·7	515	−371
2	60·1	33·8	26·3	692	59·6	26·1	33·5	1122	−430
3	64·8	28·0	36·8	1354	63·7	22·0	41·7	1739	−385
4	68·5	24·4	44·1	1945	67·4	18·8	48·6	2362	−417
5	71·0	20·9	50·1	2510	70·1	15·8	54·3	2948	−438

Mean value of $s^2-t^2 = -408{\cdot}2$. Hence, by (7),

$$n = N + (s^2-t^2)/q^2 = 39 - 408{\cdot}2/599{\cdot}2 = 38{\cdot}319,$$

and $$M/n = 1203{\cdot}7/38{\cdot}319 = 31{\cdot}41 \text{ grm.}$$

A third set of observations with $M = 1471{\cdot}91$ grm. gave $n = 47{\cdot}116$ and $M/n = 31{\cdot}24$ grm.

The mean of the three values gives $M/n = 31{\cdot}42$ grm.

Young's modulus. By (4), § 78, we have

$$E = \frac{2gl\,(4ac-r^2)}{\pi\lambda r^4}\cdot\frac{M}{n} = \frac{2\times 981\times 2{\cdot}685\times 262{\cdot}72}{\pi\times 5{\cdot}893\times 10^{-5}\times 0{\cdot}555^4}\times 31{\cdot}42$$

$$= 2{\cdot}475\times 10^{12} \text{ dyne cm.}^{-2}.$$

CHAPTER V

MATHEMATICAL DISCUSSIONS OF PROBLEMS IN SURFACE TENSION

81. The nature of surface tension. The surface separating a liquid from a gas or from another liquid is the seat of actions which cause the surface to behave in a peculiar manner. One effect of these surface actions is to cause the interface to behave as if it were a very thin contractile film. Across any (geometrical) line drawn on the surface of separation there is a definite pull of the surface layer on one side of the line upon the surface layer on the other side of the line; the amount of this pull per cm. length of the line is called the Surface Tension of the surface. For water in contact with air, at the temperature of 15° C., this pull amounts to about 73 dynes per cm. of the line, and thus the surface tension of a water-air surface is 73 dynes per cm. If we knew the thickness of the thin layer which is the seat of the surface actions, and also the distribution of the stress in different parts of this layer, we should be able to estimate the stress at each point of the layer in dynes per *square* cm. If a film of liquid be gradually diminished in thickness, no change of the surface tension is to be expected until the layers on either side of the film overlap. A film of soap solution retains its power of exhibiting surface tension unchanged until its thickness becomes reduced to about 1/60,000 cm. The thickness of the layer in which the surface forces act may be something like half this distance, or about 1/120,000 cm. If the same be true for water, the 73 dynes of pull are exerted across an area of 1/120,000 square cm. Hence the mean stress is $73 \times 120{,}000$ or $8{\cdot}76 \times 10^6$ dynes per *square* cm. or about 8·7 times the stress due to the pressure of the atmosphere.

As we have no certain knowledge of the thickness of the surface layer, or of the manner in which the stress is distributed throughout it, we have to be content with finding the total pull across a line 1 cm. long, drawn on the layer. As stated above, this total pull is known as the surface tension.

82. Surface energy. If a film occupy the area, indicated by shading in Fig. 53, between a wire FG and part of a bent wire $ABCD$, of which AB, CD are perpendicular to FG, the pull on FG due to the surface tension on the *two* sides of the film is $2Tl$ dynes at right angles to FG; here l cm. is the length of FG. If the wire FG be displaced to $F'G'$ through a distance x cm., at right angles to FG, by some agent, the work done against surface tension by the agent is $2Tlx$ ergs, provided the change of surface cause no change in the surface tension. Since the area of the film, counting both sides, has been increased by $2lx$ cm.^{2}, we see that the work done by the agent equals the product of the surface tension T and the increase of area. Thus, the energy expended by the agent in increasing the surface of the liquid by one square cm. equals T ergs. Hence, when the surface of a mass of liquid is increased by S cm.^{2}, the work spent against capillary forces in increasing the surface is ST ergs, provided that T remain unchanged.

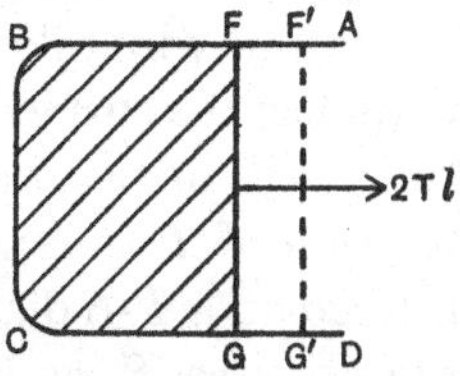

Fig. 53

When the surface is increased, the mass of liquid in the ordinary state is diminished, and the mass in the special "surface state" is increased. The energy per gramme of liquid is greater in the "surface state" than in the ordinary state; the increase of energy, which is due to the passage of liquid from the ordinary state to the surface state, accounts for the work spent in increasing the surface.

If no energy reach the film from any source other than the mechanical agent used in increasing the surface, we may speak of the surface layer as having energy of amount T ergs per cm.^{2}

The theory, so far, has been given on conventional lines. But this treatment is artificial and is felt to be unsatisfactory by students who know the elements of thermodynamics. For it can be shown that, since the surface tension diminishes with rise of temperature, energy in the particular form of heat must be supplied when the surface is increased, if the temperature of the film is to remain unchanged. We shall see, however, in § 84, that the fuller theory allows us to retain the conclusion that the film may be treated as if it had energy of T ergs per cm.^{2} of surface.

83. Application of thermodynamics. Lord Kelvin showed how thermodynamical principles may be applied to a film. The frame and movable wire of Fig. 53 play the part of the cylinder and piston of Carnot's engine. The film of liquid bounded by the frame and wire takes the place of the substance contained in the cylinder; the surface tension T and the surface S of the film take the places of the pressure and the volume of the substance. The relation of T to S during the working of the engine can be exhibited on a T-S diagram in which the ordinate represents T and the abscissa S, as in Fig. 54. At any given temperature θ, the surface tension T is independent of S,* and hence the isothermals are straight lines parallel to OS.

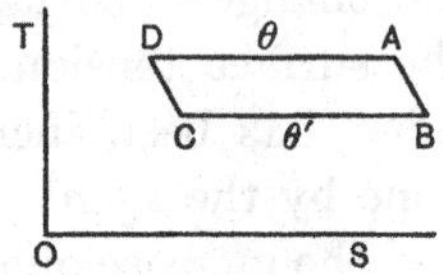

Fig. 54

Heat is communicated to or absorbed from the film only in two chambers of constant temperatures θ and θ' on the thermodynamic scale. We suppose that initially the film is at the higher temperature θ, when its surface tension is T, and that its surface, counting both sides, is S_1. The state of the film is represented by the point A on the T-S diagram. The film is stretched adiabatically, i.e. without receiving or giving out any energy in the particular form of heat, until its temperature falls to θ'. The relation between surface tension and surface during this process is shown by the adiabatic line AB. The film is now transferred to the chamber at temperature θ' and is allowed to contract at constant temperature θ', when its surface tension is T', until the point on the diagram reaches C. During this process heat Q' is given out by the film. The film is removed from the θ'-chamber and is allowed to contract adiabatically until its temperature reaches θ, the point reaching D on the isothermal AD, when the surface is S_2. Then it is put into the θ-chamber and is stretched until the point reaches A again. During this process heat Q is taken in by the film.

The total heat absorbed by the film is $Q-Q'$, and the mechanical work which has been spent upon it is represented by the area $ABCD$. If we make AD finite and $\theta-\theta'$ infinitesimal, the exact

* We suppose, of course, that the thickness of the film is always sufficient to ensure that T depends only upon the temperature and not upon S.

forms of the adiabatics will be of no consequence, and the work spent upon the film will be $(T-T')(S_1-S_2)$. The work done by the film is $-(T-T')(S_1-S_2)$.

Since DA and BC are isothermals, no heat is imparted to or taken from the frame or liquid for the purpose of changing their temperatures as the point moves along these lines. To prevent heat from passing from frame or liquid to the surface film when the representative point for the film moves along the adiabatics CD, AB, heat would have to be suitably imparted to or taken from the frame or liquid. Since CD, AB are infinitesimal, so are these amounts of heat; they are therefore negligible in the limit compared with the amounts Q and Q' which correspond to DA and BC. Moreover, it is only the *difference* in the amounts corresponding to CD and AB which concerns us.

By the Second Law of thermodynamics, the ratio of the work done *by* the film during the cycle to the heat taken in at the higher temperature, θ, equals $(\theta-\theta')/\theta$. Hence

$$\frac{-(T-T')(S_1-S_2)}{Q}=\frac{\theta-\theta'}{\theta},$$

or

$$\frac{Q}{S_1-S_2}=-\frac{\theta(T-T')}{\theta-\theta'}.$$

When we make $(\theta-\theta')$ infinitesimal, we get, in the limit,

$$\frac{Q}{S_1-S_2}=-\theta\frac{dT}{d\theta}. \quad \text{......................(1)}$$

Here $Q/(S_1-S_2)$ is the energy, in the particular form of heat, which must be supplied to keep the temperature constant when the surface of the film is increased by one square cm. at temperature θ.

The surface tension of water diminishes with rise of temperature in the manner shown in the following Table:

Temp.	T/T_{15}	Temp.	T/T_{15}	Temp.	T/T_{15}
0° C.	1·030	40° C.	0·947	80° C.	0·851
10	1·010	50	0·925	90	0·827
20	0·990	60	0·901	100	0·800
30	0·970	70	0·876		

Here T_{15} is the surface tension at 15° C. At 15° C., when $\theta=288°$ on the thermodynamic scale, T/T_{15} diminishes by 0·02 for a rise of 10°, and hence $dT/d\theta = -T_{15}/500$. Thus, by (1), we have for water at 15° C.,

$$\frac{Q}{T(S_1 - S_2)} = \frac{288}{500} = 0{\cdot}576.$$

Since $T(S_1 - S_2)$ is the mechanical work spent in stretching the film, we see that, at 15° C., the heat required to keep the temperature of the film constant is 0·576 times the mechanical work expended in stretching the film.

84. Surface energy and surface tension. In ordinary experiments, the surface film is in contact with air on both sides, as with soap films, or with air on one side and liquid on the other, as in the case of a drop of mercury on a plate, or with a liquid on both sides, as when paraffin oil floats on water. When the film is stretched, its tendency to cool will be corrected by a flow of heat to the film from its surroundings and the temperature of the film will, with sufficiently slow changes of surface, remain constant. We will consider the case of the soap film of Fig. 53. If the wire FG, of length l, be displaced by some agent through a distance x, the mechanical work done by the agent against the surface tension is $2Tlx$. By (1), § 83, heat is taken in by the film from the surroundings to amount Q, where

$$Q = \mu \,.\, 2Tlx$$

and

$$\mu = -\frac{\theta}{T} \cdot \frac{dT}{d\theta}.$$

Hence the energy of the film has been increased by

$$2Tlx(1 + \mu).$$

Since the energy of the surroundings has been diminished by $\mu \,.\, 2Tlx$, the total energy of the film and the surroundings has increased by $2Tlx$. This is exactly the amount by which the energy of the film would be increased, if the surface energy of the film were T ergs per square cm. and no heat were required to keep the temperature of the film constant when it is stretched. Since we must always regard the film in connection with its surroundings, we shall make no error if we treat the film as having surface energy T ergs per square cm. and disregard the interchange of heat between the film and its surroundings.

85. Geometry of spherical film. When a curved film is formed on a circular rim, of radius c, of a cup, by a pressure excess p, it must be a surface of revolution about the axis of the rim; the surface may be shown to be part of a sphere. Let VMO (Fig. 55) be the axis of the rim, and V the vertex of the film; the figure is a section of the system by a plane through VO. Let NMN' be the trace of a plane normal to VO, and let $MN = x$, $VM = y$. Let the tangent at N to the section make an angle θ with MN. Since the force on the film NVN' due to the surface tension, on *both* sides of the film, acting along the circle NN', in which the plane NMN' cuts the film, equals the force due to the pressure excess p acting on the circular area of radius MN, we have

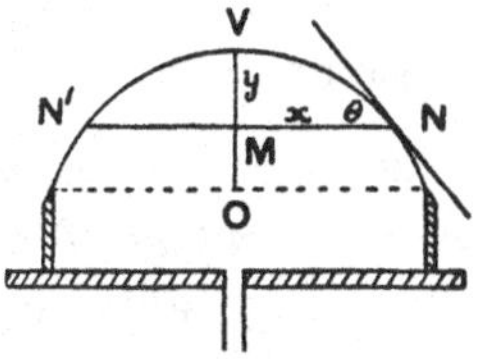

Fig. 55

$$2T \sin\theta \,.\, 2\pi x = p \,.\, \pi x^2,$$

or

$$x = (4T/p)\sin\theta = r\sin\theta,$$

where $r = 4T/p$. But $dy/dx = \tan\theta$, and hence

$$dy/dx = \sin\theta/\cos\theta = x/\sqrt{(r^2 - x^2)}.$$

Thus

$$y = -\sqrt{(r^2 - x^2)} + K.$$

Choosing the constant K so that $y = 0$ when $x = 0$, we have

$$(y - r)^2 + x^2 = r^2.$$

Hence N is on a circle of radius r whose centre lies on VO at a distance r below V. Hence the film is spherical and has radius r, where

$$r = 4T/p. \quad \text{.........................(2)}$$

If c be the radius of the rim, and h be the height VO of the vertex V above the plane of the rim, we have $h(2r - h) = c^2$, or*

$$r = (c^2 + h^2)/2h.$$

As the height, h, of the vertex of the film above the rim of the cup is gradually increased from zero to c, the radius of the rim, the radius of the bubble diminishes from infinity until it reaches c.

* On plotting r against h, we obtain an hyperbola having $h = 0$ and $r = \frac{1}{2}h$ for asymptotes and lying in the acute angle between them.

If h be further increased, the radius of the bubble increases also. Hence the pressure excess due to the film cannot exceed $4T/c$.

If h exceed c, the bubble is more than a hemisphere, and hence r also exceeds c.

86. Stability of spherical film. If a bubble of radius r, *for which h, and therefore r, is greater than c*, be connected to a source which maintains a constant pressure exceeding that of the atmosphere by $4T/r$, the bubble will be in equilibrium. The equilibrium will, however, be unstable, and any slight disturbance will cause the bubble either to swell or to contract. For, if r increase, $4T/r$ falls and then the bubble cannot cause a pressure excess as great as that due to the source, and hence the bubble will swell until it bursts. On the other hand, a decrease in r increases $4T/r$, and the bubble will drive air back to the source until h has become so much less than c that the original value of r is recovered. The bubble is then again in equilibrium, but now the equilibrium is stable.

87. General theory of stability of spherical film. If the interior of a bubble of radius r, springing from the rim of a cup of radius c, be connected to a vessel K containing air whose pressure exceeds the atmospheric pressure by $4T/r$, the film will be in equilibrium. If the pressure excess be less than $4T/c$, the film can take two forms, say A and B. The height of the vertex of the film above the plane of the ring is $r-\sqrt{(r^2-c^2)}$ for A and $r+\sqrt{(r^2-c^2)}$ for B. The volume enclosed by A increases when r diminishes, but the volume enclosed by B increases when r increases. In the case of A, an increase in r diminishes $4T/r$ and causes no change of internal pressure when the volume of K is infinite (corresponding to a source of constant pressure) and causes an increase of internal pressure when K is finite; thus whether K be finite or infinite, the film will go back to the radius r and the equilibrium will be stable.

The case of B is different. For now an increase in r diminishes $4T/r$ but increases the volume, and thus diminishes the pressure, unless the volume of K be infinite. The equilibrium is therefore unstable when K is infinite (corresponding to a source of constant pressure), but will be stable, if K be so small that the diminution of internal pressure due to increase of volume is greater than the diminution of $4T/r$.

Let v be the total volume enclosed by the film, of radius r, and the vessel K connected with it. Then there is a critical value of v —say Q—such that, if v be less than Q, the equilibrium is stable, and, if v exceed Q, the equilibrium is unstable. As a knowledge of the critical volume is essential for an intelligent appreciation of bubble problems, we will investigate the value of Q.

Let f be the volume between the bubble and the plane of the rim and let k be the remainder of the volume. Then $f+k=v$. Let y be measured from the vertex V along the normal at V towards the centre of curvature of the film and let x be the perpendicular from a point on the film upon this normal. Then we have

$$f=\int_0^h \pi x^2 dy,$$

where h is the height of V above the plane of the rim. But $x^2=y(2r-y)$, and hence

$$f=\pi\int_0^h y(2r-y)\,dy=\pi h^2(r-\tfrac{1}{3}h).$$

Since $2r=h+c^2/h$, or $2rh=h^2+c^2$, we find $dh/dr=2h^2/(h^2-c^2)$, and thus, since $dv/dr=df/dr$,

$$\frac{dv}{dr}=\pi\left\{h^2+(2rh-h^2)\frac{dh}{dr}\right\}=\pi\left\{h^2+\frac{2c^2h^2}{h^2-c^2}\right\}=\frac{\pi(h^2/c^2+1)}{h^2/c^2-1}\,.\,h^2.$$

Since $h^2=(h^2+c^2)^2/4r^2$, we can put $\frac{1}{4}(c^2/r)^2(h^2/c^2+1)^2$ for the last factor h^2, and then we find

$$\frac{dv}{dr}=\frac{\pi c^4(h^2/c^2+1)^3}{4r^2(h^2/c^2-1)}.$$

If the internal pressure be p, the external pressure is $p-4T/r$. If the volume increase from v to $v+(dv/dr)\,dr$, the new internal air pressure is

$$pv\{v+(dv/dr)\,dr\}^{-1}=p\{1-v^{-1}(dv/dr)\,dr\},$$

to the first power of dr. The new excess of the internal air pressure above the external air pressure is thus

$$\frac{4T}{r}-\frac{p}{v}\frac{dv}{dr}dr.$$

The pressure due to the film is $4T/(r+dr)$ or, to the first order,

$$(4T/r)(1-dr/r).$$

The film will return to the radius r, if $(4T/r)(1-dr/r)$ be greater than the new excess air pressure, i.e. if

$$4T/r^2 < (p/v)\,dv/dr.$$

When we use the value of dv/dr, we find that the bubble is stable if v be less than the critical volume

$$Q = \frac{3pr}{64T}\cdot\phi\left(\frac{r}{c}\right).\,U,$$

where $U = 4\pi r^3/3$, the volume of a sphere of radius r, and $\phi(r/c)$ is a function of r/c which is given by

$$\phi\left(\frac{r}{c}\right) = \frac{c^4(h^2/c^2+1)^3}{r^4(h^2/c^2-1)},$$

where $h/c = r/c + \sqrt{(r^2/c^2-1)}$. Now p is practically the atmospheric pressure, viz. 10^6 dyne cm.$^{-2}$, and we will take T for soap solution to be 30/1·12 or 26·785... dyne cm.$^{-1}$. Hence $3p/64T = 1750/l$, where l is the length of one cm. Thus

$$Q = 1750\phi(r/c).(r/l).U.$$

The values of h/c and of $\phi(r/c)$ for some values of r/c are given in the Table.

r/c	1·000	1·001	1·01	1·1	1·5	2·0	3·0	∞
h/c	1·000	1·046	1·152	1·558	2·618	3·732	5·828	∞
$\phi(r/c)$	∞	97·69	37·06	19·27	16·35	16·08	16·01	16·00

It will be seen that, as soon as r/c is appreciably greater than unity, i.e. as soon as the bubble is seriously greater than a hemisphere, $\phi(r/c)$ does not differ materially from 16 and we may put $Q = 28{,}000(r/l).U$. If $r/l = 2$, so that $r = 2$ cm., we have $U = 33·5$ c.c., and then we find that Q, the critical volume, is $28{,}000 \times 2 \times 33·5$ c.c. or 1·88 cubic *metres*.

If v exceed $28{,}000(r/l).U$, the bubble is in equilibrium with the pressure excess $4T/r$, but the equilibrium is unstable. The bubble will not, however, swell indefinitely, for Q is proportional to r^4/l and thus, unless the vessel K be enormous, the volume enclosed by K and the swollen bubble will soon become less than the critical volume corresponding to the increased radius. Hence, when equilibrium is again reached, it will be stable.

88. Relation between curvature of surface and normal force per unit area due to surface tension. Let O (Fig. 56) be a point on a curved surface. Through O we can draw on the surface two curves AOB, COD, known as lines of curvature,* which are such that the normals to the surface at points near O on AOB, COD meet in the points R_1, R_2 respectively on the normal OR_2R_1; the curves intersect at right angles in O. The points R_1, R_2 are the centres of principal curvature, and $OR_1=\rho_1$, $OR_2=\rho_2$ are the radii of principal curvature. Through A, B draw lines of curvature EK, GH, cutting AB at right angles at A, B, and through C, D lines of curvature EG, HK, cutting CD at right angles at C, D. Let

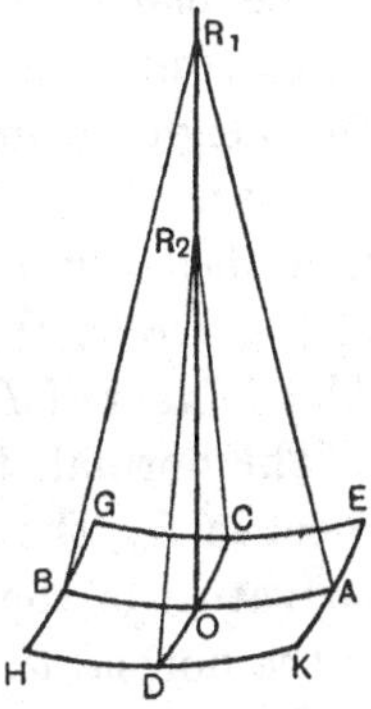

Fig. 56

$$AR_1O=BR_1O=\theta_1 \quad \text{and} \quad CR_2O=DR_2O=\theta_2.$$

Then $AB=2\rho_1\theta_1$, $CD=2\rho_2\theta_2$, and the area of the small "rectangle" is $4\rho_1\rho_2\theta_1\theta_2$.

If $EGHK$ be part of an interface, of surface tension T, the resultant pull, due to the adjoining surface, along each of the edges EK, GH is $T \,.\, 2\rho_2\theta_2$. For EK it is perpendicular to AR_1, and for GH to BR_1; in each case it is in the plane AR_1B. The resultant of the two is along OR_1 towards R_1 and is $T \,.\, 4\rho_2\theta_2 \sin\theta_1$, or $4\rho_1\rho_2\theta_1\theta_2(T/\rho_1)$, since θ_1 is small. The amount per unit area is T/ρ_1. The edges GE, HK contribute T/ρ_2 per unit area along OR_2 towards R_2. Hence, if F be the resultant force per unit area exerted by the surrounding surface upon the element, we have

$$F=T(1/\rho_1+1/\rho_2) \ldots\ldots\ldots\ldots\ldots\ldots\ldots (2a)$$

When R_1 and R_2 are on opposite sides of O, the parts of F depending on ρ_1 and ρ_2 act towards R_1 and R_2 respectively.

If $EGHK$ (Fig. 56) be an element of a soap film, and if R_1, R_2 be the centres of principal curvature for one face, they may, without appreciable error, since the film is very thin, be taken as the centres for the other face also. Hence the effects of the two surfaces will be superposed, and formula $(2a)$ will be replaced by

$$F=2T(1/\rho_1+1/\rho_2). \ldots\ldots\ldots\ldots\ldots\ldots\ldots (3)$$

* *Experimental Optics*, Chapter XI, § 269.

89. Principal curvatures of surface of revolution. Let OX (Fig. 57) be an axis and ABC a curve in a plane containing OX. If the plane AOX revolve about OX, the curve traces out a surface of revolution. Let P, P' be two neighbouring points on the curve and let PN be perpendicular to OX. When P revolves, it describes a circle of centre N. If PQ be a small arc of this circle, Q lies out of the plane of the paper. The elements PP' and PQ meet at right angles in P.

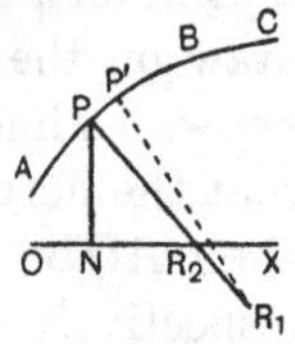

Fig. 57

The normals to the *curve* at P, P' lie in the plane POX and meet in R_1, the centre of curvature, and $PR_1 = \rho_1$, the radius of curvature of the curve. Let PR_1 cut OX in R_2. The line PR_2R_1 is the normal to the *surface* at P, for it is perpendicular both to PP' and to PQ. Since the normals to the surface at P, P' *meet* in R_1, PP' is an element of one of the two lines of curvature on the surface which pass through P, and R_1 is the corresponding centre of principal curvature. The normals to the surface at P, Q meet in R_2, and thus PQ is an element of the other line of curvature through P. Hence R_2 is the other centre of principal curvature. The radii of principal curvature at P are therefore $\rho_1 = PR_1$ and $\rho_2 = PR_2 = PN/\sin\phi$, where ϕ is the angle between PR_2, the normal at P, and the axis OX.

90. Form of meniscus in capillary tube.* When a capillary tube, i.e. a tube having a bore of small radius r, is dipped into a liquid, such as water, standing in a large dish, the liquid rises in the tube until the bottom of the meniscus attains a height h above the level of the undisturbed surface of the liquid outside the tube. If r be 0·05 cm. and g be 981 cm. sec.$^{-2}$, h will be about 3 cm. in the case of water, and thus h is large compared with r.

Let $D'OPD$ (Fig. 58) be a section of the meniscus by a plane containing the vertical axis OH of the tube. The height OH of O, the lowest point of the meniscus, above HJK, the level of the undisturbed liquid, is h. Let PUV, the normal to the curve OPD at P, cut OH in V, and let U be the centre of curvature of the curve at P. Since the surface is one of revolution about OH, PUV is the

* It has been assumed in § 90 that the liquid *rises* in the tube and that it wets the tube. Mercury *sinks* in the tube and has a finite angle of contact with the glass. The reader will easily modify the method to meet this case.

normal to the surface at P and, by § 89, $PU = R_1$ and $PV = R_2$ are the radii of principal curvature. The normal at P makes an angle ϕ with VOH. If, as we suppose to be the case, the liquid wet the tube, the tangent is vertical, and $\phi = \frac{1}{2}\pi$, at the point D, where OPD meets the tube.

If $R_2' = P'V'$ be the value of R_2 at P' near P, then

$$R_2 - R_2' = PV - P'V'.$$

But, to the first order of small quantities, $UP = UP'$, since UP is normal to the arc at P. Hence, to the first order,

$$R_2 - R_2' = UV - UV',$$

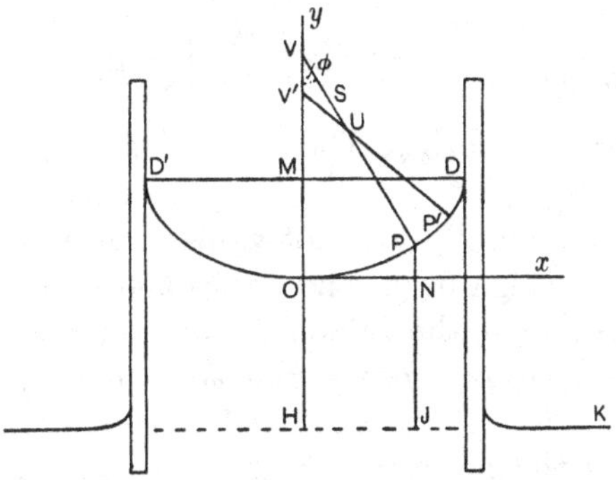

Fig. 58

and thus, in order that R_2 may be stationary for variations of ϕ, or $dR_2/d\phi = 0$, either UV must be perpendicular to OH, as at D, or U must coincide with V, as at O. Hence $dR_2/d\phi$ does not change sign as ϕ goes from 0 to $\frac{1}{2}\pi$.

The cylinder of radius $ON = x$ and height h has the volume $\pi x^2 h$. Let v be the volume traced out when the figure OPN revolves about OH. Since the pressure of the liquid at the level HJK is the same as that of the air at the same level, the apparent weight of the liquid in the volume corresponding to $OPJH$ is $g(\pi x^2 h + v)(\rho - \sigma)$, where ρ and σ are the densities of the liquid and of the air,* and this is balanced by the upward force $2\pi x T \sin\phi$ due to surface tension. Putting

$$T/\{g(\rho - \sigma)\} = a^2, \qquad \text{......................(4)}$$

we have

$$x^2 h - 2a^2 x \sin\phi + v/\pi = 0. \qquad \text{..................(5)}$$

* The variation of σ with height has been neglected.

Now $R_2 = x/\sin\phi$, and thus R is a "real" quantity. On dividing (5) by $\sin^2\phi$, we have

$$R_2{}^2 h - 2a^2 R_2 + v/(\pi \sin^2\phi) = 0,$$

or

$$R_2 = a^2 h^{-1}\{1 + \sqrt{(1-q)}\}, \quad \text{.................(6)}$$

where

$$q = vh/(\pi a^4 \sin^2\phi). \quad \text{.................(7)}$$

The positive sign has been prefixed to the square root for the reason stated below. Since R_2 is real, $|q| < 1$.

If the radius of curvature of the curve at O be b, then $R_1 = R_2 = b$ at O. When ϕ is small, $y = PN = b(1-\cos\phi) = \frac{1}{2}b\phi^2$ and $x = b\phi$. Hence, for small ϕ,

$$\frac{v}{\sin^2\phi} = \frac{1}{\sin^2\phi}\int_0^x 2\pi xy\,dx = \frac{\pi b^3}{\phi^2}\int_0^\phi \phi^3 d\phi = \tfrac{1}{4}\pi b^3 \phi^2,$$

and thus q vanishes with ϕ. Then, by (6), $R_2 = 2a^2/h$ when $\phi = 0$. Hence $b = 2a^2/h$.

Since, when $\phi = 0$, R_2 does not vanish but equals b, we must take the positive sign in (6) and consequently, for a finite value of ϕ, R_2 is less than b, the value for $\phi = 0$. Since $dR_2/d\phi$ does not change sign, $dR_2/d\phi$ is negative from $\phi = 0$ to $\phi = \frac{1}{2}\pi$, and thus R_2 diminishes from b at O to r at D and is positive throughout.

Since ON is the projection both of the arc OP and of the straight line PV, we have

$$\int_0^\phi R_1 \cos\phi\, d\phi = R_2 \sin\phi.$$

Differentiating with regard to ϕ, we find

$$(R_1 - R_2)\cos\phi = \sin\phi\, dR_2/d\phi.$$

Since $dR_2/d\phi$ is negative, R_1 is less than R_2 and thus the value of R_1 at D is less than r. At O, $R_1 = b$.

The pressure of the air at P exceeds that of the liquid at P by $g(\rho - \sigma)(h+y)$ or, by § 88, by $T(1/R_1 + 1/R_2)$. Thus, by (4), $1/R_1 + 1/R_2 = (h+y)/a^2$. If the meniscus were vertical for a finite distance above D, we should have $1/R_1 = 0$, $1/R_2 = 1/r$, and thus $1/r$ would equal $(h+y)/a^2$, or a constant would equal a variable quantity. Hence the meniscus does not extend above D.

At O, where $\phi = 0$, $1/R_1$ is positive and equal to $1/b$. If $1/R_1$ became negative as ϕ increases to $\frac{1}{2}\pi$, it would pass through the value zero before becoming negative, and R_1 would go from b to positive infinity before changing to negative infinity. But R_1 cannot do this, for, except when $\phi = 0$, R_1 is less than R_2, which is positive and does not exceed b. Hence R_1 is positive from $\phi = 0$ to $\phi = \frac{1}{2}\pi$, and thus the curve OPD is, throughout, concave upwards.

If k be the height of D above Ox,

$$k = \int_0^{\frac{1}{2}\pi} R_1 \sin\phi \, d\phi.$$

Since

$$\int_0^{\frac{1}{2}\pi} \sin\phi \, d\phi = 1,$$

we have $G > k > L$, where G and L are the greatest and least values of R_1, and thus $k < b$.

The relations obtained above are true whatever the diameter of the tube. We now consider a tube so narrow that h is large compared with r. Since $dx = R_1 \cos\phi \, d\phi$, and $dy = R_1 \sin\phi \, d\phi$, we have

$$x = \int_0^{\phi} R_1 \cos\phi \, d\phi, \qquad y = \int_0^{\phi} R_1 \sin\phi \, d\phi.$$

Since $R_1 < b$ when $\phi > 0$, we have $x < b \sin\phi$, $y < b(1 - \cos\phi)$. Hence, when $\phi > 0$,

$$v = \int_0^{x} 2\pi x y \, dx < 2\pi b^3 \int_0^{\phi} \sin\phi \, (1 - \cos\phi) \cos\phi \, d\phi$$

$$< 2\pi b^3 \left(\tfrac{1}{2}\sin^2\phi - \tfrac{1}{3} + \tfrac{1}{3}\cos^3\phi\right).$$

Hence

$$0 < q < \frac{2b^3 h}{a^4} \left\{ \frac{1}{2} - \frac{1 + \cos\phi + \cos^2\phi}{3(1 + \cos\phi)} \right\},$$

or

$$0 < q < \frac{2b^3 h}{a^4} \left\{ \frac{1}{6} - \frac{\cos^2\phi}{3(1 + \cos\phi)} \right\}.$$

As ϕ increases from 0 to $\frac{1}{2}\pi$, $\cos^2\phi/(1 + \cos\phi)$ diminishes from $\frac{1}{2}$ to zero and hence, for this range of ϕ, q lies between zero and $\frac{1}{3}b^3 h/a^4$.

By (6), $R_2 = \{1 + \sqrt{(1 - q)}\} a^2/h$; when $\phi = 0$, $R_2 = b = 2a^2/h$. Hence, as r diminishes and h, in consequence, increases, the ratio of R_2 to $2a^2/h$ for finite values of ϕ differs from 1, the value for $\phi = 0$, by an amount which we can make as small as we please by sufficiently diminishing r. The meniscus will tend to a hemispherical form as the bore of the tube diminishes, if R_1/b tend to unity for all values of ϕ as r tends to zero. Since R_1, R_2 are positive, $(R_1 R_2)^{-\frac{1}{2}}$, the geometric mean of $1/R_1$ and $1/R_2$, is less, except when $\phi = 0$, than $(h+y)/2a^2$ or $(h+y)/hb$, the arithmetic mean. Since $R_1 < b$, we have

$$1 > \frac{R_1}{b} > \frac{b}{R_2(1+y/h)^2}.$$

The greatest value of y is less than b. Since b/h tends to zero, and b/R_2 to unity, as r diminishes, R_1/b tends to unity as r diminishes, and the meniscus tends to a hemispherical form.

The volume of a hemisphere of radius r is $\frac{2}{3}\pi r^3$ and hence, when the meniscus is hemispherical, $v = \frac{1}{3}\pi r^3$ for $x = r$. Putting $\phi = \frac{1}{2}\pi$, $x = r$ and $v = \frac{1}{3}\pi r^3$ in (5), we have

$$T = g(\rho - \sigma) a^2 = \tfrac{1}{2} rh (1 + \tfrac{1}{3} r/h) g (\rho - \sigma), \quad \text{........(8)}$$

the formula generally employed.

When h/r is only moderately large, we shall still get a close approximation to v, if we replace the section of the meniscus by half an ellipse which has $2a^2/h$ for radius of curvature at O and has semi-axis $MO = k$. For an ellipse of semi-axes r and k, the radius of curvature at O is r^2/k, and thus $r^2/k = 2a^2/h$ or $k = \frac{1}{2} r^2 h/a^2$. The value of v for the ellipsoid of revolution bears the same ratio to that for a sphere of radius r as k does to r, and thus, for $x = r$,

$$v = \tfrac{1}{3}\pi r^2 k = \tfrac{1}{6}\pi r^4 h/a^2.$$

Thus, by (5), with $x = r$, $\phi = \frac{1}{2}\pi$, we have for equilibrium,

$$2a^2 r = r^2 h + \tfrac{1}{6} r^4 h/a^2,$$

or

$$12a^4 - 6rha^2 - r^3 h = 0.$$

Hence, since a^2 is positive,

$$a^2 = \tfrac{1}{4} rh \{1 + \sqrt{(1 + 4r/3h)}\}.$$

Expanding the square root as far as r^2/h^2, we have

$$a^2 = \tfrac{1}{2} rh (1 + \tfrac{1}{3} r/h - \tfrac{1}{9} r^2/h^2). \quad \text{..................(9)}$$

Poisson* and Lord Rayleigh†, by a fuller analysis, have obtained, as far as r^2/h^2,

$$a^2 = \tfrac{1}{2} rh (1 + \tfrac{1}{3} r/h - 0{\cdot}1288 r^2/h^2), \quad \text{...........(10)}$$

where 0·1288 replaces the 1/9 or 0·1111 of (9).

* Poisson, *Nouvelle Théorie*, p. 112 (1831).

† Lord Rayleigh, *Proc. Royal Soc.* Vol. 92 A, p. 189 (1916).

THEORY OF CATENOID FILM

91. Catenoid film. A soap film may be formed between the opposing plane ends AA, BB (Fig. 59) of two coaxial tubes, of circular section, in such a way that there is no difference of pressure between the two sides of the film. Hence the resultant pull, due to surface tension, across the circle in which a plane perpendicular to the axis cuts the film is obviously parallel to the axis and is independent of the position of the plane. If the form of the film be given by the revolution of the plane curve $AQKB$ about FN, which is the axis of the tubes and also the axis of x, and if $QN=y$, the pull across the circle through Q is $2T.2\pi y\cos\theta$, where T is the surface tension for each of the two faces of the film and θ is the angle between the tangent QF and the axis of x. Thus

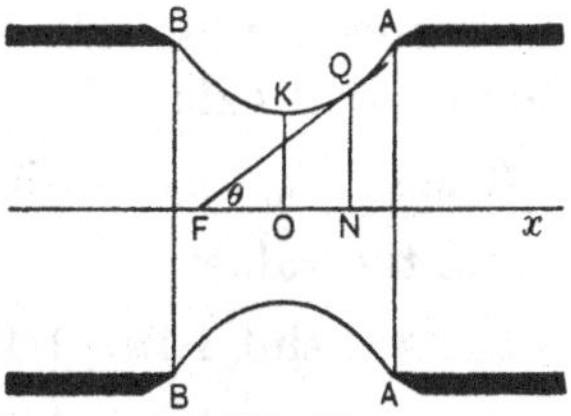

Fig. 59

$$y\cos\theta=c,$$

where the parameter c is the value of y where $\cos\theta=1$, as at K.

Hence $$y^2=c^2\sec^2\theta=c^2\{1+(dy/dx)^2\},$$

and thus $$y=c\cosh(x/c), \quad \text{.......................(1)}$$

if the origin be taken at O, so that $x=0$ when $y=OK=c$.

In EXPERIMENT 18 each tube is of radius a, and O is midway between AA and BB.

The curve (1) is a catenary having Ox as directrix; the surface formed by the revolution of the catenary about Ox is called a catenoid. Since $d^2y/dx^2=1/c$, when $x=0$, the radius of curvature of the catenary at K equals c or OK.

If $AA=BB=2a$ and $AB=l$, the point $x=\frac{1}{2}l$, $y=a$ lies on the catenary. Hence $a=c\cosh(l/2c)$ and thus

$$\frac{2a}{l}=\frac{2c}{l}\cosh\left(\frac{l}{2c}\right)=\frac{1}{n}\cosh n, \quad \text{..............(2)}$$

if $n=l/2c$.

92. Maximum length of catenoid. Let $z=n^{-1}\cosh n$. Then z is infinite when $n=0$, diminishes to finite values when n is finite and becomes infinite as n becomes infinite. A stationary value of

z occurs when $dz/dn = 0$ or when $n^{-2} \cosh n = n^{-1} \sinh n$. Since n is not zero or infinite at the stationary value of z, we have $\coth n = n$. Since $\coth n$ constantly diminishes from infinity to unity as n increases from zero to infinity, $\coth n = n$ has only one root—say n_0. Hence there is only one value of n, viz. n_0, for which $dz/dn = 0$, and for that value z has a minimum value z_0.

Then $\quad z_0 = n_0^{-1} \cosh n_0 = \tanh n_0 \cosh n_0 = \sinh n_0$. ......(2a)

From the values

$$\sinh 1{\cdot}19 = 1{\cdot}49143, \qquad \cosh 1{\cdot}19 = 1{\cdot}79565,$$

$$\sinh 1{\cdot}20 = 1{\cdot}50946, \qquad \cosh 1{\cdot}20 = 1{\cdot}81066,$$

we find

$$\coth 1{\cdot}19 - 1{\cdot}19 = 0{\cdot}013979, \qquad \coth 1{\cdot}20 - 1{\cdot}20 = -0{\cdot}000459.$$

By interpolation we find $n_0 = 1{\cdot}19968$, and $z_0 = \sinh n_0 = 1{\cdot}50888$.

For any value of z greater than z_0 there are two (and only two) values of n, since the straight line $z = h$ cuts the curve showing z in terms of n in two (and only two) points if $h > z_0$.

Thus, if $2a/l$ exceed $1{\cdot}50888$ or l/a be less than $1{\cdot}32549$, two catenaries having Ox for directrix can be drawn through A and B. The two will merge into a single catenary when $l/a = 1{\cdot}32549$. Now $a/c = (l/2c)(2a/l)$ and thus, for this catenary,

$$c/a = (1{\cdot}19968 \times 1{\cdot}50888)^{-1} = 0{\cdot}55243.$$

When l/a exceeds $1{\cdot}32549$, it is no longer possible to draw a catenary through A and B with Ox as directrix.

When $l/a < 1{\cdot}32549$, two catenoid films can be formed between the tubes. Since the film encloses no air, it will take the form of that one of the two catenoids which has the smaller surface, and this catenoid, we can show, has the larger c.

The pull across the neck of the catenoid and on either tube is $2T \,.\, 2\pi c$. If the distance between the tubes be increased by dl, the work done, viz. $2T \,.\, 2\pi c\, dl$, is expended in increasing the surface from S to $S + (dS/dl)\, dl$. Since the surface energy is $2T$ per unit area of film, the work is also $2T\,(dS/dl)\, dl$. Thus $dS/dl = 2\pi c$. Hence dS/dl is positive and is greater for the catenoid with the larger c.

When $l = 0$, the catenary (1) of larger c has $c_1 = a$ and has zero surface, and the catenary (2) of smaller c has $c_2 = 0$ and has surface $2\pi a^2$. When $S/2\pi a^2$ is plotted against l/a, as in Fig. 60, the second branch, i.e. that for the smaller value c_2 of c, starts from J on the vertical axis, where $S/2\pi a^2 = 1$, and slopes upwards. The first branch, i.e. that for the larger value c_1 of c, starts from the origin (0, 0) and also slopes upwards, but, for any value of l/a, its slope is greater than that of the second branch. Hence the diagram will be as in Fig. 60, which represents the actual values of $S/2\pi a^2$. The two branches

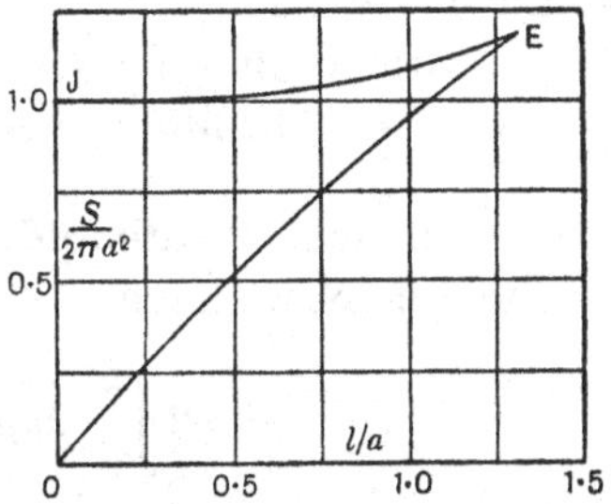

Fig. 60

will meet in E, where $l/a = 1{\cdot}32549$. At this critical value of l/a, $n = n_0$ and, by (2a), $\cosh n_0 = n_0 \sinh n_0$. If S_0 be the critical value of S, we have, by (3a), § 93,

$$S_0/2\pi a^2 = n_0 = 1{\cdot}19968.$$

The branches cannot go beyond E, since no catenoid can be formed when l/a exceeds the value 1·32549 for which the two catenoids coalesce. Thus, over the whole range of l/a from 0 to 1·32549, the catenoid of larger c has the smaller surface and is, therefore, the catenoid which will be observed in the experiment.

Hence, if a catenoid be formed between the tubes when l/a is small and if the distance, l, between the tubes be increased, the radius, c, of the neck of the film will diminish to a definite value given by $c/a = (1{\cdot}19968 \times 1{\cdot}50888)^{-1} = 0{\cdot}55243$, but beyond this it cannot go. If l/a be further increased, the catenoid film will break up into two plane films, one at the end of each tube. The surface will diminish from $1{\cdot}19968 \times 2\pi a^2$ to $2\pi a^2$.

Since $c/a = (\cosh n)^{-1}$ and $l/a = 2n(\cosh n)^{-1}$, where $n = \frac{1}{2}l/c$, we can find simultaneous values of c/a and l/a by giving n a series of values. The values in the first six columns of the Table refer to the

n	c/a	l/a	n	c/a	l/a	n	c/a	l/a
0·1	0·995	0·199	0·7	0·797	1·115	1·2	0·552	1·325
0·2	0·980	0·392	0·8	0·748	1·196	1·3	0·507	1·319
0·3	0·957	0·574	0·9	0·698	1·256	1·5	0·425	1·275
0·4	0·925	0·740	1·0	0·648	1·296	2·0	0·266	1·063
0·5	0·887	0·887	1·1	0·599	1·319	4·0	0·037	0·293
0·6	0·844	1·012				6·0	0·005	0·059

curve through the origin and E (Fig. 60); the remainder refer to the curve EJ.

The critical value of n is $n_0 = 1{\cdot}19968$ and then $c/a = 0{\cdot}55243$ and $l/a = 1{\cdot}32549$. The film is unstable if $n > n_0$.

93. Surface of catenoid. If the area of the surface of the catenoid be S, we have

$$S = 2\int_0^{l/2} 2\pi y \sqrt{\{1 + (dy/dx)^2\}}\, dx.$$

Since $\quad y = c\cosh(x/c), \qquad 1 + (dy/dx)^2 = \cosh^2(x/c),$

and hence

$$S = 4\pi c \int_0^{l/2} \cosh^2(x/c)\, dx = 2\pi c \int_0^{l/2} \{\cosh(2x/c) + 1\}\, dx$$

$$= 2\pi c\left\{\frac{c}{2}\sinh\frac{l}{c} + \frac{l}{2}\right\} = 2\pi c^2\{\sinh n \cosh n + n\},$$

since $l/2c = n$. But $c^2 = a^2/\cosh^2 n$, and thus

$$S = 2\pi a^2(\sinh n \cosh n + n)/\cosh^2 n. \quad\text{...........(3)}$$

For the critical value n_0, $\cosh n_0 = n_0 \sinh n_0$, and thus

$$S_0 = 2\pi a^2 n_0(\sinh^2 n_0 + 1)/\cosh^2 n_0 = 2\pi a^2 n_0. \quad\text{...(3a)}$$

The values of $S/2\pi a^2$ for some values of n are given in the following Table:

n	$S/2\pi a^2$	n	$S/2\pi a^2$
0	0	1·3	1·1964
0·2	0·3896	1·4	1·1879
0·4	0·7222	1·6	1·1625
0·6	0·9640	2·0	1·1053
0·8	1·1113	3·0	1·0247
1·0	1·1816	4·0	1·0023
1·1	1·1956	∞	1
$n_0 = 1{\cdot}19968$,		$S_0/2\pi a^2 = n_0 = 1{\cdot}19968$	

The catenoids with $n > n_0$ are unstable.

We have seen in § 92 that the critical value n_0 of n gives the longest catenoid. We can show that this longest catenoid is also the catenoid of greatest surface. From (3) we find

$$dS/dn = 4\pi a^2 (\cosh n - n \sinh n)/\cosh^3 n, \quad \text{.........(4)}$$

$$d^2S/dn^2 = 4\pi a^2 (-n \cosh^2 n - 3 \sinh n \cosh n + 3n \sinh^2 n)/\cosh^4 n.$$

By (4), S has a stationary value when $\cosh n = n \sinh n$, i.e. when $n = n_0$. When $n = n_0$, we find

$$d^2S/dn^2 = -4\pi a^2 n_0/\cosh^2 n_0.$$

Hence S is a maximum when $n = n_0$.

94. Solution of equation coth *n* = *n*. In § 92 it is shown that $\coth n = n$ has a single root differing little from 1·2. The value $n_0 = 1{\cdot}19968$ is as accurate as can be obtained by aid of tables of $\cosh n$ and $\sinh n$. A closer approximation is furnished by a method which depends upon the fact that $\log_e 11$ is known to a large number of decimal places.

Let $n = 1{\cdot}2 - f$, where f is a small quantity, of the order of 0·0003, which is to be found. Since

$$n = \coth n = (e^n + e^{-n})/(e^n - e^{-n}),$$

we find

$$e^{2n} = (n+1)/(n-1),$$

or

$$\exp(2{\cdot}4 - 2f) = 11(1 - 5f/11)/(1 - 5f).$$

Hence

$$2{\cdot}4 - \log_e 11 = 2f - \log_e(1 - 5f) + \log_e(1 - 5f/11).$$

Expanding the logarithms, we find

$$2{\cdot}4 - \log_e 11 = 2f + 5f\left(1 - \frac{1}{11}\right) + \frac{5^2 f^2}{2}\left(1 - \frac{1}{11^2}\right) + \frac{5^3 f^3}{3}\left(1 - \frac{1}{11^3}\right) + \dots.$$

Since $\log_e 11 = 2{\cdot}39789\,52727\,98\dots$, we have, to 10 decimal places,

$$0{\cdot}00210\,47272 = \frac{72f}{11} + \frac{1500f^2}{121} + \frac{125 \times 1330 f^3}{3 \times 1331} + \dots \quad \text{...(5)}$$

For the first approximation we neglect the terms in $f^2, f^3, \dots$, and then

$$f_1 = 11 \times 0{\cdot}00210\,47272/72 = 0{\cdot}00032\,15555.$$

This value gives $1500f^2/121 = 1{\cdot}2818 \times 10^{-6}$. We use this in (5) and neglect terms in $f^3, f^4, \dots$, and have, as a second approximation,

$$f_2 = 11 \times 0{\cdot}00210\,34454/72 = 0{\cdot}00032\,13597.$$

If we put $f=f_2$ in the terms in (5) which involve f^2 and f^3, they become $1{\cdot}2802\times10^{-6}$ and $1{\cdot}4\times10^{-9}$. Neglecting terms in f^4, ..., we have, as a third approximation,

$$f_3=11\times0{\cdot}00210\ 34456/72=0{\cdot}00032\ 13597.$$

Further approximation, limited to 10 decimal places, will not affect this result. The value

$$n_0=1{\cdot}2-f_3=1{\cdot}19967\ 86403$$

will not be in error by more than 1 or 2 in the last place.

THEORY OF CYLINDRICAL FILM

95. Unduloid film. If a cylindrical film be formed between the opposed ends of two coaxial tubes of radius a, the pressure excess within it is $2T/a$. But this is not the only film joining the tubes in which the excess is $2T/a$.

If an ellipse of major axis $2a$ and minor axis $2b$ roll on the axis Ox, a focus will trace out a "roulette" in the form of a waved curve whose complete wave-length λ equals the circumference of the ellipse. The tops of the crests and the bottoms of the troughs are at distances $a(1+e)$ and $a(1-e)$ respectively from Ox, where e is the eccentricity, and thus are at equal distances from the line $y=a$, but the points of intersection of the roulette with $y=a$ are not equally spaced; the roulette is not a *simple* sine curve. The points of inflexion of the curve lie on $y=b$ and not on $y=a$.* The distance between the two consecutive points on $y=a$ bounding a crest exceeds the distance between the points bounding a trough. When e is small, the distances are approximately $a(\pi+e-\frac{1}{4}\pi e^2+\ldots)$ and $a(\pi-e-\frac{1}{4}\pi e^2-\ldots)$ respectively. When the ellipse degenerates into a line of length $2a$, the distances are $2a\sqrt{3}$, or $3{\cdot}464a$, and $2a(2-\sqrt{3})$, or $0{\cdot}536a$. The distance between successive points of intersection of the roulette with $y=b$ is half the circumference of the ellipse and, when e is small, is $\pi a(1-\frac{1}{4}e^2-\frac{3}{64}e^4\ldots)$.

The surface formed by the revolution of the roulette about Ox is called an unduloid. For all values of e between 0 and 1, the

* If ρ be the radius of curvature of the roulette described by the focus F, $1/\rho=1/a-1/r$, where r is the distance from F to the point of contact with Ox. When the perpendicular from F on Ox is b, $r=a$ and $1/\rho=0$.

pressure excess within an unduloid film is $2T/a$, and thus there are an infinite number of unduloids for which the excess is $2T/a$, but the wave-length λ depends on e and diminishes from $2\pi a$, when $e=0$, to $4a$, when $e=1$. If an unduloid is to be formed between two tubes of radius a, it is obvious that the distance between the tubes must be equal to that between two points in which the roulette cuts the line $y=a$. Many writers have based their treatment of the stability of a cylindrical film on the properties of an unduloid, but the method depends upon comparing the cylinder of radius a with an unduloid for which e is infinitesimal and therefore cannot be applied except when the distance between the tubes is πa, $2\pi a$, ...

95 A. Cylindrical film with constant internal pressure. Many writers have described experiments to prove their statement that a cylindrical film of length l, formed on tubes of radius a, is stable so long as $l < 2\pi a$, but no mention is made of the value of $V/\pi^2 a^3$, where V is the volume contained by the film, the tubes and any communicating vessel. The discussion, in §§ 86, 87, of the stability of a spherical film may lead the reader to suspect that, when $V/\pi^2 a^3$ is *very* large, the maximum stable length is less than $2\pi a$. Theory and experiment show that, if V be infinite, or, what is the same thing, if the internal pressure excess have the *constant* value $2T/a$, the film is unstable unless $l < \pi a$.

Let the form of the film be the surface generated by the revolution of the curve

$$y = a - c \sin \pi x/l \qquad (0 < c < a)$$

about Ox, the axis of the tubes. If J be a point on the film at $x=\frac{1}{2}l$, we have $y=a-c$, and $dy/dx=0$ at J. If ρ be the radius of curvature of the curve at J, $\rho^{-1}=d^2y/dx^2=\pi^2 c/l^2$; the curve is convex to Ox at J. By §§ 88, 89, the pressure exerted by the film at J is

$$2T\{1/(a-c) - \pi^2 c/l^2\}.$$

If the pressure of the internal exceed that of the external air by $2T/a$, the film will shrink inwards at J, if

$$1/(a-c) - \pi^2 c/l^2 > 1/a,$$

or if

$$l^2 > \pi^2 a(a-c).$$

Hence the film will *begin* to shrink inwards if $l > \pi a$.

If the curve be $y = a + c \sin \pi x/l$, $(0 < c < a)$

the film will *begin* to bulge outwards when $l > \pi a$. Hence, when the pressure excess has the *constant* value $2T/a$, the cylindrical form is not stable if $l > \pi a$.

The lack of stability leads to the expectation that the film will either break up into two spherical caps or that it will bulge outwards, and the expectation is confirmed by experiment. When caps are formed, they are less than hemispheres and of radius $2a$; they are stable and each is part of an unduloid corresponding to $b = 0$ (§ 95). If l slightly exceed πa, the film can become part of an unduloid with a crest midway between the ends of the tubes, and is then stable.

We shall see, in § 96, that the cylindrical film, with *constant* pressure excess $2T/a$, is stable against *any* type of displacement, provided $l < \pi a$.

96. Theory of stability of cylindrical film. The film has initially the form of a cylinder of radius a and length l stretching between the tubes; it has surface $S_0 = 2\pi a l$. The pressure p_0 within it exceeds the atmospheric pressure P by $2T/a$. Let the film be deformed, still remaining a surface of revolution, so that $y = a + \eta$, where η is a small quantity which depends upon x and is such that $d\eta/dx$ is small. The axis of revolution is the axis of x and the origin is taken in the plane of the end of one of the tubes. Since the film joins the tubes, $\eta = 0$ when $x = 0$ and also when $x = l$. For convenience we write

$$\int_0^l \eta \, dx = F, \qquad \int_0^l \eta^2 dx = G, \qquad \int_0^l (d\eta/dx)^2 dx = H.$$

The integral G is of the second order of small quantities. The integral F is of the first order; it may, of course, be zero if η be positive for some values of x and negative for others.

If ds be the element of the arc of the plane curve $y = a + \eta$ which corresponds to dx along the axis of x,

$$ds = \{1 + (dy/dx)^2\}^{\frac{1}{2}} dx = \{1 + (d\eta/dx)^2\}^{\frac{1}{2}} dx.$$

Then, if S be the surface of the deformed film,

$$S = 2\pi \int_0^l (a + \eta) \{1 + (d\eta/dx)^2\}^{\frac{1}{2}} dx.$$

Expanding as far as terms of the second order, we have

$$S - S_0 = 2\pi F + \pi a H,$$

and thus, if E be the work spent in stretching the film,

$$E = 2T(S - S_0) = 2\pi T(2F + aH). \quad \text{............(1)}$$

If the volume enclosed by the film and any vessel communicating with it be v_0 when the film is cylindrical, and if the volume be v when the film is deformed,

$$v - v_0 = \pi \int_0^l (a+\eta)^2 dx - \pi a^2 l = 2\pi a F + \pi G.$$

To the second order, $(v - v_0)^2 = 4\pi^2 a^2 F^2$, since FG is of the third and G^2 of the fourth order.

When the volume changes from v_0 to v, the internal pressure becomes $p_0 v_0/v$; the work done by the enclosed air, during the change of volume, is

$$\int_{v_0}^{v} \frac{p_0 v_0 dv}{v} = p_0 v_0 \log \frac{v_0 + (v - v_0)}{v_0} = p_0 v_0 \left\{ \frac{v - v_0}{v_0} - \frac{1}{2}\frac{(v - v_0)^2}{v_0^2} + \ldots \right\}.$$

The work done against the external pressure is $P(v - v_0)$ or $(p_0 - 2T/a)(v - v_0)$. Hence, if W be the resultant work done by the enclosed air and the atmosphere,

$$W = \frac{2T(v - v_0)}{a} - \frac{p_0 (v - v_0)^2}{2v_0}. \quad \text{...............(2)}$$

In the second term it will suffice to write $v - v_0 = 2\pi a F$. Then

$$W = 2\pi T \left\{ 2F + \frac{G}{a} - \frac{p_0 F^2}{T(Na + l)} \right\}, \quad \text{............(3)}$$

where $Na + l = v_0/\pi a^2$, so that v_0 is the volume of a cylinder of radius a and length $Na + l$.

The first-order term $4\pi T F$ is the same in E as in W. This result was to be expected, since the cylindrical film is in equilibrium with the pressures p_0 and P. The *stability* depends on the second-order quantities G, H and F^2.

More work is needed to stretch the film than the pressures can supply, and the cylindrical film is therefore stable against the given deformation, if

$$aH > \frac{G}{a} - \frac{p_0 F^2}{T(Na + l)}. \quad \text{.....................(4)}$$

The pressures supply more work than the film can absorb, and the equilibrium will be unstable for the given deformation, if

$$aH < \frac{G}{a} - \frac{p_0 F^2}{T(Na+l)}. \qquad \text{.....................(5)}$$

Since η is free from discontinuities, we can, for any given deformation, find $c_1, c_2, \ldots$, so that, for all values of x from 0 to l,

$$\eta = c_1 \sin(\pi x/l) + c_2 \sin(2\pi x/l) + \ldots. \qquad \text{...........(6)}$$

We easily find

$$F = 2lf/\pi, \qquad G = lg^2/2, \qquad H = \pi^2 h^2/2l,$$

where

$$f = c_1 + \tfrac{1}{3}c_3 + \tfrac{1}{5}c_5 + \ldots,\ g^2 = c_1{}^2 + c_2{}^2 + \ldots,\ h^2 = c_1{}^2 + 2^2 c_2{}^2 + 3^2 c_3{}^2 + \ldots.$$

We may assume that h^2/g^2 is finite.

If, for a given value of l, the film be unstable for one particular type of deformation, we need not enquire as to its stability for other deformations. The number of possible deformations is, of course, infinite. It may help the reader if we consider a mechanical system of more than one degree of freedom. If at the origin point O on a smooth surface the planes of curvature be vertical and in the planes $y=0$ and $x=0$, and if O be the lowest point of the section of the surface by $y=0$ and the highest point in the section by $x=0$, the equilibrium of a particle at O will be stable for displacements in the plane $y=0$ but unstable for displacements in the plane $x=0$. Here we can gain stability if we restrain the particle to remain in the plane $y=0$. In the case of the film, no restraint is possible, and we conclude that, if, for a given l, the film be unstable for a deformation of a given type, it is unstable for every deformation.

When we express (4) in terms of f, g, h, we find that the film is stable against the deformation specified by (6) so long as l is sufficiently small to satisfy the inequality

$$\frac{\pi^2 a h^2}{2l} > \frac{lg^2}{2a} - \frac{4 p_0 l^2 f^2}{\pi^2 T(Na+l)},$$

or

$$l^2 < \frac{\pi^2 a^2 h^2}{g^2} + \frac{J l^3 f^2}{(Na+l) g^2}, \qquad \text{.................(7)}$$

where $J = 8p_0a/\pi^2T$. The least value of h^2/g^2 is unity. The film will therefore be stable when $l < \pi a$, if N be infinite. When N is infinite, the internal pressure is independent of l and therefore exceeds the atmospheric pressure by the constant difference $2T/a$. If N be finite, the film will be stable when $l < \pi a + q$, where $q > 0$ and, as we shall see, $q < \pi a$.

If $\eta = c_2 \sin 2\pi x/l$, so that all the c's vanish except c_2, we have $f^2/g^2 = 0$, $h^2/g^2 = 4$, and the film is unstable for this deformation, and therefore for *all* deformations, if $l > 2\pi a$, whatever the value of N. Hence $q < \pi a$.

Thus the film is stable when $l < \pi a$ and unstable when $l > 2\pi a$ and thus, when we deal with finite values of N, we need consider only those values of l for which $\pi a < l < 2\pi a$.

Using rough values, $p = 10^6$ dyne cm.$^{-2}$, $T = 27{\cdot}76$ dyne cm.$^{-1}$, and thus, with $a = 1{\cdot}37$ cm. as in § 130, $J = 8p_0a/(\pi^2T) = 40{,}000$. When N is not very large, the J-term in (7) is great unless f^2/g^2 be nearly zero, and the film is stable against this type of deformation so long as l^2 is less than $\pi^2a^2h^2/g^2 + \Omega$, where Ω is large. But true stability cannot extend beyond $l = 2\pi a$, for, when $l > 2\pi a$, the film is unstable to the deformation $c_2 \sin 2\pi x/l$. We conclude that, unless N be so great that J/N is small, the film will be stable so long as $l < 2\pi a$.*

THEORY OF SESSILE DROP

97. Sessile drops. When a quantity of mercury or other liquid is placed on a horizontal plate of glass which it does not wet, it does not spread over the plate but forms a so-called sessile drop, which, with clean mercury and clean glass, will be symmetrical about a vertical axis. A small mass of mercury gathers itself into an approximately spherical globule. As the mass is increased, the radius, c, of the girth of the drop increases, the curvature of the surface near the vertex diminishes and the departure of the drop from a spherical form increases. As c increases from zero, the height, h, of the drop, which at first increases, reaches a maximum value; this is so, for, when c/h is large, h steadily diminishes to

* The problem of finding the (large) value of N which will make the cylindrical film truly stable for a given length *between* πa and $2\pi a$ awaits solution.

a finite limiting value as c increases to infinity.* The central part of the upper surface of a large drop is found to be very nearly plane.

In an actual experiment, the radius of the drop must be finite and the corresponding theory must be used. The simple theory of a drop of infinite radius given in § 98 may serve as an introduction.

98. Introduction to theory. As the mass of mercury is increased, the girth or circle of the greatest horizontal section, corresponding to G (Fig. 61), increases in radius and in the limit may be represented by a straight line. Fig. 61 shows a part of the section of the drop by a vertical plane containing the (horizontal) normal GK to the surface of the drop at G. The horizontal axis Ox is asymptotic to the curve $WPGM$, i.e. it touches the surface of the drop at an infinite distance from G, the point where the tangent is vertical. We measure z vertically downward from Ox. If P be a point on the section at depth z below Ox, and R be the radius of curvature, PU, of the arc at P, then $1/R$ is one of the principal curvatures at P. The other principal curvature is zero, since the section of the surface by a plane perpendicular to the diagram and containing PU cuts the surface in a straight line. The excess of the pressure of the liquid at P above that of the air at P is thus (§ 88) T/R, where T is the surface tension of the liquid-air interface. At an infinite distance from G, $1/R = 0$, and thus the two pressures are $F + g\rho z$ and $F + g\sigma z$, where F is the atmospheric pressure at the level of Ox, and ρ, σ (treated as constant) are the densities of the liquid and the air. Hence

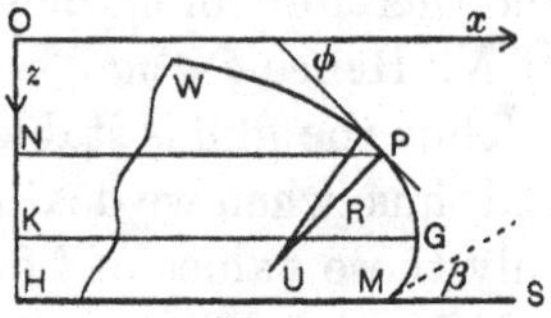

Fig. 61

$$z/a^2 = 1/R, \quad \ldots\ldots\ldots\ldots\ldots\ldots\ldots\ldots(1)$$

where

$$a^2 = T/\{g(\rho - \sigma)\}. \quad \ldots\ldots\ldots\ldots\ldots\ldots(2)$$

Let the tangent at P make an angle ϕ with Ox and let s be the arc, measured in the direction WGM, from W. Then

$$R = \frac{ds}{d\phi} = \frac{ds}{dz} \cdot \frac{dz}{d\phi} = \frac{1}{\sin\phi} \cdot \frac{dz}{d\phi}.$$

* Equation (30), § 104, shows that, when c, and therefore l, is large, h/a diminishes and tends to $2a \cos \frac{1}{2}\beta$ as l tends to infinity; here β is the angle of contact. The value of c which makes h a maximum is not known.

Hence, by (1), $z/a^2 = \sin\phi \, . \, d\phi/dz.$ ....(3)

Integrating from $z=0$ and $\phi=0$ to larger values z and ϕ, we have

$$\tfrac{1}{2}z^2/a^2 = 1-\cos\phi = 2\sin^2\tfrac{1}{2}\phi. \quad \text{..............(4)}$$

Thus $$z/2a = \sin\tfrac{1}{2}\phi. \quad \text{......................(4a)}$$

At G, $\phi=\tfrac{1}{2}\pi$, and thus, if f be the depth OK of G below Ox,

$$a^2 = \tfrac{1}{2}f^2, \quad \text{........................(5)}$$

which, by (2), gives the surface tension.

If the curve meet the plate HMS in M at the angle of contact β, $\phi=\pi-\beta$ at M. Then, if h be the height OH of the drop, we have, by (4a),

$$\cos\tfrac{1}{2}\beta = \sin\tfrac{1}{2}(\pi-\beta) = h/2a = h/(f\sqrt{2}). \quad \text{........(6)}$$

By (1), R_G, the radius of curvature at G, equals a^2/f, and thus, by (5), $R_G = \tfrac{1}{2}f$.

Since $dx/d\phi = dx/dz \, . \, dz/d\phi$, and $dx/dz = \cot\phi$, we have, by (3) and (4a),

$$\frac{dx}{d\phi} = \frac{a\cos\phi}{2\sin\tfrac{1}{2}\phi} = \tfrac{1}{2}a\left(\frac{1}{\sin\tfrac{1}{2}\phi} - 2\sin\tfrac{1}{2}\phi\right).$$

If the axis of z pass through G (*not* O), so that $x=0$ when $\phi=\tfrac{1}{2}\pi$,

$$x = a\{\log_e\tan\tfrac{1}{4}\phi - \log_e\tan\tfrac{1}{8}\pi + 2\cos\tfrac{1}{2}\phi - \sqrt{2}\}.$$

By (4a), $$z = 2a\sin\tfrac{1}{2}\phi.$$

If we take a series of values of ϕ and find the corresponding values of x and z, we can plot the curve. The values of x will be *negative*.

99. Large finite drop. Fig. 62 shows a section of the drop by a vertical plane through O, its vertex. Let $ON=z$ and $NP=x$ be the coordinates of P on the surface of the drop. The tangent at P to

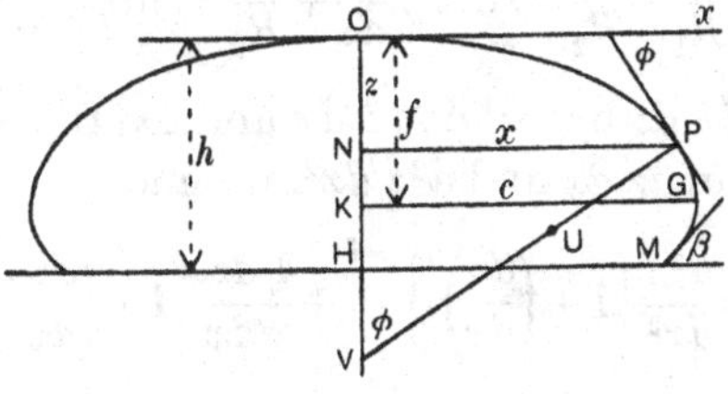

Fig. 62

the section makes an angle ϕ with Ox. Let $KG=c$ be the maximum value of x; the tangent at G is vertical. Let $OK=f$, and $OH=h$.

As c increases from zero, both f and the height h rise from zero to finite maximum values; they then decrease and tend to the finite limiting values $f=a\sqrt{2}$, $h=2a\cos\frac{1}{2}\beta$, as c tends to infinity.

The section meets the plate in M at an angle β, the angle of contact. Then $MH=l$ is the radius of the circular area of contact of the drop with the plate.

Let PU be the normal and U the centre of curvature of the *arc* at P. Let PU meet the axis OH in V, and let $PU=R_1$, $PV=R_2$. Then, by § 89, PUV is the normal to the *surface* at P, and R_1, R_2 are the principal radii of curvature at P. By § 88, the excess of the pressure of the *liquid* at P above that of the *air* at P is

$$T(1/R_1+1/R_2).$$

At O, $R_1=R_2=b$, where b is the radius of curvature of the arc OP at O. If ρ, σ be the densities of the liquid and the air, treated as constant,

$$T(1/R_1+1/R_2)=g(\rho-\sigma)z+2T/b. \quad\text{.........(7)}$$

For convenience, we write $T/\{g(\rho-\sigma)\}=a^2$; then

$$1/R_1+1/R_2=z/a^2+2/b. \quad\text{..................(8)}$$

The length a serves as a standard of magnitude. For a mercury-air surface, $T=520$ dyne cm.$^{-1}$ approximately, for mercury $\rho=13{\cdot}6$ and for air $\sigma=1{\cdot}29\times10^{-3}$ grm. cm.$^{-3}$. Hence $a=0{\cdot}20$ cm. For a water-air surface, $T=72$, $\rho=1$, $\sigma=1{\cdot}29\times10^{-3}$, and then $a=0{\cdot}27$ cm. The difference between the two values of a is slight.

If the arc $OP=s$, $1/R_1=d\phi/ds=\sin\phi\,.\,d\phi/dz$, since $dz/ds=\sin\phi$. We also have $1/R_2=1/PV=\sin\phi/x$. Then (8) becomes

$$\frac{z}{a^2}+\frac{2}{b}=\frac{1}{R_1}+\frac{1}{R_2}=\sin\phi\,.\,\frac{d\phi}{dz}+\frac{1}{R_2}=\sin\phi\,.\,\frac{d\phi}{dz}+\frac{\sin\phi}{x}. \quad\text{....(9)}$$

If we replace $d\phi/dz$ by $\cot\phi\,.\,d\phi/dx$ and express $\sin\phi$, $\cot\phi$ and $d\phi/dx$ in terms of dz/dx and d^2z/dx^2, we find

$$\frac{z}{a^2}+\frac{2}{b}=\frac{d^2z}{dx^2}\left\{1+\left(\frac{dz}{dx}\right)^2\right\}^{-\frac{3}{2}}+\frac{1}{x}\frac{dz}{dx}\left\{1+\left(\frac{dz}{dx}\right)^2\right\}^{-\frac{1}{2}}. \quad\text{...(10)}$$

100. Curvature at vertex. It is evident, on viewing objects by reflexion at the surface of an actual drop for which c exceeds one cm., that near the vertex the surface is nearly plane. Hence b/a

is large when c/a is large. We cannot, however, treat b/a as infinite, if we are dealing with a drop for which c/a is finite. We must, therefore, make an estimate of at least the order of magnitude of b/a.

Equation (9) or its equivalent (10) cannot be solved exactly, and we must be content with approximate results. We shall first find the form of that central circular patch of the surface of the drop for which $\phi < \epsilon$, where ϵ is so small that $\tan^2\epsilon$ is negligible compared with unity. We shall find that b, the radius of curvature at the vertex, is so large that, for points on the edge of the patch, the depth z, although very small compared with a, is very large compared with $2a^2/b$.

When ϕ is not greater than ϵ,

$$1/R_1 = d^2z/dx^2, \quad \sin\phi = \tan\phi = dz/dx$$

and

$$1/R_2 = x^{-1}.\sin\phi = x^{-1}.dz/dx.$$

Then, by (8), if $u = x/a$,

$$z + \frac{2a^2}{b} = a^2\left\{\frac{d^2z}{dx^2} + \frac{1}{x}\frac{dz}{dx}\right\} = \frac{d^2(z+2a^2/b)}{du^2} + \frac{1}{u}\frac{d(z+2a^2/b)}{du}. \quad \ldots(11)$$

We obtain the same result if, in (10), we neglect $(dz/dx)^2$ in comparison with unity.

The Bessel function $I_0(u)$ is defined by

$$I_0(u) = \frac{1}{\pi}\int_0^\pi e^{u\cos\psi}\,d\psi. \quad \ldots\ldots\ldots\ldots(12)$$

Hence

$$\frac{dI_0(u)}{du} = \frac{1}{\pi}\int_0^\pi \cos\psi\, e^{u\cos\psi}\,d\psi$$

$$= \frac{1}{\pi}\left[\sin\psi\, e^{u\cos\psi}\right]_0^\pi + \frac{1}{\pi}\int_0^\pi u\sin^2\psi\, e^{u\cos\psi}\,d\psi$$

$$= \frac{u}{\pi}\int_0^\pi \sin^2\psi\, e^{u\cos\psi}\,d\psi. \quad \ldots\ldots\ldots\ldots(13)$$

If we differentiate (12) twice, we have

$$\frac{d^2I_0(u)}{du^2} = \frac{1}{\pi}\int_0^\pi \cos^2\psi\, e^{u\cos\psi}\,d\psi. \quad \ldots\ldots\ldots\ldots(14)$$

From (12), (13) and (14), we find that $I_0(u)$ satisfies

$$I_0(u) = \frac{d^2 I_0(u)}{du^2} + \frac{1}{u}\frac{dI_0(u)}{du}, \quad \text{..............(15)}$$

and that, when $u=0$, $I_0(u)=1$, $dI_0(u)/du=0$, $d^2I_0(u)/du^2=\frac{1}{2}$. Since $u=x/a$, we have $u=0$ when $x=0$ and also

$$dI_0(x/a)/dx = a^{-1}.dI_0(u)/du, \qquad d^2I_0(x/a)/dx^2 = a^{-2}.d^2I_0(u)/du^2.$$

Hence, when $x=0$,

$$dI_0(x/a)/dx=0, \qquad d^2I_0(x/a)/dx^2=\tfrac{1}{2}a^{-2}.$$

By comparing (11) with (15), we see that z will satisfy (11), if

$$z+2a^2/b = C.I_0(u) = C.I_0(x/a),$$

where C is a constant. Since $z=0$ and $I_0(x/a)=1$ when $x=0$, $C=2a^2/b$. Hence

$$z=\frac{2a^2}{b}\left\{I_0\left(\frac{x}{a}\right)-1\right\}. \quad \text{..................(16)}$$

This gives $z=0$, $dz/dx=0$, $d^2z/dx^2=1/b$ when $x=0$; thus (16) satisfies all the conditions.

The function $I_0(u)$ increases rapidly as u increases. Since

$$I_0(u) > \frac{1}{\pi}\int_0^{\frac{1}{4}\pi} e^{u\cos\psi}\,d\psi,$$

and since $\cos\psi$ is not less than $1/\sqrt{2}$ over the range 0 to $\frac{1}{4}\pi$ of ψ,

$$I_0(u) > \tfrac{1}{4}e^{u/\sqrt{2}}.$$

The mathematical theory of the I_0 function shows that the ratio of $I_0(u)$ to $e^u/\sqrt{(2\pi u)}$ has the limit unity when u tends to infinity. Some values of the two functions are given below:

u	$I_0(u)$	$e^u/\sqrt{(2\pi u)}$	u	$I_0(u)$	$e^u/\sqrt{(2\pi u)}$
1	1·266	1·084	12	$1{\cdot}895\times10^4$	$1{\cdot}874\times10^4$
3	4·881	4·627	15	$3{\cdot}396\times10^5$	$3{\cdot}367\times10^5$
6	67·23	65·71	18	$6{\cdot}219\times10^6$	$6{\cdot}174\times10^6$
9	1094	1078	21	$1{\cdot}155\times10^8$	$1{\cdot}148\times10^8$

We can now obtain an estimate of $2a/b$; then, if we know a, we can find b. When c/a is large, ϕ is small until x approaches c, and will certainly be small if $x=3c/4$. Since ϕ is small when $x=3c/4$, the value of z/a given by (16) for $x=3c/4$ is a good approximation. Hence, as a good approximation,

$$2a/b = z/ap, \quad \text{......................(17)}$$

where $$p=\{I_0(3c/4a)-1\}. \quad \text{..................(18)}$$

We have not yet found z, but it is certain that z for $x = 3c/4$ is much less than a, when c/a is large. Hence $2a/b < p^{-1}$, $b > 2ap$, and $4a^2/b$, which is needed in § 102, is much less than $2a/p$. The Table gives some values of p^{-1}; the values of $2ap$ and $2a/p$ have been calculated for $a = 0{\cdot}2$ cm.

c/a	p	p^{-1}	$2ap$ cm.	$2a/p$ cm.
8	$6{\cdot}6 \times 10^1$	$1{\cdot}5 \times 10^{-2}$	$2{\cdot}6 \times 10^1$	$6{\cdot}0 \times 10^{-3}$
12	$1{\cdot}1 \times 10^3$	$9{\cdot}1 \times 10^{-4}$	$4{\cdot}4 \times 10^2$	$3{\cdot}7 \times 10^{-4}$
16	$1{\cdot}9 \times 10^4$	$5{\cdot}3 \times 10^{-5}$	$7{\cdot}6 \times 10^3$	$2{\cdot}1 \times 10^{-5}$
20	$3{\cdot}4 \times 10^5$	$2{\cdot}9 \times 10^{-6}$	$1{\cdot}4 \times 10^5$	$1{\cdot}2 \times 10^{-6}$
24	$6{\cdot}2 \times 10^6$	$1{\cdot}6 \times 10^{-7}$	$2{\cdot}5 \times 10^6$	$6{\cdot}4 \times 10^{-8}$

Since the ratio of $2a/b$ to z/a equals p^{-1}, we see that, if $c/a > 12$, $2a/b$ is negligible compared with the value of z/a at a point where $x = 3c/4$.

We cannot calculate the exact value of b, the radius of curvature at the vertex of the drop, but we know that b exceeds $2ap$. When $a = 0{\cdot}2$ cm. and $c = 4$ cm., we have $c/a = 20$. The Table shows that, in this case, b exceeds $1{\cdot}4 \times 10^5$ cm. or $1{\cdot}4$ kilometres.

101. Circle of contact. The equilibrium of the drop, of volume v, requires that

$$\pi l^2 \{gh(\rho - \sigma) + 2T/b\} = gv(\rho - \sigma) + 2\pi l T \sin\beta.$$

If $v = \pi h c^2 \lambda^2$, λ cannot be much less than unity, when c/a is large. Neglecting $2T/b$ and writing $T/\{g(\rho - \sigma)\} = a^2$, we obtain a quadratic

$$l^2 - 2l \sin\beta \,.\, a^2/h - \lambda^2 c^2 = 0.$$

Hence $$l/a = \sin\beta \,.\, a/h + (c^2\lambda^2/a^2 + \sin^2\beta \,.\, a^2/h^2)^{\frac{1}{2}},$$

for l/a is positive. Hence $l > \lambda c$ and thus l/c is little less than unity.

102. Form of drop where x is comparable with c. On integrating (9) with respect to z from $z = 0$ to any greater value not exceeding h, we obtain

$$\tfrac{1}{2}z^2/a^2 + 2z/b = \tfrac{1}{2}(z/a^2)(z + 4a^2/b) = 2\sin^2\tfrac{1}{2}\phi + Q(\phi), \quad \text{...(19)}$$

where $$Q(\phi) = \int_0^z \frac{dz}{R_2} = \int_0^\phi \frac{\sin\phi}{x} \cdot \frac{dz}{d\phi}\, d\phi. \quad \text{............(20)}$$

If v be the volume of the drop,

$$\int_0^h \pi x^2 dz = v. \qquad \text{.......................(21)}$$

When $z = h$, $\qquad dz/dx = -\tan\beta. \qquad$ ....(22)

An exact solution of (19) subject to (20), (21) and (22) has not so far been found. The values found for a and β by the following method of approximation become more accurate as c, the maximum radius of a horizontal section of the drop, increases.

At the vertex O (Fig. 62), $R_2 = b$, and at G, $R_2 = c$. If the normal at P' meet OH in V', R_2 will have the same value at P' as at P if $P'V' = PV$. If the two normals meet in W and if P' be very close to P, $P'W - PW$ is of the second order of small quantities, since WP is a normal. As P' tends to P, W tends to U, the centre of curvature of the arc at P. Hence $P'V' = PV$, if $WV' = WV$, which requires, in the limit, that UV be perpendicular to OH. Hence R_2 has a stationary value only when UV is horizontal, and thus R_2 has no stationary value *between* O and G. Since R_2 diminishes from b to c as P goes from O to G, the stationary value of R_2 at G must be a minimum. Hence $KG = c$ is the minimum value of R_2. Hence $Q < z/c$, and accordingly Q diminishes to the limit zero as c/a increases.

The depth $OK(=f)$ approximately equals $a\sqrt{2}$ (see § 98) and thus, if $a = 0{\cdot}2$ cm., $f = 0{\cdot}28$ cm. The Table of § 100 shows that, if $c > 2{\cdot}4$ cm. or $c/a > 12$, $2a/p < 3{\cdot}7 \times 10^{-4}$ cm. Hence we can take J (Fig. 63), where $z = k$, $x = q$, $\phi = \epsilon$, so that k is (1) small compared

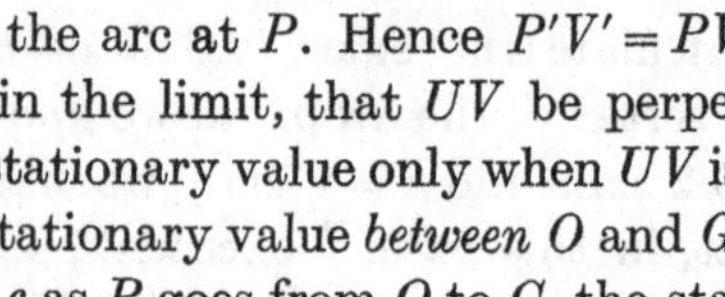

Fig. 63

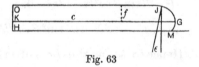

with f or with the larger h, and yet is (2) large compared with the very small $4a^2/b$, which is much smaller than $2a/p$. Since the nature of the curve $OJGM$ allows us to take J so that q/c is nearly unity without violating (1) or (2), we shall definitely choose J so that q nearly equals c.

Since R_2 diminishes steadily from b at O to the smaller, but still very large, $R_2(\epsilon)$ at J, $Q(\epsilon) < k/R_2(\epsilon)$. Thus $Q(\epsilon)$ is entirely negligible compared with f^2/a^2, which approximates to 2, or with the larger h^2/a^2. Now

$$Q(\phi)=\int_0^k \frac{dz}{R_2}+\int_k^z \frac{dz}{R_2}=Q(\epsilon)+\int_k^z \frac{dz}{R_2}.$$

Hence, when z is large compared with k, we may neglect $Q(\epsilon)$ and write

$$Q(\phi)=\int_k^z \frac{dz}{R_2}=\int_\epsilon^\phi \frac{\sin\phi}{x}\cdot\frac{dz}{d\phi}\,d\phi. \qquad \text{.........(23)}$$

To find an approximate value of $Q(\phi)$, which is small, being less than $(z-k)/c$, for a large *finite* value of c/a, we first find the relation between z and ϕ when c/a is infinite and Q and $4a^2/b$ are zero. Then, in accordance with the general principles of approximation, that relation is used in (23) to give $Q(\phi)$.

When c/a is infinite, $Q(\phi)=0$, $1/b=0$ and $1/R_2=0$, and hence (19) and (9) give respectively

$$z=2a\sin\tfrac{1}{2}\phi \quad \text{and} \quad d\phi/dz=z/(a^2\sin\phi).$$

Hence $$dz/d\phi=a\cos\tfrac{1}{2}\phi. \qquad \text{.....................(24)}$$

Using (24) in (23), we have, for values of ϕ large compared with ϵ,

$$Q(\phi)=2a\int_\epsilon^\phi \frac{\sin\frac{1}{2}\phi\cos^2\frac{1}{2}\phi}{x}\,d\phi.$$

Since $$d(1-\cos^3\tfrac{1}{2}\phi)/d\phi=\tfrac{3}{2}\sin\tfrac{1}{2}\phi\cos^2\tfrac{1}{2}\phi,$$

we have $$Q(\phi)=G(\phi)+J(\phi),$$

where $$G(\phi)=\frac{4a}{3}\left[\frac{1-\cos^3\frac{1}{2}\phi}{x}\right]_\epsilon^\phi, \quad J(\phi)=\frac{4a}{3}\int_q^x \frac{1-\cos^3\frac{1}{2}\phi}{x^2}\,dx.$$

In place of $1-\cos^3\frac{1}{2}\phi$, $A-\cos^3\frac{1}{2}\phi$ might have been used, where A is any constant, but it is convenient to put $A=1$, for $1-\cos^3\frac{1}{2}\phi=0$, when $\phi=0$. Since ϵ is small, $1-\cos^3\frac{1}{2}\epsilon=3\epsilon^2/8$. Hence, since $x=q$ when $\phi=\epsilon$,

$$G=\frac{4a}{3}\left\{\frac{1-\cos^3\frac{1}{2}\phi}{x}-\frac{3\epsilon^2}{8q}\right\},$$

or, since ϵ^2 is of the second order and q is of the same order as c,

$$G=4a(1-\cos^3\tfrac{1}{2}\phi)/3x.$$

Since $1-\cos^3\frac{1}{2}\phi$ increases with ϕ, when $0<\phi<\pi$,

$$J<(4a/3)(1-\cos^3\tfrac{1}{2}\phi)\int_q^x \frac{dx}{x^2}$$
$$<G(x-q)/q.$$

Thus J/G is small, for $x-q$ is small compared with c and therefore with q. Hence we may put $Q(\phi)=G$. Since $4a^2/b$ is negligible compared with z when ϕ exceeds ϵ, we may, on the left side of (19), neglect $2z/b$ in comparison with $\frac{1}{2}z^2/a^2$. Then we have

$$\tfrac{1}{2}z^2/a^2=2\sin^2\tfrac{1}{2}\phi+4a(1-\cos^3\tfrac{1}{2}\phi)/3x. \quad\text{.........(25)}$$

Equation (25) is a valid approximation for all values of ϕ large compared with ϵ.

103. Determination of surface tension. The values of z and x when $\phi=\frac{1}{2}\pi$, viz. f and c, are easily measured, and hence (25), with $\phi=\frac{1}{2}\pi$, is used for finding a^2. We have, with $\lambda=\frac{1}{3}(\sqrt{8}-1)$,

$$\tfrac{1}{2}f^2/a^2=1+a(4-\sqrt{2})/3c=1+\sqrt{2}\,.\,\lambda a/c. \quad\text{......(26)}$$

We see that $\frac{1}{2}f^2/a^2$ is a little greater than unity. If we put $2a^2/f^2=1-\eta$, η will be small and positive. If $\mu=\lambda f/c$, μ is positive, and we have $\sqrt{2}\,.\,\lambda a/c=\mu\sqrt{(1-\eta)}$. Then, by (26),

$$\eta=\mu(1-\eta)^{\frac{3}{2}}, \quad\text{........................(27)}$$

or
$$\eta^3+(\mu^{-2}-3)\eta^2+3\eta-1=0.$$

This cubic equation in η has one real and positive root. It has two other roots, which, if real, are negative. If we write

$$\eta_1=\mu,$$

and $\eta_2=\mu(1-\eta_1)^{\frac{3}{2}},\quad \eta_3=\mu(1-\eta_2)^{\frac{3}{2}},\dots\quad \eta_n=\mu(1-\eta_{n-1})^{\frac{3}{2}},\dots,$

then η is the limit to which η_n tends as n tends to infinity. If $\mu<0{\cdot}1$ the approximation proceeds rapidly.

We can exhibit η in terms of μ. If we expand $(1-\eta)^{\frac{3}{2}}$, we have

$$\eta=\mu(1-\eta)^{\frac{3}{2}}=\mu(1-\tfrac{3}{2}\eta+\tfrac{3}{8}\eta^2+\tfrac{1}{16}\eta^3+\dots).$$

The first approximation is $\eta=\mu$. For the second, we put μ for η in the term $\frac{3}{2}\eta$. Then $\eta=\mu(1-\frac{3}{2}\mu)$. For the third, we put μ for η in $\frac{3}{8}\eta^2$ and $\mu(1-\frac{3}{2}\mu)$ for η in $\frac{3}{2}\mu$. Then

$$\eta=\mu\{1-\tfrac{3}{2}\mu(1-\tfrac{3}{2}\mu)+\tfrac{3}{8}\mu^2\}=\mu\{1-\tfrac{3}{2}\mu+\tfrac{21}{8}\mu^2\}.$$

If we take two further steps in the process, we find

$$\eta = \mu\{1 - \tfrac{3}{2}\mu + \tfrac{21}{8}\mu^2 - 5\mu^3 + \tfrac{1287}{128}\mu^4\}. \quad \text{.........(28)}$$

Since $\lambda = \frac{1}{3}(\sqrt{8} - 1) = 0{\cdot}6094757...$, and $\mu = \lambda f/c$, we have

$$\begin{aligned} a^2 &= \tfrac{1}{2}f^2(1-\eta) \\ &= \tfrac{1}{2}f^2\{1 - 0{\cdot}6095\nu + 0{\cdot}5772\nu^2 - 0{\cdot}5943\nu^3 + 0{\cdot}6899\nu^4 \\ &\qquad - 0{\cdot}8456\nu^5\}, \quad \text{......(29)} \end{aligned}$$

where $\nu = f/c$.

By § 99, the surface tension is given by $T = ga^2(\rho - \sigma)$.

104. Determination of angle of contact. At M (Figs. 62, 63), $\phi = \pi - \beta$, $z = h$ and $x = l$. Then, by (25),

$$\tfrac{1}{2}h^2/a^2 = 2\cos^2\tfrac{1}{2}\beta + 4a(1 - \sin^3\tfrac{1}{2}\beta)/3l. \quad \text{.........(30)}$$

If we put $\sin\frac{1}{2}\beta = s$, $(4a^2 - h^2)/4a^2 = \omega^2$, $2a/3l = \gamma$,

in (30), we have $$s^2 = \omega^2 + \gamma - \gamma s^3. \quad \text{....................(31)}$$

If $\gamma < s^2/(1-s^3)$, ω^2 is positive and ω is real. But $0 < s < 1/\sqrt{2}$, and hence $s^2/(1-s^3)$ is less than $2/(4-\sqrt{2})$ or $0{\cdot}773$. The theory requires, for the validity of the approximations, that γ be small—certainly less than $0{\cdot}773$—and this condition is satisfied in the experiments, and thus ω^2 is positive. Hence $\omega^2 + \gamma$ is positive, and the cubic for s has one real and positive root. It has two other roots, which, if real, are negative. If we write

$$s_1{}^2 = \omega^2,$$

and $$s_2{}^2 = \omega^2 + \gamma - \gamma s_1{}^3, \ \ldots \quad s_n{}^2 = \omega^2 + \gamma - \gamma s_{n-1}{}^3, \ \ldots,$$

then s^2 is the limit to which $s_n{}^2$ tends as n tends to infinity. If $\gamma < 0{\cdot}1$ the approximation is rapid.

We can exhibit s^2 in terms of γ. The first approximation to s^2 is ω^2. The second is $\omega^2 + \gamma(1-\omega^3)$. The third is got by using the second in s^3 and expanding as far as γ^2. Continuing the process, we find, as far as γ^3,

$$s^2 = \omega^2 + k\gamma - \tfrac{3}{2}\omega k\gamma^2 - \tfrac{3}{8}(k/\omega)(1 - 7\omega^3)\gamma^3, \quad \text{......(32)}$$

where $k = 1 - \omega^3$.

When s^2 has been found, we have $\cos\beta$, for

$$\cos\beta = 1 - 2\sin^2\tfrac{1}{2}\beta = 1 - 2s^2. \quad \text{............(33)}$$

CHAPTER VI

EXPERIMENTS ON SURFACE TENSION

105. Soap solution. The following method of preparing a soap solution for surface tension experiments is based on the recipe given by W. Watson. To make up 1 litre of solution, place 750 grm. of distilled water in a bottle fitted with a glass stopper and add 18·75 grm. of pure fresh sodium oleate in small shavings. Shake the bottle vigorously until the mixture jellies, and leave the bottle in a dark place for a month, shaking it occasionally. Then add 315 grm. or 250 c.c. of glycerine and shake the bottle until the glycerine and the solution are well mixed. Leave the bottle in a dark place for four or five months. The sodium oleate will then have dissolved completely and the bottle will contain clear liquid only. Add two or three drops of strong ammonia to every litre. Two litres of solution is suitable for a "Winchester quart."

When the solution is made in this way, the oleate should be put into the water six months before the solution is required. A quicker, but more wasteful, method is to add the glycerine as soon as the sodium oleate has been well shaken with the water and subsequently to draw off with a siphon all the clear liquid formed when the mixture has stood for a week. A good deal of the oleate will be still undissolved at the end of a week and the solution will consequently be weaker than when it is prepared by the slower method. The surface tension will not be appreciably affected, but the films formed of the weaker solution will have shorter lives than those formed of the stronger solution.

The solution should be kept in the dark and will improve with keeping. It should not be warmed or filtered or be exposed to the air for any length of time.

Solution which has been poured out of the stock bottle for use in experiments should, as far as possible, be kept covered to exclude dust. If, after the experiment, the used solution be put into a bottle provided with a stopper, it will serve for a few days, but it should not, on any account, be returned to the *stock* bottle.

105 A. Cleaning of glass surfaces.* Alkaline solutions have a tendency to attack glass, which is a complex silicate mixture, with the formation of soluble silicates. Thus, the inner surface of a glass bottle containing a solution of caustic soda in water soon appears to be etched. Long before the attack has proceeded far enough for an appreciable amount of silica to be removed, sodium ions have penetrated a small distance into the glass, and, as they are held there by quasi-chemical forces, they are not removed at all readily by mere washing of the surface with water. Consequently, a glass surface which has been treated with caustic alkali is unsuitable for use in experiments on the surface tension of water. Even though the alkali have been diluted by rinsing the surface many times with water, the film of water on the surface will probably still contain enough alkali to cause the value of the surface tension, as found in the experiment, to be less than the value for pure water.

The quickest way to remove any adsorbed alkali is to treat the glass with a dilute acid; the concentration of uncombined alkali in the acid solution in contact with the glass is then as low as possible, with the result that the rate of diffusion of alkali out of the glass is a maximum.

For removing greasy matter from glass, a soap solution is usually more effective than caustic soda or caustic potash. Adsorption of alkali from soap solution is relatively small, but, even when soap is used as a detergent, it is advisable to rinse the surface several times with water, then with dilute acid and again with water.

Experiment 12. **Measurement of surface tension of liquid by capillary tube.**

106. Method. The surface tension of the surface separating a liquid from the air may be found from observations on the rise of the liquid in a vertical tube, which has a bore 1 mm. or less in diameter—a so-called capillary tube. When such a tube DK (Fig. 64), open at both ends, has its lower end immersed in liquid contained in a pan G, the liquid rises in the tube and is bounded by a meniscus DOD'. The height of O, the lowest point, above

* Section 105 A is due to Dr Guy Barr, of the National Physical Laboratory.

the surface of the liquid in the pan, depends upon the bore of the tube near O, the densities of the liquid and of the air, the surface tension of the liquid-air surface and the angle of contact between the liquid and the tube. We shall suppose that the liquid wets the tube, i.e. that the angle of contact is zero.

In the neighbourhood of the glass tube DK and of the side of the pan G, the surface of the liquid in the pan is disturbed by the action of surface tension, but, if the diameter of G be a few cm., the greater part of the surface of the liquid in G will not differ appreciably from a horizontal plane. Let FF' be the level of this plane part of the surface.

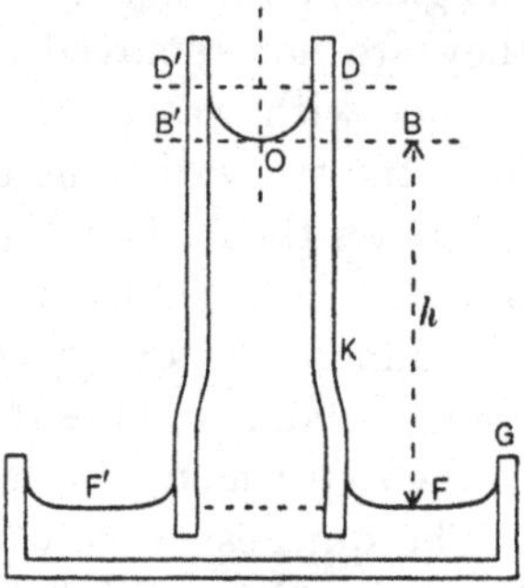

Fig. 64

Let the horizontal plane BB' touch the meniscus at its lowest point O. The form of the tube below BB' is, as we shall see, of no consequence, but, for a short distance above BB', the tube is supposed of uniform circular section with its axis vertical. Under these conditions, the meniscus will be a surface of revolution about the vertical axis and its edge will be in a horizontal plane DD'. The internal radius of the tube near O is r cm. and the height BF of O above FF' is h cm. The densities of the liquid and of the air near the tube are ρ and σ grm. cm.$^{-3}$ respectively. If v be the small volume of liquid in the tube above the level BB', the volume of the air between DD' and BB' is $\pi r^2 k - v$, where $k = D'B'$.

Let the atmospheric pressure at level F be P. Then the pressure of the liquid at O is $P - g\rho h$ and is independent of the form of the tube; the pressure of the air at D is $P - g\sigma(h+k)$, if the change of σ with height be neglected.

Since the liquid *wets* the glass, the surface of the liquid is vertical at the level DD'. The total upward force on the liquid due to surface tension is therefore $2\pi rT$.

We now consider the equilibrium of the matter—partly liquid and partly air—contained between the tube and the planes BB', DD'. Acting vertically upwards on it are the force $2\pi rT$, due to surface tension, and the force $\pi r^2(P - g\rho h)$, due to pressure of the liquid on BB'. Acting vertically downwards are the force $\pi r^2\{P - g\sigma(h+k)\}$, due to atmospheric pressure on DD', the weight $g\rho v$ of the volume v of liquid and the weight $g\sigma(\pi r^2 k - v)$ of the volume $\pi r^2 k - v$ of air.

Since there is equilibrium,

$$2\pi r T + \pi r^2(P - g\rho h) = \pi r^2\{P - g\sigma(h+k)\} + g\rho v + g\sigma(\pi r^2 k - v),$$

or
$$T = \tfrac{1}{2}g(\rho - \sigma)(hr + v/\pi r). \quad \text{.................}(1)$$

It is generally assumed that, when r is small compared with h, the meniscus is a hemisphere. The volume of a hemisphere of radius r is $\frac{2}{3}\pi r^3$ and that of a cylinder of radius r and length r is πr^3. The difference is v, and thus $v = \frac{1}{3}\pi r^3$. Then (1) becomes

$$T = \tfrac{1}{2}g(\rho - \sigma)r(h + \tfrac{1}{3}r). \quad \text{.................}(2)$$

If the density of the air be neglected in comparison with that of the liquid, we have

$$T = \tfrac{1}{2}g\rho r(h + \tfrac{1}{3}r). \quad \text{....................}(3)$$

If r/h be not very small, the meniscus is no longer part of a sphere. In § 90 it is shown that, in this case, (2) should be replaced by

$$T = \tfrac{1}{2}g(\rho - \sigma)rh(1 + \tfrac{1}{3}r/h - \tfrac{1}{9}r^2/h^2). \quad \text{...........}(4)$$

A word may be said about the case when the horizontal section of the bore is an ellipse of major and minor axes $2b$, $2c$. Let h be the height BF in Fig. 64. Then, if the minor axis be so small that c/h is small, the volume of liquid above BB' is negligible compared with πbch. We then have, by § 112,

$$Ts = \pi g(\rho - \sigma)bch,$$

where s is the perimeter of the section. Now $s = 4bE_1(k)$, where $E_1(k)$ is the Second Complete Elliptic Integral of modulus k, and $k = \sqrt{(1 - c^2/b^2)}$. Thus

$$T = \tfrac{1}{4}\pi g(\rho - \sigma)ch/E_1(k).$$

Dale's Mathematical Tables give $E_1(\sin\theta)$, where $\sin\theta = k$, and show that $E_1(k)$ gradually diminishes from $\frac{1}{2}\pi$ or $1{\cdot}5708\ldots$, when $k = 0$, to unity, when $k = 1$. When c/b is nearly unity and, consequently, k is small,

$$E_1(k) = \tfrac{1}{2}\pi\left\{1 - \frac{1}{1}\left(\frac{1}{2}\right)^2 k^2 - \frac{1}{3}\left(\frac{1\times 3}{2\times 4}\right)^2 k^4 - \ldots\right\}.$$

When c/b is small, k is nearly unity and, if $k'^2 = 1 - k^2$, k' is small. Then

$$E_1(k) = 1 + \tfrac{1}{2}k'^2\left(\log_e \frac{4}{k'} - \frac{1}{1\times 2}\right) + \ldots.$$

107. Preparation of the tube. The bore of the tube should be very nearly uniform, for then very little error is introduced if we fail to cut the tube, for the purpose of measuring its radius, exactly at the edge of the meniscus. A suitable tube may be drawn from a piece of glass tube, with fairly thick walls, about 0·8 cm. in diameter. If necessary, the tube is cleaned first with nitric acid and then with water (see § 105 A); it is then dried by passing warm air through it.* The middle of the tube is heated in a blow-pipe flame large enough to soften 2 or 3 cm. of the tube. One end of the tube is held in each hand, and the tube is continually turned about its axis so as to ensure uniform heating. A little care is needed to rotate both ends at the same rate, when the glass is soft. When the glass is quite soft, the tube is removed from the flame. As soon as it is out of the flame, the two ends are drawn slowly apart. If they be drawn apart too quickly, the resulting tube will be too narrow for the experiment. A very small pull is required to draw the ends apart while the glass is still hot. But when, by gentle pulling, the central part of the heated portion has been drawn down to nearly the desired diameter, the pull can be gradually increased, since the central portion stiffens as it cools. After the central part has solidified, as it does very quickly, a continued extension draws the glass off from the two ends, which, on account of their bulk, retain a high temperature much longer than the central part. The pull may be maintained till the glass has cooled so much that no further extension can be produced. If this process be carefully carried out, a tube will be produced 20 to 40 cm. in length, with an external diameter of 1 to 2 mm., the diameter being very nearly uniform over the whole length of the tube. By rotating the tube on its axis while it is pressed between finger and thumb, it is easy to detect any devia-

* The chief enemy of surface tension is grease and, as this is destroyed when the tube is heated in the flame, not much is gained by the cleaning process. Of course an obviously dirty tube should be cleaned.

tion of the outer cross-section from a circular form. If the cross-section differ widely from a circle, the tube should be rejected. The thick ends are now cut off, and, if the tube is not to be used for some time, the ends are sealed up to keep out dust.

The tube should be of such a bore that the liquid will rise from 2 to 5 cm.

For cutting the tube, a commercial "glass knife," or a triangular file ground smooth so as to have three sharp edges, is convenient. A still better cutter can be made from a piece of hack-saw. The teeth are ground off and the blade is ground to a knife edge. It is then hardened in mercury and is finished on a stone. The sharp edge of the knife is drawn with a light pressure across the tube, the edge being perpendicular to the axis of the tube. If the knife be in good condition, a slight scratch will result. The tube is now held in the two hands, between the thumb and the forefinger of each hand. If the hands be held with the palms upwards, the side of one hand being pressed against the side of the other, it is easy to apply a pull along the tube. If the tube have been properly scratched by the knife, it will now break in two, the surface of the fracture being a plane perpendicular to the axis of the tube. This operation should be practised till a tube can be cut properly at any desired point.

108. Measurements. The capillary tube is attached to a millimetre scale by indiarubber bands, so that the tube lies over the divisions on the scale. The scale and tube are held vertical by a clip. The lower ends of the scale and of the tube dip into the liquid in a beaker. The diameter of the beaker should not be less than 4 cm. The tube should lie approximately along the axis of the beaker, for, unless this be so, the part of the tube below the liquid will, owing to refraction, appear to be not continuous with the part of the tube above the liquid. If the eye be placed slightly below the level of the liquid, the under side of the surface of the liquid will, owing to total internal reflexion, appear bright and silvery. As the eye is raised, this silvery surface appears more and more foreshortened, till, when the eye is at the level of the liquid, the silvery surface just disappears. The reading of the scale corresponding to the line in which the plane, so ascertained, cuts the scale is then taken. The level of the *bottom* of the meniscus in the tube is also read off on the scale.

Care should be taken that the beaker and the scale are clean; they may be cleaned by soap and water. Care must be taken to remove the soap by thorough rinsing with tap water. If a liquid other than water or a solution of a salt in water be used in the experiment, the beaker and scale are dried after the rinsing. If water be used, it should be freshly drawn from the tap, as the surface soon becomes contaminated when exposed to the air. There is no advantage in using distilled water which has been kept for days in a bottle. Before a reading is taken, the beaker may be raised a few mm., so as to cause the liquid to rise a little in the tube and to ensure that the tube is properly wetted. Unless the meniscus come back to its old position, after the beaker has been raised and lowered, the tube is dirty, and no accurate result will be obtained.

When the readings have been taken, the tube is removed from the scale, and is cut as nearly as possible at the point at which the meniscus stood.

The temperature of the air should be noted, for this determines the temperature of the meniscus in the capillary tube.

The diameter of the tube is measured by a travelling microscope. One of the two parts of the tube is held with its axis vertical in a suitable clip. The base of the microscope and the clip holding the tube should be firmly supported so that there may be no danger of relative movement. Some students, who are without any "mechanical sense," are content, and perhaps love, to use books as supports. The microscope, which has a cross-wire in the focal plane of the eye-piece, is focused upon the upper end of the tube. This is the end which has been formed by cutting the tube. The cross-wire must be set perpendicular to the direction of motion of the microscope. The micrometer screw is then adjusted so that the cross-wire appears as a tangent to the inside of the tube first at one end of a diameter and then at the other. The difference of the two micrometer readings gives the internal diameter of the tube. The setting of the cross-wire is facilitated by suitable illumination, which may be secured by placing below the tube a plane mirror arranged to reflect light upwards.

As the tube may not be *exactly* circular in section, two diameters at right angles are measured. It is convenient if the clip be so arranged that it can be turned through a right angle; the tube must not be removed from the clip. Two or three measurements of each diameter should be taken, the position of the tube being slightly changed after each pair of readings, in order to get independent settings of the micrometer for each measurement of the diameter, with a view to detecting any misreading of the micrometer.

109. Practical example.

G. F. C. Searle used water at 21° C. Beaker stood on horizontal slab and was raised about 1 cm. and was then replaced on slab before each reading of meniscus.

Reading of plane of undisturbed surface of water, 2·01 cm.

Five readings of meniscus were: 6·50, 6·52, 6·51, 6·51, 6·51. Mean 6·510 cm. Hence $h = 6{\cdot}51 - 2{\cdot}01 = 4{\cdot}50$ cm.

Readings of travelling microscope gave for one diameter:

2·3063 − 2·2397 = 0·0666, 2·3052 − 2·2386 = 0·0666.

Mean = 0·06660 cm.

For perpendicular diameter, readings gave

0·0670, 0·0667. Mean 0·06685 cm.

Mean radius $= r = \frac{1}{4}(0{\cdot}06660 + 0{\cdot}06685) = 0{\cdot}03336$ cm. Hence

$$h + \tfrac{1}{3}r = 4{\cdot}511 \text{ cm.}$$

Density of water at 21° C. $= \rho = 0{\cdot}9980$. Density of air at 21° C. and pressure of 76·36 cm. of mercury $= \sigma = 1{\cdot}206 \times 10^{-3}$. Hence

$$\rho - \sigma = 0{\cdot}9968 \text{ grm. cm.}^{-3}.$$

The term $\frac{1}{9}r^2/h^2$ in (4) equals $6{\cdot}1 \times 10^{-6}$ and is negligible. By (2),

$$T = \tfrac{1}{2}g(\rho - \sigma)\,r\,(h + \tfrac{1}{3}r) = 0{\cdot}5 \times 981 \times 0{\cdot}9968 \times 0{\cdot}03336 \times 4{\cdot}511$$
$$= 73{\cdot}58 \text{ dyne cm.}^{-1}.$$

Experiment 13. Measurement of surface tension of soap solution by torsion balance.

110. Method. Let $ABCD$ (Fig. 65) be a rectangular frame of thin wire, the plane of the rectangle being vertical. If the frame be dipped into a soap solution and then be partially withdrawn, so that the horizontal surface of the solution cuts the frame at E and F, a film will be formed, which will fill the area $ABFE$. This film will pull the frame downwards. If the surface tension of the solution be T dynes

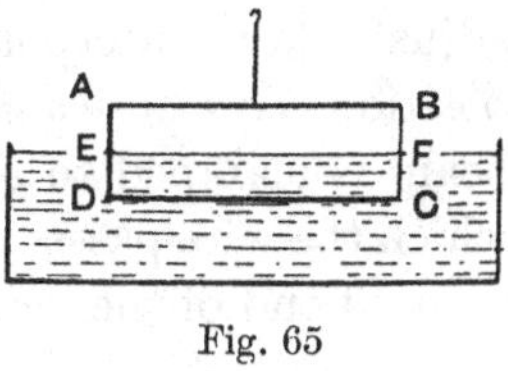

Fig. 65

per centimetre,* and if the distance EF be l cm., the downward pull on the frame will be $2Tl$ dynes, *each* surface of the film contributing Tl dynes. If the downward pull be measured, the surface tension can be calculated.

The pull of the soap film is easily measured by aid of the simple torsion balance shown in Fig. 66. This was designed, in conjunction with Messrs W. G. Pye and Co., as a more convenient form of the apparatus originally constructed at the Cavendish Laboratory.

Fig. 66

The base of the balance is a tripod stand furnished with a levelling screw. From this stand rises an adjustable vertical rod carrying a stiff metal frame, across which is stretched a torsion wire; the tension of the wire can be adjusted by a screw. A double-ended beam is attached to the wire. The long arm of the beam is pointed to serve as an index, and this index point moves near a vertical scale divided to millimetres; the short arm carries an adjustable counterpoise for bringing the beam to a horizontal position. To secure a good connexion of the beam to the wire, the beam is clamped to a short metal tube of small bore, through which the wire passes and to which the wire is soldered. Near the pointed end of the beam is cut a small notch which serves to de-

* So many elastic constants are expressed in dynes *per square centimetre* that students easily fall into the error of stating surface tensions in dynes per square cm.

fine the position of a hook supporting a small scale pan. Below the scale pan hangs the wire frame on which the film is formed. This frame is about 8 cm. in length and 3 cm. in height.

The sensitivity of the torsion balance depends upon the thickness of the torsion wire; by using wires of different diameters a wide range of sensitivity can be covered. For the present purpose, it is convenient to use a torsion wire such that one decigramme in the scale pan gives a deflexion of about 2·5 mm.

A beaker containing soap solution is placed so that the frame dips into it, and the height of the torsion balance is adjusted so that, when the frame is drawn down by the film, the film is one or two centimetres in height. There must be enough solution in the beaker to allow the frame to be completely immersed when necessary.

The measurements are taken as follows: The end of the balance arm is depressed so as to immerse the frame completely. The arm is then allowed to rise, when a film will be formed between the emergent part of the frame and the horizontal surface of the solution. The solution will then begin to drain off from the wire and from the film itself, and the scale reading of the index will change slightly, but will reach a steady value after a short time. This steady reading is recorded. The reading must be taken to $\frac{1}{10}$ mm.; a short-focus lens mounted in a clip may be used to give the necessary optical assistance.

The film is now broken. The arm then rises; it is brought back to the position it had, when the film was intact, by placing a mass m grm. in the pan. If necessary, the exact mass required is obtained by interpolation from two scale readings, the mass being a little too large for one reading and a little too small for the other. A "rider" may also be used to obtain the fine adjustment. If its mass be r grm., its effect is equivalent to that of a mass rx/d grm. placed in the pan, where d and x are the distances from the axis of the torsion wire to the notch and to the point of suspension of the rider. Since the volume of the frame below the surface of the solution is the same as when the film was intact, the upward thrust of the solution is the same in each case. The vertical force, due to surface tension, on the vertical members of the frame is also the same in each case (see § 112). Hence the weight, mg

dynes, of the mass in the pan is equal to the force which was exerted by the film.* Thus $mg = 2Tl$, or

$$T = \tfrac{1}{2}mg/l \text{ dynes per cm.,} \quad \ldots\ldots\ldots\ldots\ldots\ldots (1)$$

where l is the distance EF (Fig. 65) between the points where the frame cuts the solution. This is the shortest distance between the two circles in which the plane of the surface intersects the two wires.

After each pair of readings, the height of the torsion balance may be slightly changed, or a small mass may be placed in the pan. In this way a number of independent readings may be obtained.

111. Practical example.

Width of frame $= l = 8{\cdot}00$ cm. Temperature about 19° C.

Reading with film unbroken	Readings with film broken 0·40 grm. in pan	Readings with film broken 0·45 grm. in pan	m
cm.	cm.	cm.	grm.
3·83	3·79	3·85	0·433
3·87	3·82	3·88	0·442
3·90	3·85	3·90	0·450
4·00	3·94	4·00	0·450

Mean value of $m = 0{\cdot}444$ grm.

Hence, by (1),

$$T = \frac{mg}{2l} = \frac{0{\cdot}444 \times 981}{2 \times 8{\cdot}00} = 27{\cdot}22 \text{ dynes per cm.}$$

EXPERIMENT 14. **Measurement of surface tension of liquid by torsion balance.**

112. Introduction. If a solid body dip into a liquid which wets it, the liquid rises up the sides by capillary action and consequently exerts a downward force upon the body. Even though the sides of the body be vertical, the line on its surface separating the wetted from the unwetted part will not, in general, be horizontal. An extreme case occurs when two glass plates, which touch each other along a vertical line and include a small angle, dip into water. The boundary between wetted and unwetted glass on each plate is then approximately a rectangular hyperbola with horizontal and vertical axes, the vertical axis being the line of

* Strictly, the downward force exerted by the load m is not mg dynes but $mg(1 - \sigma/\beta)$, where the density of the air is σ and of the loading mass (probably brass) is β. The accuracy of the experiment is not, however, high enough to require the application of the correction.

contact of the plates. When the body is a cylinder of circular section with its axis vertical, the boundary is, by symmetry, a horizontal circle.

Provided that the body have vertical sides, the vertical component of the force exerted on it by surface tension is easily calculated. Let PQ, $P'Q'$ (Fig. 67) be two neighbouring vertical lines on the surface of the body, let QQ' be horizontal and let PP' be an element of the curve forming the boundary between the wetted and the unwetted surface. Since QQ' cuts PQ, $P'Q'$ at right angles, it is an element, ds, of the curve in which the body is cut by a horizontal plane. If PP' make an angle θ with QQ', $PP' \cos\theta = QQ' = ds$. The pull of the liquid-air surface on the body along PP' is in the vertical plane $PQQ'P'$ and is normal to PP'. If T be the surface tension of the liquid-air surface, the pull upon PP' is $T.PP'$, of which the vertical component is $T.PP' \cos\theta$ or $T.ds$. Hence, if F be the resultant vertical force exerted on the body by surface tension, we have simply

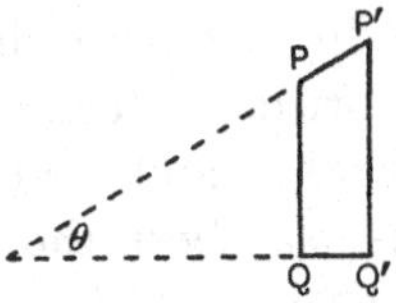

Fig. 67

$$F = T.s, \qquad \text{.........................(1)}$$

where s is the perimeter of the curve in which a horizontal plane cuts the surface of the body.

Let M be the mass of the body and A the area of a horizontal section; we suppose the body to have horizontal plane ends. The density of the liquid is ρ and of the air is σ, and the pressure of the atmosphere at the level of the undisturbed surface of the liquid is P. Let the lower end of the body be at depth h below this level and the upper end at a height k above it. Then, if R be the upward force which must be applied to keep the body in equilibrium, we have, by (1),

$$R + A(P + g\rho h) = Mg + Ts + A(P - g\sigma k). \qquad \text{......(2)}$$

If $M'g$ be the "apparent weight" of the body in air,

$$M'g = Mg - g\sigma(h + k)A.$$

Thus, by (2), $$R = M'g + Ts - ghA(\rho - \sigma). \qquad \text{..............(3)}$$

If $h = 0$, $$R = M'g + Ts. \qquad \text{......................(4)}$$

113. Method. The torsion balance of § 110 may be used to determine the surface tension of water or any other transparent liquid which will not form persistent films. A thin rectangular glass plate is held in a clip (Fig. 66) by which it is suspended below the scale pan, the plate taking the place of the wire frame. The glass slips sold for microscope slides are convenient. The plate is adjusted in the clip so that, when it is suspended, its lower edge is horizontal. The plate is then allowed to dip into the liquid in the beaker, and the balance is raised until the lower edge of the plate is exactly in the plane of the undisturbed part of the surface of the liquid. A fine adjustment for height is provided by the levelling screw in the base of the instrument; it may be more convenient to mount the beaker on a table whose height can be adjusted by a screw.

Since the lower edge of the plate has been adjusted to the level of the liquid, the downward pull exerted by the plate upon its clip is, by § 112, $M'g + Ts$, where $M'g$ is the "apparent weight" of the plate in air and s is the perimeter of a horizontal section of the plate. If the length of this section be l and the width be c cm., $s = 2(l+c)$ cm.

The scale reading of the torsion-arm index is taken; the beaker of liquid is then removed and the plate is dried by filter-paper. A mass of m grm. is then placed in the pan to bring the index to the first reading. The downward force exerted by this load is $mg(1-\sigma/\beta)$ dynes, where σ is the density of the air and β is the density of the load m; the correcting factor $1-\sigma/\beta$ does not differ appreciably from unity. The downward force exerted by the plate is $M'g$ and thus, since the total downward force is the same in the two cases, the force Ts is equivalent to that exerted by the load m. Hence $Ts = mg(1-\sigma/\beta)$, and thus

$$T = \frac{mg(1-\sigma/\beta)}{2(l+c)} \text{ dynes per cm.} \quad \text{............(5)}$$

If the mass m be of brass, $\beta = 8{\cdot}5$. Since for air $\sigma = 1{\cdot}3 \times 10^{-3}$ under ordinary conditions, $\sigma/\beta = 1{\cdot}5 \times 10^{-4}$. Hence σ/β may be neglected in comparison with unity. Thus we may write

$$T = \tfrac{1}{2}mg/(l+c). \quad \text{........................(6)}$$

If water be used, the surface should be free from grease. The beaker should be cleaned with soap and should be filled with water freshly drawn from the tap. The glass plate should also be cleaned with soap. The beaker, plate and clip should be well rinsed with tap water to remove the soap. Caustic alkali is to be avoided, as even a small quantity has a considerable effect in lowering the surface tension (see § 105 A).

114. Practical example.

G. F. C. Searle used water at 21° C. Length of glass plate $= l = 7{\cdot}55$, thickness $= c = 0{\cdot}146$. Hence $2(l+c) = 15{\cdot}392$ cm.

Reading of torsion arm index, when lower edge of plate was in plane of undisturbed surface of water, was 0·25. When plate was out of water and adhering water had been removed by filter-paper, the reading 0·25 was obtained when a rider of copper, of mass $m_0 = 1{\cdot}777$ grm., was placed on torsion arm at distance 8·03 cm. from axis of wire. Distance of point of suspension of plate from axis of wire was 12·65 cm. Hence

$$\text{effective mass of rider} = m_0 \times 8{\cdot}03/12{\cdot}65 = 1{\cdot}128 \text{ grm.}$$

Density of copper $= \beta = 8{\cdot}89$. Density of air at 21° C. and pressure of 76·36 cm. of mercury $= \sigma = 1{\cdot}206 \times 10^{-3}$. Hence $\sigma/\beta = 1{\cdot}36 \times 10^{-4}$ and is negligible in comparison with unity.

By (6),

$$T = mg/2(l+c) = 1{\cdot}128 \times 981/15{\cdot}392 = 71{\cdot}89 \text{ dyne cm.}^{-1}.$$

Experiment 15. **Measurement of surface tension of soap solution by thread method.**

115. Method. On the horizontal arm QR (Fig. 68) of a bent glass rod PQR slide two rings A, B. Through eyes on these rings passes a thread whose ends are attached to a glass rod CD. By adjusting the rings and the thread, the distances AB and CD can be made equal, and the rod can be made to hang with its axis horizontal. The points C, D should be at equal distances from the corresponding ends of the rod. The distance AC should be three or four times the distance AB.

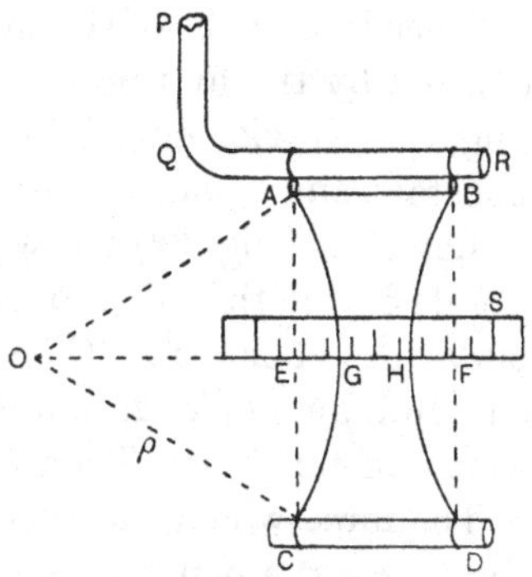

Fig. 68

If the whole system be dipped into soap solution and be then withdrawn, a film will be formed in the area bounded by the thread and the lower rod CD. The parts AC, BD, which may be treated as independent threads, now take the form of curves AGC, BHD, which we shall show are arcs of circles. The vertical part PQ of the bent rod is fixed in a clamp, a horizontal scale S is placed close to the film, and the distance GH, between the two points on the threads where the tangents are vertical, is determined from the scale readings of the threads. The film is then broken and the scale readings of the threads are again taken. Before the first pair of readings is taken, as much as possible of the solution adhering to the lower rod is removed by filter-paper so that the supported mass may be as nearly as possible that of the rod alone.

Let the distance EF (Fig. 68) between the threads when they are vertical be b cm. and let the minimum distance GH between the curved threads be c cm. Let the mass of the rod CD be m grm.; the mass of the threads, of the film and of the air displaced by the rod CD may be neglected. Let the tension of the threads at G and H be N dynes.

The weight of the part of the system below a horizontal plane through G, H is mg dynes, and this is supported by the stresses which act across the plane. The force due to the film is $2Tc$ dynes, since there are *two* faces to the film, and the force due to the threads is $2N$ dynes. Hence

$$2Tc + 2N = mg. \quad \text{.......................(1)}$$

Since the weight of the threads is negligible and since the force exerted by the film on any element of either thread is at right angles to the element, it follows that the tension of each thread is constant and equal to N dynes.

Let P, Q (Fig. 69) be two neighbouring points on the thread, and let the radius of curvature of the arc PQ be ρ cm. Let PQ subtend an angle 2θ radians at O, the centre of curvature. Then $PQ = 2\rho\theta$.

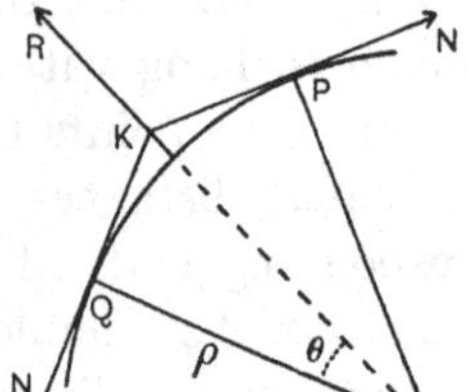

Fig. 69

The force which the other parts of the thread exert upon PQ is $2N \sin\theta$ in the direction KO, where K is the point of intersection of the tangents at P and Q. The resultant force R, which the film exerts on PQ, is in the direction OK and lies between $2T.PQ$ and $2T.PQ\cos\theta$, or between $4T\rho\theta$ and $4T\rho\theta\cos\theta$ dynes. Since the force due to the thread balances the force due to the film, N lies between

$$2T\rho\theta/\sin\theta \quad \text{and} \quad 2T\rho\,(\theta/\sin\theta)\cos\theta.$$

When θ approaches zero, the limit of $\theta/\sin\theta$ is unity and that of $\cos\theta$ is also unity. Hence

$$N = 2T\rho \text{ dynes.} \qquad \text{.......................(2)}$$

Since N and T are constant, ρ also is constant, and thus the threads form arcs of circles.

By (1) and (2), $\quad T(2c+4\rho) = mg,$

or $\quad T = mg/(2c+4\rho). \quad$ ....(3)

The radius of curvature, ρ, of the threads must now be found. Let $x = \frac{1}{2}(b-c)$, so that (Fig. 68) $x = EG = HF$. Then, if $AC = h$ cm., we have, as an exact result,

$$x(2\rho - x) = \tfrac{1}{4}h^2,$$

or $\quad \rho = \frac{1}{8}h^2/x + \frac{1}{2}x. \quad$ ....(4)

The vertical distance $h = AC$ is measured *while the film is unbroken.*

The radius ρ can be found, by an *approximate* formula, when the length, l cm., of the arc AGC, i.e. the length of each thread, is known. Thus, if $AOG = \phi$, we have $\phi = l/2\rho$. But

$$\frac{x}{\rho} = 1 - \cos\phi = \frac{\phi^2}{2} - \frac{\phi^4}{24} + \ldots = \frac{l^2}{8\rho^2}\left(1 - \frac{l^2}{48\rho^2} + \ldots\right).$$

Hence $\quad \rho = \dfrac{l^2}{8x}\left(1 - \dfrac{l^2}{48\rho^2} + \ldots\right).$

The first approximation is $\rho = l^2/8x$. Using this in the term $l^2/48\rho^2$, we have, as a second approximation,

$$\rho = \frac{l^2}{8x}\left(1 - \frac{4}{3}\frac{x^2}{l^2} + \ldots\right) = \frac{l^2}{8x} - \frac{x}{6} + \ldots. \qquad \text{.........(5)}$$

For the third approximation, we have

$$\rho = \frac{l^2}{8x} - \frac{x}{6} - \frac{16x^3}{45l^2}. \qquad \text{.....................(6)}$$

When ρ has been found by (4), (5) or (6), T is found by (3).

116. Practical example.

Mr C. E. Simmons found distance EF between threads when vertical $=b=1{\cdot}68$ cm. Minimum distance GH between threads when curved $=c=1{\cdot}15$ cm. Vertical distance AC when film is unbroken $=h=7{\cdot}40$ cm. Mass of glass rod $CD=m=2{\cdot}94$ grm.

Hence $x=\frac{1}{2}(b-c)=0{\cdot}265$ cm., and, by (4),

$$\rho=\frac{h^2}{8x}+\frac{x}{2}=\frac{7{\cdot}40^2}{8\times 0{\cdot}265}+\frac{0{\cdot}265}{2}=25{\cdot}83+0{\cdot}13=25{\cdot}96 \text{ cm.}$$

By (3), $$T=\frac{mg}{2c+4\rho}=\frac{2{\cdot}94\times 981}{2{\cdot}30+103{\cdot}84}=27{\cdot}17 \text{ dynes per cm.}$$

Unfortunately, the temperature was not recorded.

EXPERIMENT 16. **Measurement of surface tension of soap film by viscosity potentiometer.**

117. Method. The pressure excess due to a spherical soap film may be measured by aid of a device which, by an electrical analogy, we shall call a "viscosity potentiometer". Air from a gasometer G (Fig. 70) flows through two tubes AB, CD, which

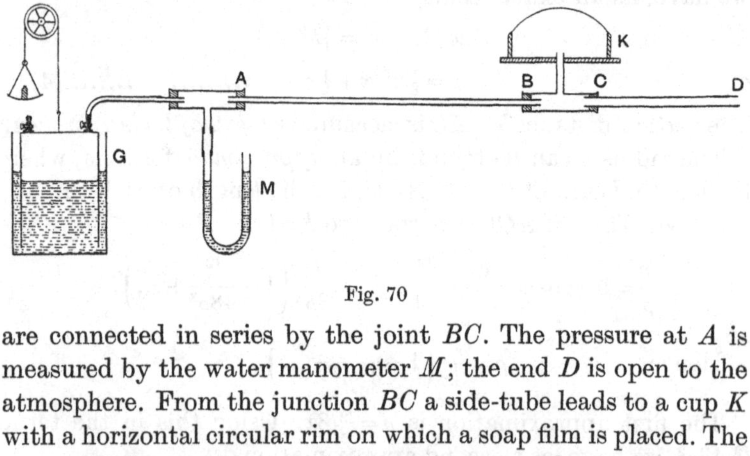

Fig. 70

are connected in series by the joint BC. The pressure at A is measured by the water manometer M; the end D is open to the atmosphere. From the junction BC a side-tube leads to a cup K with a horizontal circular rim on which a soap film is placed. The excess of the pressure at A above the atmospheric pressure is very small compared with the latter, and the temperature is, by assumption, constant along the tubes from A to D. Thus we may treat the air in the flow tubes as a viscous fluid of constant density. There is then a fall of pressure along each tube, and, for a given flow of air, the fall of pressure in either tube is proportional to the length of the tube and inversely proportional to the fourth power of its internal radius, provided the flow be so slow that stream-line motion exists. The excess of the pressure in the cup K, above that of the atmosphere at the same level, causes the film to rise above the rim and to take the form of part of a spherical surface. From the distance, h cm., of the highest point of the film above the plane of the rim and from the radius, c cm., of the rim, the radius, r cm., of the spherical surface is deduced. Thus $h(2r-h)=c^2$, or

$$r=\tfrac{1}{2}(c^2+h^2)/h. \quad \text{......................(1)}$$

If the tube CD be l_1 cm. in length and have a uniform internal radius of a_1 cm., if l_2, a_2 be corresponding quantities for the tube AB, and if the pressure excess at B be p and that at A be P, then

$$\frac{P-p}{p}=\frac{a_1^4}{l_1}\cdot\frac{l_2}{a_2^4}. \quad \text{......................(2)}$$

Hence
$$p=P\{1+l_2a_1^4/(l_1a_2^4)\}^{-1}. \quad \text{..................(3)}$$

From (3), the value of p can be found in terms of P, the pressure excess observed on the gauge M. It is convenient to arrange the tubes so that P is about 100 times p. If ρ be the density of the liquid in the gauge, σ be the density of the air and z be the difference of levels, then $P=g(\rho-\sigma)z$, or, with sufficient accuracy,

$$P=g\rho z. \quad \text{........................(4)}$$

The internal radii of the tubes may be found by means of mercury. For accurate work they should be calibrated.

If b_1, b_2 be the radii of the tubes as found by filling the tubes with mercury, and if F_1, F_2 be the calibration corrections obtained by the method of § 162, we must use F_1/b_1^4 and F_2/b_2^4 in place of $1/a_1^4$ and $1/a_2^4$.

When r, the radius of the bubble, and p, the pressure excess due to the bubble, are known, the surface tension is (§ 85) given by

$$T=\tfrac{1}{4}rp=zr.(\tfrac{1}{4}g\rho).\{1+l_2a_1^4/(l_1a_2^4)\}^{-1}. \quad \text{........(5)}$$

The gasometer G is formed of a cylindrical can about 16 cm. in diameter and 24 cm. in height. Part of its weight is supported by a string which passes over a ball-bearing pulley and carries a pan and weights. By varying these weights, the pressure excess in the gasometer can be adjusted. The can is furnished with two gas-fitter's taps, as shown in Fig. 70. The lower rim of the can is loaded with lead so that the equilibrium of the can is stable when the can is floating with its axis vertical. The can is connected to the joint A by a piece of flexible rubber tube. Since the walls of the can are thin, the pressure excess diminishes only very slowly as the can sinks in the cistern.

The internal radii of the flow tubes must not be too small. With tubes of small radii, the flow of air is so small that a considerable time elapses before the film reaches its full height, and there is a danger that the film may break before the necessary measurements can be made.

The time required for the bubble to reach its full height can be shortened by closing the end D of the flow tube with the finger until the full height has been nearly attained.

A series of measurements may be taken by varying the counterpoise of the gasometer, and determining r, the radius of the spherical film, in each case. The value of zr, where z cm. is the difference of level of the manometer columns, is found in each case, and the mean of these values is used in calculating the surface tension. Since the pressure in the gasometer slowly diminishes, the observations for r and z should be made as nearly as possible simultaneously.

118. The bubble holder. The details of the arrangement on which the spherical film is formed and measured are shown in Fig. 71. A brass plate 8·5 cm. in diameter is carried by a tripod. Into a central opening in the plate is soldered a short vertical tube communicating with the horizontal tube shown in Fig. 71; the latter tube is connected to the joint BC (Fig. 70). To prevent soap solution from draining into the tubes, the vertical tube projects about 1 cm. above the plate (see Fig. 72). A ring rests upon the plate. The upper end is bevelled so that the edge lies on the internal

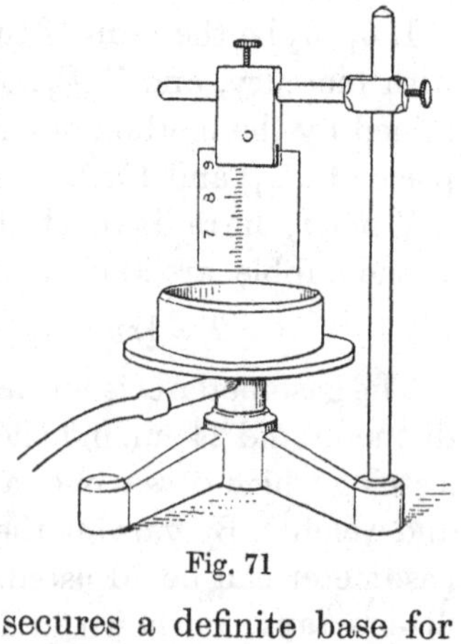

Fig. 71

cylindrical surface of the ring; this secures a definite base for the spherical film. The joint between the ring and the plate is made air-tight by a little of the soap solution. This arrangement allows a number of rings of different radii to be used and also allows the ring to be adjusted on the table. A rod rising from the tripod base is furnished with an adjustable horizontal arm which carries a clamp holding a glass scale divided to millimetres. It is convenient to adjust the scale and ring so that, when the spherical film is formed, the lower edge of the scale may be as near as possible to the highest point of the film. There is then little error due to parallax, if the film and scale be observed through a telescope set at the same level as the top of the film; the film and the scale will both be sufficiently in focus at the same time, if the distance of the telescope, of low power, from the film be 2 or 3 metres.

The film should be illuminated by light which is directed towards the telescope and along its axis. If a flame or an electric lamp be used, it should be at such a distance that the film and the air it encloses are not affected by the heat. A piece of ground glass may be placed between the lamp and the film. In an alternative method, light from the sky is reflected towards the telescope by a plane mirror. When the light falls properly on the telescope, the object glass will form by reflexion an image of the source of light, and this image may be seen by an observer looking towards the telescope, if his eye be not far from the axis of the telescope.

When a film is to be formed on the ring, one edge of a thin plate of metal is dipped into the solution, and then the wetted edge of the plate is gently drawn across the ring from one side to the other. The film will at once break unless the plate, during the whole motion, touch the ring at *two* points, so that there may always be a complete metallic boundary formed by the edge of the plate and the edge of the ring. Success will be more difficult to attain if the plate be held stiffly than if it be drawn along by a single finger, which also acts as the third point of support.

The position of the zero of the glass scale relative to the plane of the rim of the ring is found by placing a steel scale of known width on the rim with the plane of the scale vertical and then observing the reading of the upper edge of the steel scale on the glass scale.

An alternative method of determining the height of the vertex of the film is to use a horizontal microscope provided with a vertical motion and to "set" the microscope first on the vertex of the film and then on the rim of the ring; the plane of the edge of the ring must be horizontal. The difference of the readings gives the distance h.

119. Note on design of apparatus. In Fig. 70 the flow tubes are horizontal, an arrangement which is convenient since they can rest on the table. The apparatus becomes more compact, if the tubes AB, CD be fixed to a vertical board so that B and C are vertically above A and D respectively; the joint BC will need modification. The manometer and the pulley may be mounted upon the board or its base. There are now differences of pressure due to differences of height, but an examination on the lines of § 158 shows that, if the temperature of the tubes be equal to that of the atmosphere near them, the pressure in the bubble exceeds the atmospheric pressure just outside the bubble by an amount which, for a given difference of manometer levels, may be taken, with ample accuracy, to be the same as if the tubes were horizontal. The density of air diminishes, and its viscosity increases,* with rise of temperature. Thus the resistance due to viscosity increases on two counts as the temperature rises and some approach to uniformity of temperature along the flow tube is important.

120. Practical example.

Mr H. F. Kenyon found:

Length of tube CD (Fig. 70) $= l_1 = 53{\cdot}66$ cm.

Mass of mercury filling tube $= 69{\cdot}939$ grm. Density $= 13{\cdot}56$ grm. cm.$^{-3}$.

(Radius of $CD)^2 = a_1^2 = 69{\cdot}939/(\pi \times 13{\cdot}56 \times 53{\cdot}66) = 3{\cdot}060 \times 10^{-2}$ cm.^{2}

Hence

$$a_1 = 1{\cdot}749 \times 10^{-1}\,\text{cm.}, \quad a_1^4 = 9{\cdot}361 \times 10^{-4}\,\text{cm.}^4, \quad a_1^4/l_1 = 1{\cdot}744 \times 10^{-5}\,\text{cm.}^3$$

Length of tube $AB = l_2 = 101{\cdot}95$ cm. Mass of mercury $= 15{\cdot}964$ grm.

(Radius of $AB)^2 = a_2^2 = 15{\cdot}964/(\pi \times 13{\cdot}56 \times 101{\cdot}95) = 3{\cdot}676 \times 10^{-3}$ cm.^{2}

Hence

$$a_2 = 6{\cdot}063 \times 10^{-2}\,\text{cm.}, \quad a_2^4 = 1{\cdot}351 \times 10^{-5}\,\text{cm.}^4, \quad a_2^4/l_2 = 1{\cdot}325 \times 10^{-7}\,\text{cm.}^3$$

By (2), § 117, $$\frac{P-p}{p} = \frac{a_1^4}{l_1} \cdot \frac{l_2}{a_2^4} = \frac{1{\cdot}744 \times 10^{-5}}{1{\cdot}325 \times 10^{-7}} = 131{\cdot}6.$$

Hence $p/P = 1/132{\cdot}6 = 7{\cdot}541 \times 10^{-3}$.

Radius of rim of cup $= c = 2{\cdot}90$ cm.

* The viscosity of liquids *diminishes* with rise of temperature.

The manometer scale was in inches; differences have been reduced to centimetres.

Difference of water levels, z	Height of vertex above rim, h	Radius of bubble, $r=(c^2+h^2)/2h$	zr
cm.	cm.	cm.	cm.^{2}
3·23	0·99	4·742	15·32
4·85	1·96	3·125	15·16
5·18	2·51	2·930	15·18

Mean value of $zr = 15{\cdot}22$ cm.^{2}

Since $P = 981 \times \rho z$, where ρ = density of water = 1 grm. cm.$^{-3}$ approximately, we have $Pr = 981 zr$. Hence, by (5), § 117,

$$T = \tfrac{1}{4}rp = \tfrac{1}{4}(p/P) \, . \, Pr = \tfrac{1}{4} \times 7{\cdot}541 \times 10^{-3} \times 981 \times 15{\cdot}22 = 28{\cdot}15 \text{ dyne cm.}^{-1}.$$

Sad to say, the temperature was not noted.

EXPERIMENT 17. Measurement of surface tension of soap film by buoyancy method.

121. Method. The principle of this experiment was suggested by the plan adopted by Mr J. D. Fry in the calibration of his Micromanometer;* it depends upon the difference of density

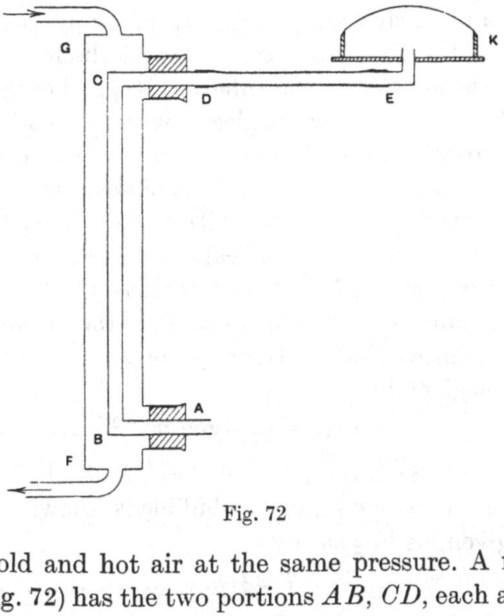

Fig. 72

between cold and hot air at the same pressure. A metal tube $ABCD$ (Fig. 72) has the two portions AB, CD, each a few centimetres long, at right angles to the main portion BC, which is about one metre in length. This bent tube is surrounded by a

* J. D. Fry, *Philosophical Magazine*, April 1913, p. 494.

second tube FG, which is used as a steam jacket for heating the tube $ABCD$. The parts AB and CD are horizontal; the part BC may be set at any desired angle to the vertical. The inner tube passes out of the steam jacket through rubber bungs. The opening D is connected by a horizontal rubber tube DE to the cup K, on which a bubble is to be formed. The bubble holder described in § 118 is used. The end A remains open to the atmosphere. In the theory we suppose that the tubes AB, CD are very fine. Since AB is *horizontal*, there is no difference of pressure between A and B in spite of the difference of temperature, and similarly for CD.

Let the temperature of the atmosphere in the neighbourhood of the apparatus be t_1° K., i.e. t_1° on the Kelvin or "absolute" scale, and let the temperature of the air in the tube BC be t_2° K. On the centigrade system, the Kelvin temperature exceeds the ordinary temperature by 273°. Let P dyne cm.$^{-2}$ be the atmospheric pressure at the level of A. Let the density of air at normal pressure p_0 and normal temperature t_0° K. be ρ_0 grm. cm.$^{-3}$, let the density of the air in the atmosphere at the level of A be ρ_1 and that of the air at B in the tube BC be ρ_2. Let the height of D above A be z cm. If we neglect the very small changes of density due to differences of level, the pressure in the atmosphere at the level of CD is $P - g\rho_1 z$,* and the pressure in the tube at C is $P - g\rho_2 z$. If the horizontal tube DE be long enough to ensure that the bubble stand K is not heated by the steam jacket, the fall of pressure between E and the bubble is, to all the accuracy required, the same as that which occurs in the atmosphere over the same difference of level.† Hence p, the pressure excess within the bubble, is given by

$$p = gz(\rho_1 - \rho_2) \text{ dyne cm.}^{-2}, \qquad \text{.................. (1)}$$

where $\rho_1 = \rho_0 P t_0/(p_0 t_1)$, $\quad \rho_2 = \rho_0 P t_0/(p_0 t_2)$.

When the radius, r cm., of the bubble is known, the surface tension is given, as in § 85, by

$$T = \tfrac{1}{4} r p. \qquad \text{.......................... (2)}$$

* The exact expression is $Pe^{-g\rho_1 z/P}$ or $P - g\rho_1 z + \frac{1}{2}g^2\rho_1^2 z^2/P - \ldots$.

† It will be easily seen that no special care need be taken to ensure that those parts of the tube DE *which are at atmospheric temperature* shall be in the same horizontal plane as CD.

122. Practical details. The bends at B and C are formed as in Fig. 73. The end of the long tube is soldered into a block of brass. The short side-tube screws into the block; the joint is made tight by a flange and a leather washer. The short tube must be stout to withstand the strain involved in making the joint tight. The side-tubes are screwed into the blocks after the long tube has been placed within the steam jacket. Steam is supplied by a small boiler; the waste steam from the jacket should be led away clear of the apparatus. The inner tube is about 0·8 cm. in diameter and the tube forming the steam jacket is about 2·5 cm. in diameter.

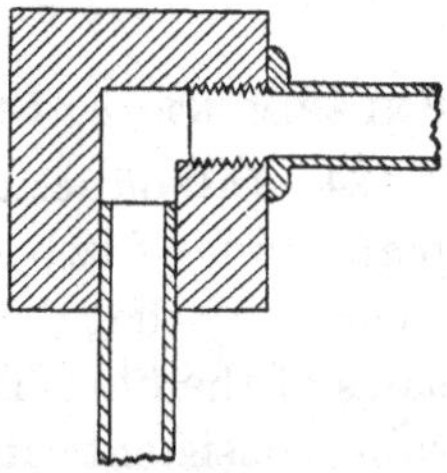

Fig. 73

The bubble is formed on the ring, and the observations for its radius are made, just as in §§ 117, 118. The temperature of the surrounding air is read on a thermometer.

If, as suggested by Mr J. D. Fry, the tube $ABCD$ be turned about CD as a horizontal axis, the difference of level, z, between D and A can be changed, and thus the pressure excess in the bubble can be varied. We have $z = AD \cos\theta$, where θ is the inclination of the plane ACD to the vertical.

To ensure that the interior of the tube $ABCD$ is dry, air may be blown through it while it is heated.

123. Practical example.

The following results were obtained by G. F. C. Searle and A. J. Berry.
Height of heated column $= z = 95{\cdot}8$ cm. Barometric height $= 76{\cdot}53$ cm.
Temperature of atmosphere $= t_1 = 273 + 19{\cdot}5 = 292{\cdot}5°$ K.
Temperature of steam $= t_2 = 273 + 100{\cdot}2 = 373{\cdot}2°$ K.
Density of air at normal pressure and temperature

$$= \rho_0 = 1{\cdot}293 \times 10^{-3} \text{ grm. cm.}^{-3}.$$

Hence
$$\rho_1 = 1{\cdot}293 \times 10^{-3} \times \frac{76{\cdot}53}{76{\cdot}00} \times \frac{273}{292{\cdot}5} = 1{\cdot}2152 \times 10^{-3} \text{ grm. cm.}^{-3},$$

$$\rho_2 = 1{\cdot}293 \times 10^{-3} \times \frac{76{\cdot}53}{76{\cdot}00} \times \frac{273}{373{\cdot}2} = 0{\cdot}9524 \times 10^{-3} \text{ grm. cm.}^{-3}.$$

Thus $p = gz(\rho_1 - \rho_2) = 981 \times 95{\cdot}8 \times 0{\cdot}2628 \times 10^{-3} = 24{\cdot}70$ dyne cm.$^{-2}$.
The values of distance, h, of vertex of bubble from plane of rim for five bubbles were 1·09, 1·10, 1·09, 1·10, 1·10 cm. Mean $h = 1{\cdot}096$ cm.

Radius of rim $= c = 2{\cdot}90$ cm. Mean radius of bubble (see (1), § 117)

$$= r = (c^2 + h^2)/2h = 4{\cdot}385 \text{ cm}.$$

Hence $\quad T = \tfrac{1}{4}rp = \tfrac{1}{4} \times 4{\cdot}385 \times 24{\cdot}70 = 27{\cdot}07$ dynes per cm.

EXPERIMENT 18. **Study of catenoid film.**

124. Introduction. A soap film is formed between the opposing plane ends of two equal and coaxial tubes of circular section in such a way that there is no difference of pressure between the two sides of the film. The greatest distance between the ends of the tubes consistent with continued existence of the film is determined.

125. Apparatus. The apparatus used in this EXPERIMENT and in EXPERIMENT 19 is shown diagrammatically in Figs. 74, 75. The tubes A, B are of equal diameter. The tube A slides in a pair of V's formed in a cradle C, and is held in the V's by a strip of spring brass E, the requisite pressure being obtained by milled-headed screws, as indicated in Fig. 74. The tube B can slide in V's in the cradle D. The two cradles C, D are soldered to the tube F, which slides on the vertical rod G springing from the heavy tripod H.

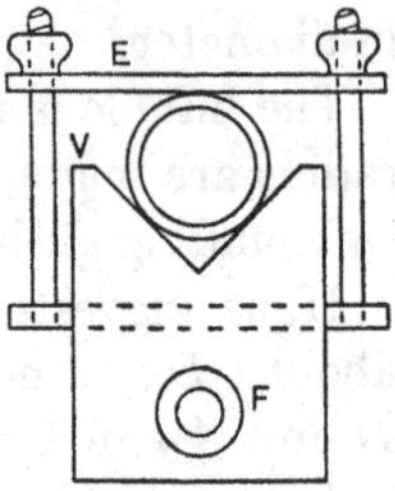

Fig. 74

The lower cradle rests upon a cam K carried by a shaft passing through the block J, which is adjustable on the rod G. By turning the cam (by a suitable handle), the tube F with its cradles can be raised and lowered through the range of the cam. The shaft of the cam is fitted with a nut and spring washer, by which the friction may be made great enough to ensure that the cam does not turn under the weight of the cradles, etc.

The tubes A, B are provided with side-tubes L, N, which, in the present experiment, are joined by flexible rubber tubes to a T-piece T fitted with a rubber mouth-piece M.

The lower end of B can dip, when necessary, into water contained in a beaker Q, which, in Fig. 75, is shown standing on a block R.

126. Method. The lower end of the tube B dips into the water in the beaker and so is closed. The length of the gap between the tubes A, B (Fig. 75) is made approximately equal to their radius.

With the mouth-piece M open to the air, plane films are formed on A and B by drawing across the ends of the tubes a metal plate wetted with soap solution. The observer then applies a very gentle pressure by breathing out air against the end of M, and so blows the films into spherical bubbles. With a little care, these may be caused to coalesce into a single film bridging the gap

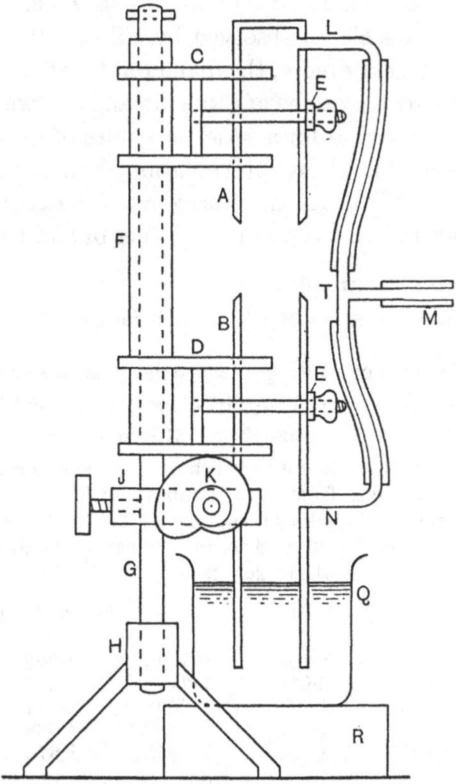

Fig. 75

between the tubes. When the observer ceases to apply pressure, the film will become a catenoid, as indicated in Fig. 59, since the internal and external pressures are now equal. If the gap be cautiously lengthened, by sliding one of the tubes in its **V**'s, the waist of the film narrows and a stage is soon reached when the waist collapses and the film breaks up into two plane films on the ends of A and B.

When the gap only slightly exceeds the critical length, the collapse of the film is very gradual, and accordingly the observer must wait half a minute or more before he concludes that the film formed across any particular gap is stable. When the greatest gap for which the film is stable has been found, its length is measured by a graduated wedge (see Fig. 76) or other means.

If a horizontal scale ruled on glass be placed close to the film and the film and scale be observed by aid of a telescope from a distance of 2 metres or more, the diameter, $2c$, of the waist can be measured for various values of l, the distance between the tubes. The values of c/a found for a series of values of l/a may be compared with those found, by interpolation, from the Table in § 92 for those values of l/a. As an alternative, a series of values of a may be deduced from the observed values of l and c as in § 127.

127. Practical example.

Mr J. G. Wilson used tubes of internal diameter $2a = 2{\cdot}74$ cm. and $a = 1{\cdot}37$ cm.

Five measurements of limiting distance between tubes gave

$l = 1{\cdot}794, 1{\cdot}783, 1{\cdot}766, 1{\cdot}778, 1{\cdot}774$ cm. Mean $l = 1{\cdot}779$ cm.

By theory of § 92, $l = 1{\cdot}32549a = 1{\cdot}816$ cm.

Diameter, $2c$, of neck was measured by telescope and scale. Objective of telescope was 198 cm. from axis of tubes and 200 cm. from scale. If difference of scale readings be D, then $2c = D \times 198/200$. Three readings for D were made for each l; second column gives mean values. By theory, $2c \cosh (l/2c)$ equals $2a$; as above $2a = 2{\cdot}74$ cm.

l cm.	D cm.	$2c$ cm.	$l/2c$	$\cosh (l/2c)$	$2c \cosh (l/2c)$
1·536	2·190	2·168	0·7085	1·262	2·736
1·601	2·097	2·076	0·7712	1·312	2·724
1·661	1·997	1·977	0·8402	1·374	2·717
1·745	1·777	1·759	0·9920	1·534	2·698
1·770	1·617	1·601	1·1055	1·676	2·683

By § 92, minimum value of $2c$, the diameter of neck,

$$= 2a \times 0{\cdot}55243 = 1{\cdot}514 \text{ cm.}$$

Experiment 19. Study of cylindrical film.

128. Method. The apparatus is that described in § 125. The side-tubes L, N (Fig. 75) are now joined by a single rubber tube so that the pressure in A is the same as that in B; if more convenient, the mouth-piece M is closed by a screw clip. As in §§ 96, 126, the distance between the tubes A and B is denoted by l.

In the first part of the experiment, the volume enclosed by the film, the tubes A, B and their auxiliaries is less than $30\pi a^3$. Hence the length $Na+l$, which appears in (7), § 96, is less than $30a$. For this small value of $Na+l$, the maximum stable length of the film is $2\pi a$.

At first the vessel Q stands on the table and the tube B does not reach to the water. With AB roughly equal to $2a$, where a is the internal radius of the tubes, plane films are formed on the ends of A and B as in EXPERIMENT 18. The vessel Q is then lifted so that the water rises above the end of B and a block R of suitable height is placed beneath Q. The increased pressure in the tubes blows the two films into two bubbles which can be caused to coalesce into a single film bridging the gap.* The film may be made cylindrical by sliding the upper tube A in its V's, a movement which rapidly changes the volume enclosed between the tubes, the film and the surface of the water. Since the length l of the cylindrical film, as thus formed, is much less than the critical length $2\pi a$, we must, for the stability test, lengthen the film but yet keep it cylindrical, and must therefore maintain the pressure excess at the value $2T/a$. If the level of the water in Q were constant, the volume of the geometrical cylinder of radius a extending from the end of A to the water would be unchanged by a downward movement of B, and so the film bridging AB would remain cylindrical. Actually, the level of the water rises slightly as B descends, but the consequent change of volume is too small to cause the film to deviate appreciably from the cylindrical form. When l has been increased to $4a$ or $5a$, the cradle carriage may be raised or lowered by the cam so as to make the film as nearly cylindrical as can be judged. If the gap be then further increased by sliding B downwards in its V's, the lower part of the film will bulge out slightly and the upper part will contract. The film is now very sensitive, and a slight movement of the air due to a flick of the hand will cause the film to vibrate up and down. The vibrations are slower as the critical length is approached. A very small increase in AB will now cause the film to break up into two

* If this single film bulge outwards, the pressure excess within it will exceed $2T/a$ and consequently it will be in the form of an unduloid generated by an ellipse of major axis greater than $2a$.

spherical caps of equal radius, the cap on B being greater and the cap on A less than a hemisphere.

Why the cap on B is the greater is, so far, unexplained.

When the maximum length of AB has been found, the gap is measured. The measurement may be made by aid of a distance-piece XY (Fig. 76) of suitable length. One end of a stout tube is turned down so that it fits easily into the lower film tube B. At X the tube terminates in a plane perpendicular to the axis. The distance between the shoulder at Y and the plane end at X is found by calipers. The distance between X and the film tube A is measured by aid of a graduated wedge, as indicated in Fig. 76.

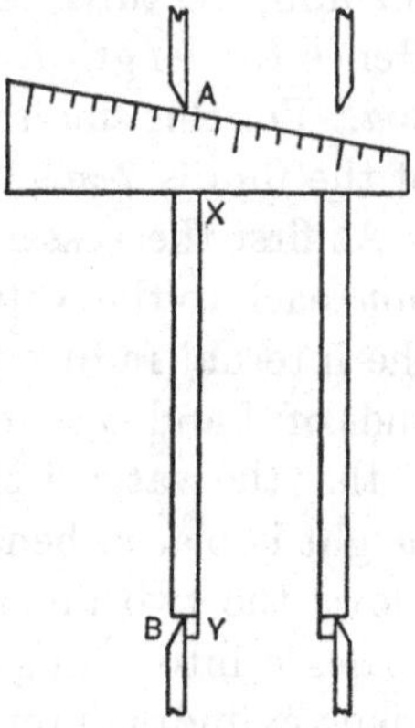

Fig. 76

129. Spherical caps left by collapsing cylindrical film. Since the pressure exerted by the film (perhaps 40 dyne cm.$^{-2}$) is very small compared with the atmospheric pressure (10^6 dyne cm.$^{-2}$), and since the length of the supporting tubes is only a few times the maximum length, $2\pi a$, of the cylindrical film, the volume of the contained air does not differ appreciably from its volume at the same temperature and at atmospheric pressure. Hence, the volume of the enclosed air after the cylindrical film has broken up into two spherical caps may be considered as equal to its volume before the transformation.

The flexible tube LTN (Fig. 75) connecting the tubes A, B ensures that all the enclosed air is at the same pressure, and thus the two spherical caps are of equal radii. A symmetrical, but unstable, configuration would be produced, if the cylindrical film broke up into two spherical caps of radius R and height k. If V be the volume of a cap, viz. the volume between the cap and its plane base on the end of the tube, $V = \frac{1}{3}\pi k^2(3R-k)$. Since the volume between the ends of the tubes A, B, when instability is reached, is $2\pi^2 a^3$, $V = \pi^2 a^3$, and thus V is greater than $\frac{2}{3}\pi a^3$, the volume of a hemisphere of radius a. Hence each cap is greater than a hemisphere. Since the base of the cap is of radius a, $R = \frac{1}{2}(a^2 + k^2)/k$. With this value of R, we find

$$\pi^2 a^3 = V = \tfrac{1}{6}\pi k (3a^2 + k^2).$$

Hence $$k^3 + 3a^2 k - 6\pi a^3 = 0.$$

This cubic equation has only one real root, viz. $k/a = 2{\cdot}2884$. Then $R/a = 1{\cdot}3627$. The surface of each cap is $2\pi Rk$ and of both is $4\pi Rk$ or $39{\cdot}189a^2$. The surface of the cylinder of length $2\pi a$ is $4\pi^2 a^2$ or $39{\cdot}478a^2$. Hence the surface of the two caps is a little less than that of the cylinder.

If the radius of one cap become greater than R, that of the other will become less than R, and then the pressure due to the second cap will exceed that due to the first cap. Thus the first cap will expand and the second contract. This process will continue until the caps reach another equilibrium configuration, this time stable, in which each cap is of radius r, one cap being greater and the other less than a hemisphere. Since the tubes are of equal radii, the two caps together form a complete sphere of radius r, and enclose a volume $\frac{4}{3}\pi r^3$. This equals the volume, $2\pi^2 a^3$, enclosed by the cylindrical film, and hence $r^3 = \frac{3}{2}\pi a^3$, or $r = 1{\cdot}6765a$. If h be the distance of the vertex of a cap from its base, which is of radius a, $h = r \pm \sqrt{(r^2 - a^2)}$, and thus for the lower cap $h/a = 3{\cdot}0221$ and for the upper cap $h/a = 0{\cdot}3309$.

Before the film separates into two parts, the lower half bulges outwards, and thus the two caps are different at birth. The lower cap (on B) is initially, and, therefore, must remain, the greater.

The total surface of the two caps is $4\pi r^2$ or $35{\cdot}320a^2$. The surface of the cylinder was $39{\cdot}478a^2$. Hence the surface of the caps is (considerably) less than that of the cylinder, as must be the case since the cylinder actually breaks up into the two caps.

The diameter $2r$ of the lower cap may be measured by aid of a horizontal scale placed close to the film, the readings being made with a telescope at a distance of two or three metres. The distance from vertex to base of each of the caps may be measured by a telescope and a vertical scale.

129 A. Cylindrical film with constant internal pressure. An electrically driven fan is mounted approximately in the plane of one end of a brass tube, about 7·5 cm. in diameter. A tube joined to the other end of the brass tube, and fitted with a tap, leads to M (Fig. 75). The lower end of the tube B is, throughout, below the surface of the water. The fan is started, and then (plane) films are formed on the ends of the tubes. The films, under the pressure due to the fan, become spherical caps. With a suitable distance between A and B, the caps unite, as in § 128, to form a single film bridging the gap AB. This film is now made cylindrical by adjusting the speed of the fan. The fan is then kept running at this speed. The distance AB is gradually increased, and the greatest value of l for stability is determined; it will, by § 96, be nearly πa (*not* $2\pi a$), since the pressure excess has the *constant* value $2T/a$.

If l be a little too great for stability, a narrow waist develops and ultimately the film breaks up into two spherical caps, less than hemispheres, and of radius $2a$. It may be verified that the work done by surface tension, on account of the diminution in area, is $8{\cdot}28\pi a^2 T$, and that the work done against the excess pressure of $2T/a$ is $5{\cdot}73\pi a^2 T$. The latter is the smaller, as it must be, since the change actually occurs.

129 B. Unduloid film. The apparatus of Fig. 75, if slightly modified, may be used for an experiment on an unduloid film. A disk, with a shoulder, fits into the top of the tube B; the joint is sealed with soap solution. Soldered to the disk, and passing coaxially through it, is a short tube D, of comparatively small diameter, which projects above the disk. The lower end of B dips into water, as in Fig. 75. The tube M is connected with the blowing apparatus of § 129 A.

Films are formed on A and D. By a suitable pressure, the films are blown out until the A-film envelops the end of D. A film of small area will still cover the end of D, but this may be broken if the water in the lower part of B be made to oscillate by inserting some object, e.g. a finger, into the submerged end of B. There remains a single unduloid film, of axial length l, bridging the gap between A and D. The pressure and the length l are then adjusted so that, as judged by eye, the unduloid film has vertical tangents at A and D, and the length l of the gap is measured. This length is u, where $2u$ is one *complete* wave-length of the unduloid; by the theory of the unduloid, $2u$ is the circumference of the generating ellipse. If $2c$ be the major axis of the ellipse,

$$u = 2cE_1(\sin\gamma), \quad\text{......................}(1)$$

where $\sin\gamma = e$, and e is the eccentricity of the ellipse. Values of $E_1(\sin\gamma)$, the Second Complete Elliptic Integral of modulus $\sin\gamma$, are given in Dale's Mathematical Tables.

If f, h be the radii of A and D, and if the pressure excess be $2T/c$, we have $f+h=2c$, $f-h=2ce$, and hence

$$e=(f-h)/(f+h).$$

The observed value of l is compared with the value of u as given by (1).

130. Practical example.

Method of § 129.

Mr J. G. Wilson used tubes with internal radius $a = 1{\cdot}37$ cm.

Four measurements of distance between tubes for maximum stable length of cylindrical film gave

8·375, 8·382, 8·453, 8·340 cm. Mean 8·388 cm.

Theoretical maximum length $= 2\pi a = 8{\cdot}608$ cm.

Spherical caps left by cylindrical film. Diameter and height of cap on tube B and height of cap on A were measured by telescope and scale. Objective of telescope was 200 cm. from scale and 203 cm. from axis of tubes. Factor 203/200 was applied to differences of scale readings to give distances entered below.

Observations of diameters of caps, greater than hemispheres, left on B by four cylindrical films gave

4·588, 4·588, 4·537, 4·578 cm. Mean $2r = 4{\cdot}572$ cm.

By theory of § 129, $2r = 2 \times 1{\cdot}6765a = 4{\cdot}594$ cm.

Observations of heights h_A and h_B of caps on A and B respectively gave

$h_A = 0{\cdot}416$, 0·467, 0·416, 0·386 cm. Mean $h_A = 0{\cdot}421$ cm.
$h_B = 4{\cdot}090$, 4·050, 4·040, 4·040 cm. Mean $h_B = 4{\cdot}055$ cm.

By theory,

$$h_A = 0{\cdot}3309a = 0{\cdot}453 \text{ cm.}, \quad h_B = 3{\cdot}0221a = 4{\cdot}140 \text{ cm.}$$

The sum $h_A + h_B$ equals diameter of sphere of which caps are part.

By observation, $h_A + h_B = 0{\cdot}421 + 4{\cdot}055 = 4{\cdot}476$ cm.

By theory, $h_A + h_B = 2r = 4{\cdot}594$ cm.

Method of § 129*A*.

Mr C. H. Garrett used tubes of radius $a = 1{\cdot}37$ cm. The rate of collapse, when the film was very slightly too long for stability, was very slow.

Films which lasted for two minutes without contracting appreciably were considered stable. Measurements of the maximum stable length for four films gave

4·14, 4·14, 4·20, 4·20 cm. Mean 4·17 cm.

Theoretical maximum stable length $= \pi a = 4{\cdot}304$ cm.

Unduloid film. Method of § 129*B*.

Mr C. H. Garrett used tubes of radii $f = 1{\cdot}37$, $h = 0{\cdot}47$ cm. Hence $2c = f + h = 1{\cdot}84$ cm. and

$$\sin\gamma = e = (f-h)/(f+h) = 0{\cdot}90/1{\cdot}84 = 0{\cdot}4891.$$

Hence $\gamma = 29°\ 17'$, and, by Tables, $E_1(\sin\gamma) = 1{\cdot}472$. Then, by (1), § 129B,

$$u = 2cE_1(\sin\gamma) = 1{\cdot}84 \times 1{\cdot}472 = 2{\cdot}71 \text{ cm.}$$

For tubes of radii $f = 1{\cdot}37$, $h = 0{\cdot}25$ cm., $2c = 1{\cdot}62$ cm., $\gamma = 43°\ 44'$, $E_1(\sin\gamma) = 1{\cdot}362$; thus

$$u = 1{\cdot}62 \times 1{\cdot}362 = 2{\cdot}21 \text{ cm.}$$

Observations of length of gap when tangents at A and D were vertical gave, for the two cases,

2·75, 2·90, 2·85, 2·73. Mean $l = 2{\cdot}808$ cm.

and 2·40, 2·40, 2·35. Mean $l = 2{\cdot}383$ cm.

EXPERIMENT 20. **Measurement of surface tension of mercury by Quincke's sessile drop method.**

131. Introduction. Mercury is poured on to a horizontal glass plate from a bottle fitted with a fine delivery tube, until a drop at least 4 cm. in diameter be formed. The drop will be about 0·35 cm. above the plate at its highest point. If the pouring be done judiciously and if the mercury and the plate be clean, a nearly circular drop will be obtained. Fig. 77 shows approximately the

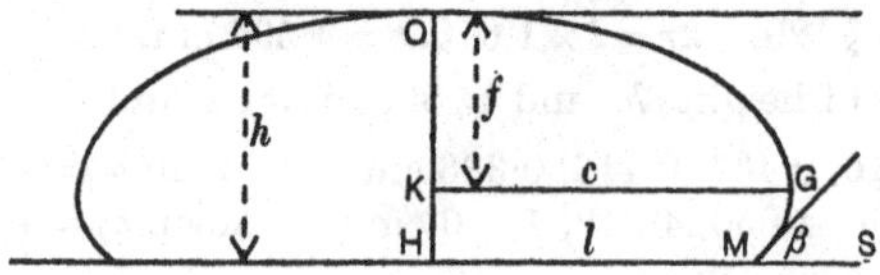

Fig. 77

section of such a sessile drop by a vertical plane through the vertex O. The central part of the upper surface is exceedingly nearly plane. It is shown in § 100 that, if $KG > 4$ cm., the radius of curvature at O of the curve OG exceeds 1·4 kilometres. As we go from O along the arc, the curve gradually falls and at G has a vertical tangent. The curve meets the surface HMS of the plate at the angle of contact β. A horizontal plane through G cuts the drop in a circle of radius c. The thickness OH is h, the depth OK of G below O is f and the radius HM is l.

By § 103, the surface tension T of the interface between the mercury, of density ρ, and the air, of density σ, is given by

$$T = a^2 g(\rho - \sigma) = \tfrac{1}{2} f^2 g(1-\eta)(\rho-\sigma), \quad \text{...........(1)}$$

where η is the positive root of the cubic

$$\eta = \mu(1-\eta)^{\frac{3}{2}}, \quad \text{........................(2)}$$

and $$\mu = 0{\cdot}6095 f/c.$$

When a has been found, the angle of contact, β, is, by § 104, given by

$$\cos\beta = 1 - 2s^2, \quad \text{........................(3)}$$

where s is the positive root of the cubic

$$s^2 = \omega^2 + \gamma(1-s^3), \quad \text{....................(3a)}$$

and $$\omega^2 = 1 - h^2/4a^2, \quad \gamma = 2a/3l.$$

132. Method. The glass plate L (Fig. 78) is carried by a table standing in a dish B, which catches any mercury spilled from the plate. The dish is fixed with plasticene to a levelling stand U,

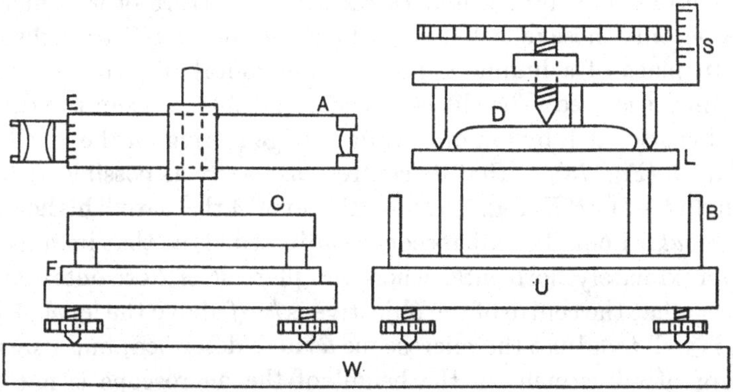

Fig. 78

which rests on a suitable base W, such as a block of stone or an iron slab. By aid of a level, the surface of L is made horizontal. The drop of mercury is indicated by D.

A microscope AE is used to determine the level of the G-line, i.e. the line (corresponding to G (Fig. 77) on the edge of the drop) along which the tangent plane to the surface is vertical. The microscope carriage C can be moved about on the glass plate F, which is made horizontal by the levelling screws of the stand carrying F. The spherometer S is used to measure the thickness of the drop and the depth of the G-line below the vertex.

The height of the G-line is found optically. If a horizontal ray PG (Fig. 79) meet the surface of the mercury at G, where the tangent plane is vertical, the reflected ray GQ will also be horizontal. If P be a luminous point and if from P a beam of small solid angle fall on the surface near G, the reflected rays pass through two focal lines, one vertical, the other horizontal. These focal lines intersect QG, or QG produced, and thus the horizontal focal line is at the same level as the G-line. Hence, if a microscope, with its axis along the horizontal line GQ, be adjusted, by movement along that line, to view the horizontal focal line, the reading of that line on the micrometer scale in the eye-piece E (Fig. 78) will be the same as if the G-line itself had been viewed.

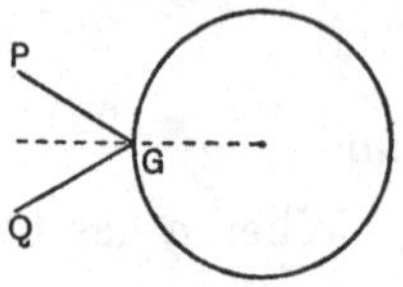

Fig. 79

The rays from a luminous point at P actually spread out in all directions, but, on account of the small aperture of the microscope, only those reflected rays which are close to GQ are utilised.

In place of a luminous point, a horizontal slit with a lamp behind it is used. The slit is placed about 30 cm. from the drop and is adjusted, by aid of a straight-edge resting on the levelled plate L (Fig. 78), so that its centre is as nearly as possible at the height $h-f$ or HK (Fig. 77) above the level of the plate. This height is *about* 0·1 cm. The microscope is adjusted (1) so that its axis is approximately horizontal when the plate F is horizontal, and (2) so that the centre of the objective is $h-f$ above the level of L. It is well to adjust the microscope as here described, but a small error of adjustment in the height of the microscope is not so serious as a small error in the adjustment of the height of the slit.

When the microscope carriage is suitably adjusted on the plate F (Fig. 78), the "image" of the slit will be seen as a fine horizontal line of light crossing the micrometer scale. The reading of this line on the scale is taken. If the line do not disappear when an obstacle is placed in front of the slit, it is due to some lamp in the room. If the slit be somewhat too high, the horizontal line of light will appear double. This effect is discussed in § 133.

If the microscope carriage be turned in azimuth on the plate F so that the edge of the drop is in focus, it will be seen that the scale reading found for the focal line may well be taken as the reading of the point on the curve where the tangent is vertical, as judged by eye. But the accuracy with which this point can be located is so low that the vertical tangent method cannot be considered as a check on the focal line method.

The thickness, h, of the drop is measured by the spherometer. The tip of the screw is adjusted to touch (1) the vertex of the drop and (2) the glass plate; the difference of the readings gives h.

When the eye-piece micrometer reading of the horizontal focal line has been found, the microscope is turned a little in azimuth so that the line of sight clears the drop. The spherometer is then placed so that the tip of the screw is in focus, and the screw is adjusted so that the reading of its tip on the eye-piece scale is identical with that of the focal line. Care must be taken that the microscope be not displaced relative to its carriage during these operations. The difference of spherometer readings* for the setting just made and the setting for the vertex gives f.

Since a little practice is needed in the various operations, a preliminary series of adjustments and *readings* should be made with a drop which is afterwards discarded. The observer should repeat the adjustments and the readings three or four times, so that he may be able to make them quickly and with certainty. Speed is necessary, because the mercury surface is soon contaminated by the air. When the observer has gained skill, the plate L is cleaned afresh, and a new drop of mercury is put upon it. The tips of the three feet and of the screw of the spherometer should be free from grease. Care should be taken to avoid touching the new drop with more than the tip of the spherometer screw. The radius of the drop is limited by the distance between the feet of the spherometer. If the spherometer be clamped to a disk

* The vertical scale at the side of the spherometer head is generally difficult to read and is sometimes unreliable because the disk may have been bent by a fall. The observer should check the readings on the vertical scale by *counting* the number of turns of the disk.

provided with three feet, of sufficient length, set on a circle of 5 cm. radius, and with four holes to accommodate the sphero-meter legs, larger drops can be used.

133. Optical discussion. The doubling of the "image" of the slit may be treated by aid of the caustic surfaces to which rays reflected at the drop are tangents. The general idea will be understood from the consideration of the simple case, in which the reflecting surface is a horizontal cylinder of radius r; the axis of the cylinder is perpendicular to the (vertical) plane of Fig. 80 and meets it at A. The luminous point P and the axis of the microscope M are in the plane of the figure, and the axis of M passes

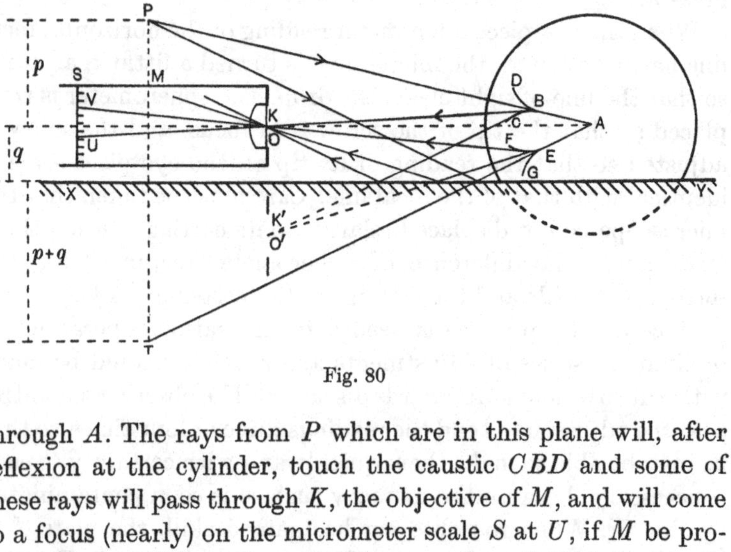

Fig. 80

through A. The rays from P which are in this plane will, after reflexion at the cylinder, touch the caustic CBD and some of these rays will pass through K, the objective of M, and will come to a focus (nearly) on the micrometer scale S at U, if M be properly adjusted. We say "nearly," because the reflected rays come, not from a single point, but from points lying along a small arc of the caustic.

Let T be the image of P by reflexion at the plane XY, the upper surface of the glass plate on which the drop rests. If P be above XY, some rays from P are reflected at XY, and these rays meet the cylinder just as if they had come from T. After reflexion at the cylinder, these rays touch the caustic FEG and some will come to a focus (nearly) at V. Hence, in place of a single image of

P at U, two images will be seen at U and V. Since the microscope is inverting, the lower of the two images corresponds to rays which fall directly from P upon the cylinder.

When the angles made with XY by PA and TA are small, and when the centre O of the objective is at, or nearly at, the same level as A, the points U, V very nearly coincide with the images of the cusps B, E of the caustics CBD, FEG respectively. When PA is large compared with r, the radius of the cylinder, $AB = AE = \frac{1}{2}r$, approximately. If the height of P above A be p and the height of A above XY be q, the depth of T below A is $p + 2q$. If the horizontal distance from P to A be d, the angle between PA and XY is p/d and that between TA and XY is $(p+2q)/d$. Since the height of A above XY is q, the heights of B, E are $q + \frac{1}{2}rp/d$, $q - \frac{1}{2}r(p+2q)/d$ respectively. The mean is $q - \frac{1}{2}rq/d$. This differs from q, and consequently we cannot use the mean of the readings for U and V to give the true height of A.

If P move towards the glass plate, the small patch on the plate, where the rays, which enter the objective to form the image V, are reflected, moves away from the cylinder. If the plate be comparatively small and if PT be sufficiently small, the reflexion necessary for the formation of V will not occur. Then, of V and U, only U is seen. In any case, V will disappear, if a piece of paper, of sufficient size, be laid on the plate between the slit and the drop. If the slit be set accurately to the height of the line on the drop—the G-line—where the tangent plane is vertical, and if V do not exist, the reading on the micrometer scale of the single image U will correspond to the G-line.

Some of the direct rays from P which are reflected at the cylinder will strike XY and will suffer a second reflexion. If K' be the image in XY of the aperture K, those rays, which before reflexion at XY pass (in direction) through K', will, after that reflexion, pass through K. When these rays enter the microscope, they form a second image of (practically) B. When P is so high above XY that it gives rise to the caustic FEG and the image V, some of the rays from this caustic will, after reflexion at XY, enter the microscope and form a second image of (practically) E. These two images will disappear, if a piece of paper, of sufficient size, be laid on the glass plate between the drop and the microscope.

134. Practical example.

In an experiment by Mr G. W. Hart, the drop had diameter $2c = 3{\cdot}78$ cm. An estimate of radius, l, of circle of contact gave $l = 1{\cdot}8$ cm.

Reading on micrometer scale of microscope of "image" of slit, or G-line (§ 132), corresponding to line on edge of drop where tangent plane is vertical, was 1·35 divs.

Spherometer reading for glass plate (zero reading) $= -0{\cdot}0430$ cm. Reading when screw tip was seen at 1·35 divs. on scale $= 0{\cdot}0232$ cm. Reading when tip touched vertex of drop $= 0{\cdot}3063$ cm. Then

$$h = 0{\cdot}3063 + 0{\cdot}0430 = 0{\cdot}3493 \text{ cm.}, \qquad f = 0{\cdot}3063 - 0{\cdot}0232 = 0{\cdot}2831 \text{ cm.},$$

$$h^2 = 0{\cdot}12201 \text{ cm.}^2, \qquad \tfrac{1}{2}f^2 = 0{\cdot}04007 \text{ cm.}^2, \qquad f/c = 0{\cdot}2831/1{\cdot}89 = 0{\cdot}1498,$$

and (see § 103) $\mu = 0{\cdot}6095 f/c = 0{\cdot}09130.$

Then, with $\eta_1 = \mu = 0{\cdot}09130,$

we have $\eta_2 = \mu(1-\eta_1)^{\frac{3}{2}} = 0{\cdot}07909, \quad \eta_3 = \mu(1-\eta_2)^{\frac{3}{2}} = 0{\cdot}08068,$

$$\eta_4 = 0{\cdot}08048, \qquad \eta_5 = 0{\cdot}08050, \qquad \eta_6 = 0{\cdot}08050.$$

The series (28), § 103, gives

$$\eta = 0{\cdot}09130\{1 - 0{\cdot}13695 + 0{\cdot}02188 - 0{\cdot}00380 + 0{\cdot}00070\} = 0{\cdot}08050.$$

With $\eta = 0{\cdot}08050$, we have, by § 103 or § 131,

$$a^2 = \tfrac{1}{2}f^2(1-\eta) = 0{\cdot}04007 \times 0{\cdot}9195 = 0{\cdot}036844 \text{ cm.}^2,$$

and $a = 0{\cdot}19195$ cm.

Then $h^2/4a^2 = 0{\cdot}8279, \quad 2a/3l = 0{\cdot}0711.$

Density of mercury at 15° C. $= \rho = 13{\cdot}5585$ grm. cm.$^{-3}$. Density of air at 15° C. and 760 mm. pressure $= \sigma = 0{\cdot}001225$ grm. cm.$^{-3}$. Hence $\rho - \sigma = 13{\cdot}557$ grm. cm.$^{-3}$, with sufficient accuracy.

For surface tension, by § 103 or § 131,

$$T = a^2 g(\rho - \sigma) = 0{\cdot}036844 \times 981{\cdot}3 \times 13{\cdot}557 = 490{\cdot}3 \text{ dyne cm.}^{-1}.$$

For angle of contact, by (33), § 104,

$$\cos\beta = 1 - 2s^2.$$

With $\omega^2 = 1 - h^2/4a^2 = 0{\cdot}1721$, $\gamma = 2a/3l = 0{\cdot}0711$, and $\omega^2 + \gamma = 0{\cdot}2432$, we have, § 104, $s_1^2 = \omega^2 = 0{\cdot}1721$, and

$$s_2^2 = \omega^2 + \gamma - \gamma s_1^3 = 0{\cdot}2381, \qquad s_3^2 = \omega^2 + \gamma - \gamma s_2^3 = 0{\cdot}2349,$$

$$s_4^2 = 0{\cdot}2351, \qquad s_5^2 = 0{\cdot}2351.$$

The series (32), § 104, gives

$$s^2 = 0{\cdot}1721 + 0{\cdot}06602 - 0{\cdot}00292 - 0{\cdot}00015 = 0{\cdot}2351.$$

With $s^2 = 0{\cdot}2351$, we have

$$\cos\beta = 1 - 2s^2 = 0{\cdot}5298.$$

Hence $\beta = 58° \, 0{\cdot}5'.$

The value found for T is lower than that found by observers who are able to take special care to secure cleanliness. In the experiment, purified mercury was used, but it had been exposed to the air in its bottle for some days since it was distilled.

EXPERIMENT 21. **Surface tension of interface between two liquids.**

135. Introduction. When two liquids A, B, which do not mix, are in contact, their interface possesses surface tension. In general, A will be somewhat soluble in B and B in A. When the liquids have been sufficiently long in contact, the whole of A will be saturated with B, forming the solution A', and the whole of B will be saturated with A, forming the solution B'. In many cases the two solutions A' and B' are distinct and have a definite interface when they are in contact. In the immediate neighbourhood of the interface, the saturation of A with B and B with A will take place very quickly, and thus, if we pour A into B, the surface tension effects we observe *at the interface* will be due to the saturated solutions A' and B'. We may denote the surface tension of the interface by T_{AB}.

Pairs of solutions, which do not mix, may be divided into two classes. Let a vertical rectangular plate, which is wetted by each solution, be suspended so that its upper horizontal edge is in the lighter solution A' and its lower edge is in B'. Then, with a pair of solutions of the first class, the interface rises near the plate, but with a pair of the second class it falls.

The surface tension of the interface between the two solutions can be measured by suitable methods.

136. Methods for pair of solutions of first class. The two solutions A', B', of which A' is the lighter, are contained in a vessel. A thin vertical rectangular plate $CEFD$ (Fig. 81), seen in

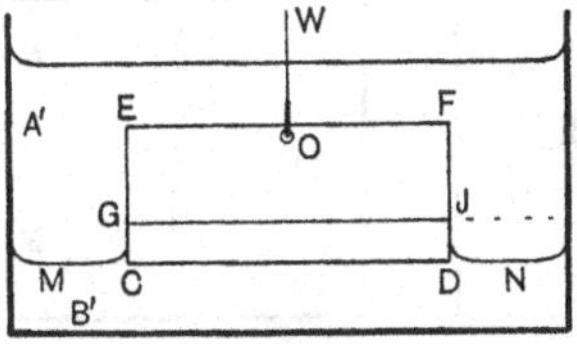

Fig. 81

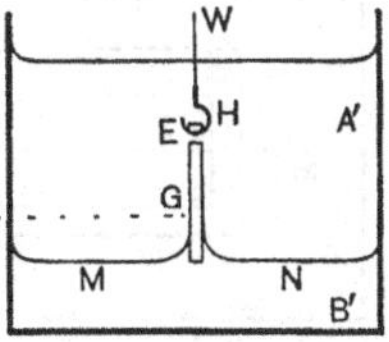

Fig. 82

section in Fig. 82, is suspended by a fine wire W and a *small* hook H, which passes through a *small* hole O in the plate, and the lower edge CD of the plate is horizontal; the hook must be completely submerged. Initially, the plate is adjusted so that CD is above and very close to MN, the plane interface between the solutions A' and B'. The gap between the edge and the interface can, in practice, be made as small as $\frac{1}{4}$ mm. The pull on the wire is $g(m-v\rho_A)$, where m and v are the mass and volume of the plate and ρ_A is the density of the solution A'. If the plate be lowered, so that it is partially immersed in B', and then be lifted again to its initial position, the solution B' will, if suitable, stick to the plate and, in consequence, the interface between A' and B' will rise up to the line GJ on the plate and will take the form indicated in Figs. 81, 82. If the area of the horizontal section of the vessel be sufficiently great, the change of level of the plane part of the interface is negligible. Then the pressures along CD and EF are the same as before, and thus the buoyancy is the same, viz. $gv\rho_A$. If the interface between A' and B' be vertical where it meets the plate, the pull due to surface tension is $T_{AB}.l$, where l is the perimeter of a horizontal section of the plate (see § 112). Thus, the pull on the wire W is increased from $g(m-v\rho_A)$ to $g(m-v\rho_A)+T_{AB}.l$. The difference $T_{AB}.l$ can be measured by a suitable torsion balance.

If the solutions A', B' differ little in density, the effect of surface tension in raising the interface near the plate will be very conspicuous and the error due to the small gap between the lower edge of the plate and the interface in the initial position will be negligible.

Instead of a plate, we may use a frame $CEFD$ (Fig. 83). In its initial position, the frame hangs with its lower edge CD slightly

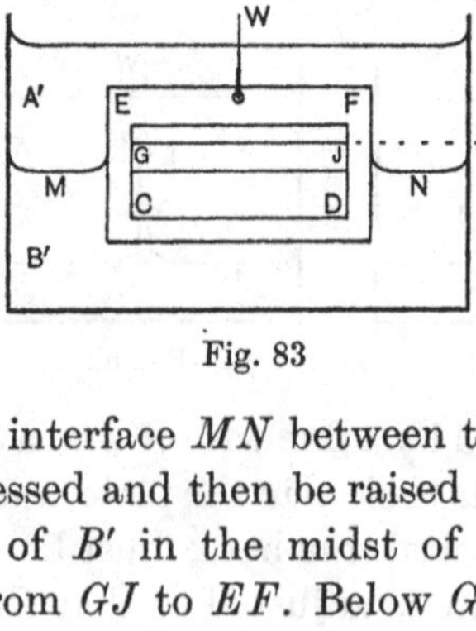

Fig. 83

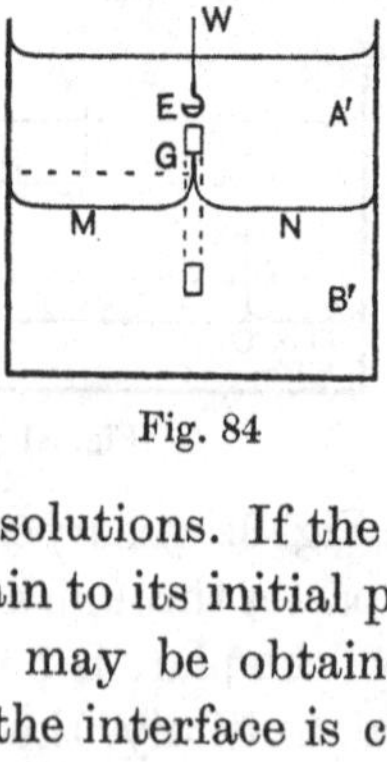

Fig. 84

below the interface MN between the solutions. If the frame be now depressed and then be raised again to its initial position, a thin film of B' in the midst of A' may be obtained which extends from GJ to EF. Below GJ the interface is curved, as indicated in Fig. 84. The pull on EF is $2T_{AB}.EF$, and the pull on the vertical limbs of the frame is $T_{AB}.s$, where s is the sum of the perimeters of the sections of those limbs by a horizontal plane; the total is $T_{AB}(2EF+s)$. If the film be broken and the frame be maintained at the same height, the pull will be $T_{AB}.s$. The difference is $T_{AB}.2EF$, and this we can measure. The lower horizontal limb is not necessary in this case, but it stiffens the frame and, by its weight, it adds to the stability of the frame. It *is* necessary when the film is *below* MN, as in Figs. 87, 88.

137. Methods for pair of solutions of second class. So far we have supposed that, when a vertical plate passes through the interface, the lower solution *rises* as in Fig. 82; this occurs when water rests upon aniline. When paraffin oil rests on water, the opposite effect occurs, and we must then arrange to make the *upper* edge of the plate the effective edge. Since the "plate" is suspended from above, it takes the form shown in Fig. 85. As a

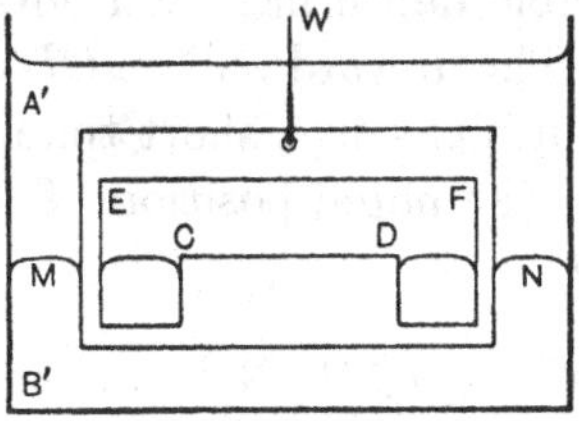

Fig. 85

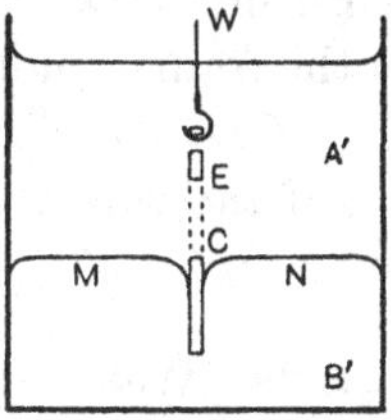

Fig. 86

preliminary, the plate is immersed so that CD is so far below MN that CD does not disturb the interface. The plate is then raised to an initial position in which CD is only just below MN. If the plate be slightly raised, so that CD actually touches MN, the interface will be depressed as in Fig. 86. The pull on the wire W is now less, by $T_{AB}.l$, than when the plate was in its initial position; here l is the perimeter of a horizontal section of the "plate" CD. The interface will be disturbed by the vertical limbs of the "plate," but, if they be of small section and the gaps between them and CD be considerable, the force exerted on CD by the interface will be practically the same as if the vertical limbs were absent.

When we wish to obtain a film of A' in the midst of B', we use a simple frame $CEFD$ (Fig. 87). The frame is gradually lowered from a position in which both the horizontal limbs are in the upper solution A'. When CD, the upper edge of the lower limb, is only a little below MN, the interface is tangential to the plane of the limb where it meets it, just as in Fig. 86. If the frame be still further lowered, a film will make its appearance between CD and

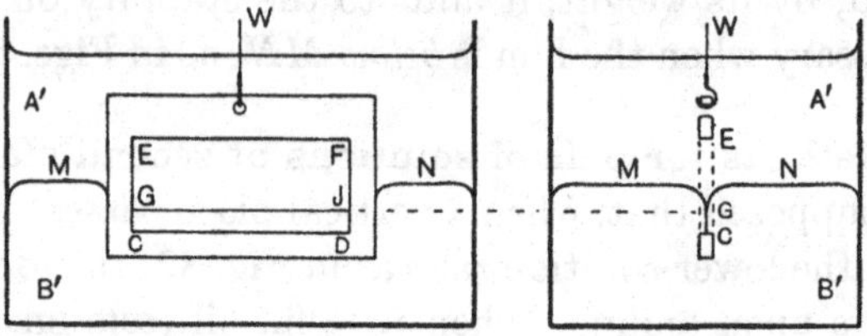

Fig. 87 Fig. 88

a line GJ at a higher level. The distance CG may be increased up to a few mm., but the life of the film is short when CG is considerable. When a film of appreciable depth has been obtained, the position of the frame is noted. The upward pull on CD due to the film is $2T_{AB} \,.\, CD$. The film will break in a short time, and then the tension of the wire W, for the noted position of the frame, will be increased by $2T_{AB} \,.\, CD$.

138. Liquids. Water and aniline ($C_6H_5.NH_2$) at temperatures below 60° C. are a pair of liquids of the first class (§ 135). They have nearly the same density. At 4° C. aniline is the denser, but, with rise of temperature, its density diminishes more rapidly than that of water and at about 70° C. the densities are equal. At higher temperatures aniline floats on water. Approximate values of the densities ρ_A and ρ_W of aniline and of water are given in the Table.

Temperature	10°	30°	50°	70°	90° C.
ρ_A	1·030	1·013	0·996	0·978	0·960
ρ_W	1·000	0·996	0·988	0·978	0·965

When aniline and water are in contact, they form two solutions A' and W'. At 20° C., the solution A' contains 4·5 grm. of water in 100 grm. of solution and does not differ in appearance from pure aniline. The other, W', contains 3·2 grm. of aniline in 100 grm. of solution and is transparent and of a slightly yellow colour. At 80° C., the numbers become 7·0 and 4·5 respectively.

At any given temperature, the densities of A' and W' naturally differ somewhat from those of pure aniline and pure water.

Ordinary paraffin oil and water are a pair of liquids of the second class. The density of paraffin oil is about 0·8; it therefore floats on water.

Aniline in the presence of air is affected by exposure to light and the interfacial tension for aniline and water is liable to change with time. The products of oxidation, which are formed under the influence of light when aniline is exposed to the air, have a dark red colour. When paraffin oil has rested on water for a day or two, a film of scum forms at the interface and the surface tension of the interface is much diminished.

For accuracy, the conditions need more efficient control than is possible in an elementary class. The student may consider his work satisfactory, if he obtain consistent results by readings taken within the space of an hour or two.

Care should be taken to prevent any aniline from coming into contact with books, for a mere trace will stain the paper pink.

139. Torsion balance. A torsion balance is used to determine the surface tension. The pull due to surface tension might be found by aid of an adjustable rider as in EXPERIMENTS 13 and 14, but, for the present experiments, it is convenient to modify the balance shown in Fig. 66. One end of the torsion wire is fixed to the frame of the instrument, as in Fig. 66. The other end is fixed to a horizontal shaft turning in a bearing attached to the frame. The wire is as nearly as possible coaxial with the shaft. The shaft carries a circular head divided to degrees; the readings are taken against a pair of fixed index points. The position of the pointed end of the torsion arm is observed with reference to a fixed vertical scale, as in Fig. 66, or, better, by aid of a fixed horizontal microscope furnished with a micrometer eye-piece. A hook of thin wire rests in a V-notch in the arm near its point, and from this hook the frames and plates used in the experiment are suspended by very fine wire.

When the balance is to be calibrated, the point of the arm is brought to a definite reading on the vertical scale, and the head is read. Then a known load of m grm. is suspended from the hook, and the torsion head is turned, through an angle θ, until the reading of the arm is the same as before. Unless draughts can be excluded during the calibration, the position of the arm will be unsteady. This difficulty is evaded if, during the calibration, a small horizontal damping vane be suspended from the arm, by a very fine wire, so that it is immersed in a single liquid. If, when an unknown load m' is substituted for m, the head has to be turned through θ' from its initial position to bring the arm back to *its* initial position, we have

$$m'g = mg \,.\, \theta'/\theta. \quad \text{.........................(1)}$$

In the experiment, a vertical pull due to surface tension takes the place of the weight $m'g$ of the load m'.

Care is needed in the construction and use of the balance. It is essential that the ends of the fine torsion wire, of phosphor-bronze about 0·03 cm. in diameter, should be securely held. The ends are soldered into holes along the axes of two brass cylinders about 0·25 cm. in diameter and 2 cm. in length. One cylinder is held by a set-screw in a socket bored in the shaft and the other in a socket in the frame. A transverse cut, 0·2 cm. wide and deep enough to meet the hole, is made in each cylinder about 0·5 cm. from the end where the wire enters it. This device promotes the flow of solder along the hole and ensures a sound joint. A short cylinder, also with a cut, is soldered to the wire near its centre and to it is clamped the torsion arm. The total length of the wire may be about 20 cm.

It is important that the wire shall not be overstrained by twisting. A pair of stops prevents the arm from turning through more than about 10°; by another pair of stops and a pin projecting from the torsion head, the head is prevented from turning through more than 40° or 50°. The cylinders at the ends of the wire are so secured in their sockets that the wire, though under considerable tension, has no unnecessary torsion.

Since the stops limit the amount of torsion available, the counterpoise must be adjusted on the arm to suit each plate or frame.

140. Experimental details. The plates and frames are made of sheet nickel 0·03 cm. in thickness. If the sheet be hammered on a smooth anvil, it becomes satisfactorily stiff. Each plate or frame is accurately cut so that, when it is suspended by the hole in it, the edges are nearly vertical and horizontal. The final adjustment is made by aid of a plate of glass which has been levelled. The plate of Fig. 81 is suspended from a fine horizontal needle passing through the hole, and is brought near the glass. If the edge CD be not parallel to the glass and D be lower than C, a little is filed off the edge FD. The edges EC, FD are made perpendicular to CD and EF is made parallel to CD. In the frames of Figs. 83, 85, 87, CD is made parallel to the lower edge of the frame, and that edge is made to hang parallel to the glass by filing one of the vertical limbs.

The plate or frame is hung from the arm by a wire. Since this wire passes through the upper surface of the upper solution and is, therefore, subject to surface tension effects which are apt to be variable, its diameter should not exceed 0·01 cm. in order that these effects and their *variations* may be small; a small hook of stouter wire is soldered to each end of the fine wire.

The two solutions are contained in a beaker. Each forms a layer two or three cm. deep. The beaker is carried by a table which is raised or lowered by a screw. The adjustments required in the four cases described in §§ 136, 137 are as follows:

Solutions of first class. (i) The plate is at first totally immersed in the upper solution A'. The table is then raised, and, when the lower edge CD of the plate is very near the interface MN, the reading of the arm on the micrometer scale is noted. The observer then views the arm through the microscope while he slowly raises the table farther. A sudden falling of the arm indicates that MN has reached CD. If CD have been made accurately horizontal, the contact is sharply defined. The arm is then brought back to the noted position by turning the torsion head. To obtain a steady reading, it may be necessary to allow a little time for the lower solution to drain off from the *upper* part of the plate.

If a load of m grm. be balanced by a torsion θ, and if ϕ be the difference of readings just found, we have

$$T_{AB} = \frac{mg \, . \, \phi}{2(CD+t)\,\theta}, \quad \text{......................(2)}$$

where t is the thickness of the plate.

(ii) The frame is at first totally immersed in the lower solution B'. The head and the table are then adjusted, so that the frame is partly in the upper solution with a film of B', in the midst of A', filling the upper part of the frame. The observer can assure himself of the existence of the film by looking through it in a horizontal direction. He will be able to see any object which is at the same level as the film. If the refractive index of one liquid differ from that of the other, this test shows that the two interfaces are parallel. For water and aniline the indices are 1·33 and 1·59 respectively. When a film has been obtained, the readings of the head and of the micrometer are taken. After the film has broken, the head is turned to bring the arm to the same micrometer reading. If ϕ be the difference of the head readings,

$$T_{AB} = \frac{mg \, . \, \phi}{2EF \, . \, \theta}. \quad \text{......................(3)}$$

Solutions of second class. (i) The "plate" is at first so deeply immersed that the edge CD is below the interface MN. The table is lowered until MN nearly touches CD, and the micrometer reading is noted. The table is then cautiously lowered farther; the contact between MN and CD is indicated by a sudden rising of the arm. The reading of the head is taken. The head is then turned to allow the "plate" to sink so far that the micrometer reading is recovered, and the reading of the head is taken again. Then

$$T_{AB} = \frac{mg \, . \, \phi}{2(CD+t)\,\theta}. \quad \text{......................(4)}$$

(ii) The frame is at first totally in the upper solution. The head and the table are then adjusted so that the frame is partly in the lower solution, with a film of the upper solution filling the lower part of the frame. The readings of the head and of the micrometer are taken. When the film has broken, the head is turned through ϕ to bring the arm to the same micrometer reading. Then

$$T_{AB} = \frac{mg \, . \, \phi}{2CD \, . \, \theta}. \quad \text{......................(5)}$$

140 A. Effect of change of surface of plate. When, as in this experiment, a plate is used, a factor additional to the two solutions themselves is introduced. The behaviour of the solutions with respect to the plate depends upon the state of its surface. A person zealous for cleanliness may think that he will improve the plate by removing grease from it by caustic soda, but his zeal is misplaced. If the plate be polished with very fine emery paper and then be rubbed with a rag moistened with paraffin oil, normal values for the surface tension between water and the oil may be found. If the plate be now treated with caustic soda, the surface tension, as found by the experiment, will fall to a low value. It will be found difficult to remove the caustic soda from the plate. The molecular change wrought in the surface of a metal by caustic soda or by other liquids, such as acids, may be persistent.* Vigorous polishing with very fine emery paper will, however, generally restore the plate to the condition it had before the caustic soda was applied.

These difficulties may render the experiment unsatisfactory to students who expect a very definite "result" in every experiment. Others, who are content to see things as they are, will find enough to interest them.

141. Surface tension of solutions in contact with air. The surface tension of the upper solution in contact with air may be measured by aid of a frame or a plate as in EXPERIMENTS 13, 14. To obtain a sample of the lower solution, the two solutions are poured into a funnel whose tube is fitted with a rubber tube closed by a pinch cock. After the solutions have stood so long that they are separate, the cock is opened and a little solution is allowed to flow out into a beaker so as to wash out the tube. Then, into another vessel, is drawn off so much of the lower solution as is re-

* A striking example of the change of state of the surface of a metal is furnished by a chemical experiment. If a clean plate of iron be dipped into a solution of copper sulphate in water, copper is rapidly deposited on the iron. If the plate be completely immersed in dilute nitric acid before it is placed in the solution, no copper is deposited from the solution. If, however, while the plate is in the solution it be struck with a glass rod, the molecular layer formed by the action of the nitric acid is ruptured at the point of impact, and deposition of copper starts at that point and rapidly spreads over the whole plate. We are thus warned that, if even a small part of the surface of the plate used in the surface tension experiment be "bad," it may affect the whole surface.

quired for the measurement of the surface tension of the interface between it and air.

The difference between the two surface tensions thus found is, according to some theories, equal to the surface tension of the interface between the two solutions.

The surface tension of water saturated with aniline will be found to be much less than that of pure water.

142. Practical example.

Mr K. G. Tupling used mass $m = 0{\cdot}1183$ grm. for calibration of torsion balance. With three different initial settings of head, angular displacements balancing m were 18·45°, 18·40°, 18·55°; each was derived from readings of *both* pointers. Hence $\theta = 18{\cdot}47°$.

Solutions of first class. Aniline and water.

(i) *Plate method.* Length of plate $= CD = 7{\cdot}00$; thickness $= t = 0{\cdot}03$ cm. Readings 0·9° and 179·8° of pointers give position of head when edge CD was in water but very near interface MN. Micrometer reading 54·0 is that observed at instant when MN met CD.

CD very near MN.	Micrometer 54·0.	Head	0·9°,	179·8°
CD in aniline.	54·0.		14·5,	193·5
		Torsion	13·6,	13·7

Mean torsion to balance surface tension $= \phi = 13{\cdot}65°$. By (2),

$$T = \frac{mg\,.\,\phi}{2\,(CD+t)\,\theta} = \frac{0{\cdot}1183 \times 981 \times 13{\cdot}65}{2\,(7{\cdot}00 + 0{\cdot}03) \times 18{\cdot}47} = 6{\cdot}10 \text{ dyne cm.}^{-1}.$$

Two other experiments gave $T = 6{\cdot}08$, $6{\cdot}06$ dyne cm.$^{-1}$.

(ii) *Frame method.* Internal width of frame $= EF = 6{\cdot}00$ cm.

Film acting.	Micrometer 43·2.	Head	25·1°,	203·9°
Film broken.	43·2.		12·9,	191·7
		Torsion	12·2,	12·2

Mean torsion to balance surface tension $= \phi = 12{\cdot}2°$. By (3),

$$T = \frac{mg\,.\,\phi}{2EF\,.\,\theta} = \frac{0{\cdot}1183 \times 981 \times 12{\cdot}2}{2 \times 6{\cdot}00 \times 18{\cdot}47} = 6{\cdot}39 \text{ dyne cm.}^{-1}.$$

Two other experiments gave $T = 6{\cdot}34$, $6{\cdot}39$ dyne cm.$^{-1}$.

Mr Tupling measured surface tension of interface between air and (i) water saturated with aniline (upper solution) and (ii) aniline saturated with water. Internal width of frame was 1·50 cm. Mean torsions in (i) and (ii) were 22·50° and 20·22°, corresponding to surface tensions of 47·1 and 42·4 dyne cm.$^{-1}$ respectively. The difference 4·7 is not far from 6·23, the mean of values for interface between solutions.

Solutions of second class. Water and paraffin oil.

(i) *Plate method.* Length of effective edge (Fig. 85) of plate
$= CD = 1{\cdot}91$ cm.; $t = 0{\cdot}03$ cm.

CD in water near oil.	Micrometer 41·5.	Head	34·9°,	213·7°
CD in oil.	41·5.		7·2,	186·0
		Torsion	27·7,	27·7

Mean torsion to balance surface tension $= \phi = 27{\cdot}7°$. By (4),

$$T = \frac{mg\,.\,\phi}{2(CD+t)\,\theta} = \frac{0{\cdot}1183 \times 981 \times 27{\cdot}7}{2(1{\cdot}91 + 0{\cdot}03) \times 18{\cdot}47} = 44{\cdot}9 \text{ dyne cm.}^{-1}.$$

A second experiment gave $T = 44{\cdot}6$ dyne cm.$^{-1}$.

(ii) *Frame method.* Internal width of frame $= CD = 1{\cdot}94$ cm.

Film acting.	Micrometer 61·5.	Head	0·7°,	179·7°
Film broken.	61·5.		29·3,	208·3
		Torsion	28·6,	28·6

Mean torsion to balance surface tension $= \phi = 28{\cdot}6°$. By (5),

$$T = \frac{mg\,.\,\phi}{2CD\,.\,\theta} = \frac{0{\cdot}1183 \times 981 \times 28{\cdot}6}{2 \times 1{\cdot}94 \times 18{\cdot}47} = 46{\cdot}3 \text{ dyne cm.}^{-1}.$$

Two other experiments gave $T = 46{\cdot}0$, $46{\cdot}0$ dyne cm.$^{-1}$.

For interface between air and water saturated with paraffin oil Mr Tupling found 67·5; for interface between air and paraffin oil saturated with water he found 25·7. The difference is 41·8 dyne cm.$^{-1}$.

143. Form of interface.

Let the densities of the upper and lower liquids be σ_1 and σ_2, and let the surface tension of the interface be T. Let the interface rise up against an

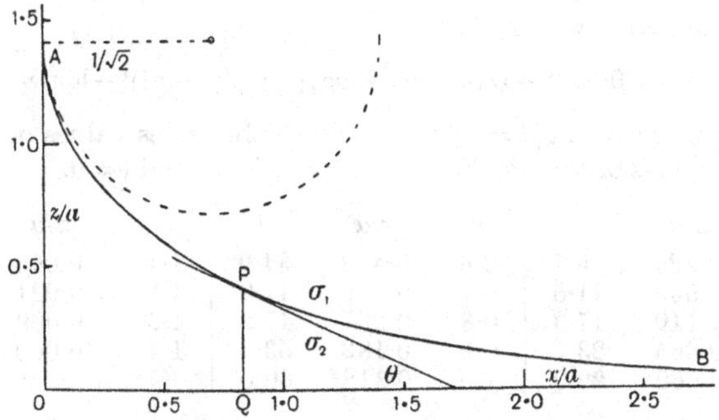

Fig. 89

infinitely wide vertical plate; we take its face as the plane of yz. The axis Oz is vertical. The plane Oxy is that of the undisturbed interface. Fig. 89 is a section of the interface by the plane Oxz. The scale of the curve depends upon the length a which is defined by (7) below. It is better to plot z/a

against x/a than to plot z against x. The relation between z and x is independent of y and thus, at each point of the interface, one of the principal curvatures is zero. Let P be any point on the surface. The pressure in the upper liquid at P is $p-g\sigma_1 z$ and in the lower liquid at P it is $p-g\sigma_2 z$; here p is the pressure in the plane Oxy. The difference $g(\sigma_2-\sigma_1)z$ is balanced by the normal stress T/ρ due to the surface tension, where ρ is the radius of curvature of the curve APB at P. Hence

$$z\rho = a^2, \quad \text{.................................(6)}$$

where

$$a^2 = T/\{g(\sigma_2-\sigma_1)\}. \quad \text{.........................(7)}$$

Thus

$$z = \frac{a^2}{\rho} = a^2 \frac{d^2 z}{dx^2}\left\{1+\left(\frac{dz}{dx}\right)^2\right\}^{-\frac{3}{2}}.$$

If we multiply each side by $2dz/dx$, we can integrate. Since $dz/dx=0$ when $z=0$, we have

$$2a^2 - z^2 = 2a^2\{1+(dz/dx)^2\}^{-\frac{1}{2}} = 2a^2\cos\theta, \quad \text{...........(8)}$$

where $\tan\theta = -dz/dx$, and θ is the slope of the surface at P.

If, as we suppose, Oz be a tangent to the curve at A, $\cos\theta=0$ at A and then $z/a=\sqrt{2}$ at A; the curve does not go above A. By (6), the radius of curvature at A is $\rho=a/\sqrt{2}$.

Solving (8) for dx/dz, which is *negative* when $0<z<a\sqrt{2}$, we find

$$\frac{dx}{dz} = \frac{z}{\sqrt{(4a^2-z^2)}} - \frac{2a^2}{z\sqrt{(4a^2-z^2)}}.$$

Hence

$$x = C - \sqrt{(4a^2-z^2)} + a\{\log_e[2a+\sqrt{(4a^2-z^2)}] - \log_e z\}.$$

Since $x=0$ when $z=a\sqrt{2}=a\times 1{\cdot}4142\ldots$,

$$C = a\{\sqrt{2} - \log_e(1+\sqrt{2})\} = 0{\cdot}5328a.$$

If we write $z=na$, we have

$$x/a = 0{\cdot}5328 - \sqrt{(4-n^2)} + \log_e\{2+\sqrt{(4-n^2)}\} - \log_e n. \quad \text{......(9)}$$

From (8) we have $\sin\frac{1}{2}\theta = \frac{1}{2}z/a = \frac{1}{2}n$. The Table gives values of x/a and θ for some values of $n=z/a$. For $n=0$, x is infinite and $\theta=0$.

n	x/a	θ	n	x/a	θ	n	x/a	θ
0·1	2·224	5·7°	0·6	0·499	34·9°	1·1	0·068	66·7°
0·2	1·536	11·5	0·7	0·370	41·0	1·2	0·031	73·7
0·3	1·140	17·3	0·8	0·267	47·2	1·3	0·009	81·1
0·4	0·865	23·1	0·9	0·183	53·5	1·41..	0·000	90·0
0·5	0·660	29·0	1·0	0·118	60·0			

Fig. 89 has been drawn from these numbers. The circle of curvature at A is shown.

When x/a is large, n^2 or $(z/a)^2$ is very small compared with 4. If we neglect the difference between $\sqrt{(4-n^2)}$ and 2, we have

$$x/a = 0{\cdot}5328 - 2 + \log_e 4 - \log_e(z/a) = -0{\cdot}0809 - \log_e(z/a). \quad \text{...(10)}$$

Hence, for large values of x/a,

$$z/a = e^{-0\cdot0809} \times e^{-x/a} = 0\cdot9223 \times e^{-x/a}.$$

It is not necessary to go to very small values of z/a in order to get good agreement between the values of x/a given by (10) and by (9). To three decimal places we have the following results:

z/a	x/a by (10)	x/a by (9)
0·05	2·915	2·915
0·10	2·222	2·224
0·20	1·529	1·536

CHAPTER VII

EXPERIMENTS ON VISCOSITY

144. Introduction. The "perfect fluid" of theoretical hydrodynamics is such that, whether its parts be at rest or be in motion relative to one another, the stress across any geometrical surface drawn in the fluid is always normal to the surface. In all actual fluids, whether liquid or gaseous, the stress is normal to the geometrical surface when the parts are at rest relative to one another, i.e. when the fluid moves as a rigid body, but, when the parts are in relative motion, there is, in general, a tangential stress across the surface. Such a fluid is said to be viscous, and the measure of the viscous quality is called the viscosity of the fluid.

To lead up to a definition of viscosity, we consider the very simple case in which the fluid flows parallel to a fixed straight line Ox (Fig. 90) in a fixed plane, which we take as the plane of xy. Let v be the velocity of the fluid at x, y, z and let v depend only upon z. Let PN be perpendicular to the plane Oxy and let $PN = z$. If PQ be parallel to Ox, PQ is a line of flow and a plane containing PQ and parallel to Oxy—call it the plane PQ—is a plane of flow. Then the law which is obeyed by a viscous liquid is expressed by the statement, formulated by Newton, that a tangential stress acts across the plane PQ proportional to the change of velocity per unit distance measured perpendicular to the plane, i.e. the stress is proportional to the velocity gradient dv/dz. The direction of the stress is such that it tends to stop the motion of the fluid on one side of PQ relative to the fluid on the other side of PQ. When dv/dz is positive at the plane PQ, the stress tends to decrease the forward velocity of the fluid on that side of the plane for which z is greater than PN and tends to increase the forward velocity of the fluid on the other side of the plane.

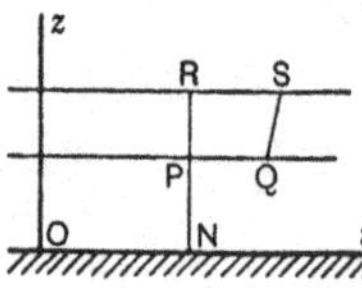

Fig. 90

Since the stress is proportional to dv/dz, we can write

$$\text{Tangential stress} = \text{Force per unit area} = \eta \, . \, dv/dz \text{....(1)}$$

The quantity η is called the coefficient of viscosity, or, more simply, the viscosity of the fluid; it is clearly the tangential stress per unit velocity gradient.

In the C.G.S. system, η is the stress in dynes per square cm. per unit velocity gradient, the unit gradient being such that the change of velocity per cm. is one cm. per sec.

The unit of viscosity on the C.G.S. system has been called a Poise in honour of Poiseuille, who made many investigations in this subject. If unit velocity gradient of one cm. sec.$^{-1}$ per cm. produce in a liquid a stress of η dyne cm.$^{-2}$, the viscosity is η poises.

The simple case of relative motion just considered has led us to a definition of viscosity which can be applied when the velocity of every part of the fluid is parallel to a fixed straight line. That it needs modification before it can be applied to more general cases a simple example will show. Suppose that a tube of circular section containing liquid rotates about its axis, which is fixed, with uniform angular velocity ω, and that each part of the fluid has angular velocity ω about the axis. The fluid moves as if it were rigid and there is obviously no shearing stress. The velocity of the fluid at distance r from the axis is $r\omega$, and hence the velocity gradient is $d(r\omega)/dr$ or ω. This would, according to the definition given above, involve a shearing stress $\eta\omega$ over the surface of the cylinder of radius r, a result which is clearly untrue.

A little consideration shows that, in the case illustrated in Fig. 90, we are not so much concerned with the change of velocity per unit distance as with the distortion suffered by the fluid per unit time. Let R be on PN, where NP is parallel to Oz, and let $RN = z + dz$, where dz is very small; let RS be parallel to PQ. If the velocity of the fluid along PQ be v, that along RS is $v + (dv/dz)\,dz$. The particles which at time t are at R and P will, at time $t + dt$, be at S and Q, where $RS = \{v + (dv/dz)\,dz\}\,dt$ and $PQ = v\,dt$. The small angle θ between QS and PR is $(RS - PQ)/dz$ or $(dv/dz)\,dt$, and this change of direction has come about in time dt. The particles which lay on the lines RS and PQ at time t lie on those lines at $t + dt$. Hence the fluid has in time dt been sheared through the angle $(dv/dz)\,dt$, and thus the rate of shearing is dv/dz.

We now amend the definition and write

$$\text{Tangential stress} = \eta \times \text{Rate of shearing.} \quad \ldots\ldots(2)$$

In the case of the rotating cylinder, there is clearly no shearing and there is no tangential stress.

On the C.G.S. system the unit rate of shearing is one radian per second.

In the general case, we may refer the velocity of the liquid to fixed axes Ox, Oy, Oz, the components of the velocity at x, y, z being u, v, w. The resultant rate of shearing depends upon the six quantities

$$du/dx, \qquad dv/dy, \qquad dw/dz,$$

$$(dw/dy + dv/dz), \qquad (du/dz + dw/dx), \qquad (dv/dx + du/dy).$$

In the simple cases with which we shall deal, the manner in which the shearing occurs will be evident, and it will be unnecessary to employ the general equations.

145. Stresses in a viscous fluid due to shearing. The stresses in a viscous fluid deserve fuller examination than was given in § 144. We shall confine our attention to the very simple case in which the fluid moves parallel to a fixed straight line Ox drawn on a fixed plane Oxy of infinite extent. We shall suppose that the velocity v at distance z from Oxy depends only upon z and is independent of x and y and of the time t.

The tangential stress on a plane at distance z from Oxy is $\eta\, dv/dz$ per unit area and on a plane at distance $z + dz$ it is

$$\eta \left\{ \frac{dv}{dz} + \frac{d}{dz}\left(\frac{dv}{dz}\right) dz \right\}.$$

Since the momentum of the fluid contained between the two planes is constant, the two stresses are equal, and thus $d^2v/dz^2 = 0$. If T be the tangential stress, $\eta\, dv/dz = T$, and thus $v = v_0 + Tz/\eta$, where v_0 is the velocity in the plane for which $z = 0$. If the plane Oxy be the surface of a stationary solid body, $v_0 = 0$, since the fluid does not slip at the bounding surface.

Let the four planes $z = z_0$, $z = z_0 + l$, $x = x_0$, $x = x_0 + l$ cut the plane Oxz in the square $PRR'P'$ (Fig. 91), each side being of length l. These planes enclose a prism of square section and of infinite length, but it is obvious that we may confine attention to a length

l of the prism. At time t certain particles of fluid lie on the faces of this cube. At time $t+dt$ they lie in corresponding positions on the faces of the parallelepiped of section $QSS'Q'$, where

$$PQ = P'Q' = v\,dt,$$

and $$RS = R'S' = (v + l\,dv/dz)\,dt.$$

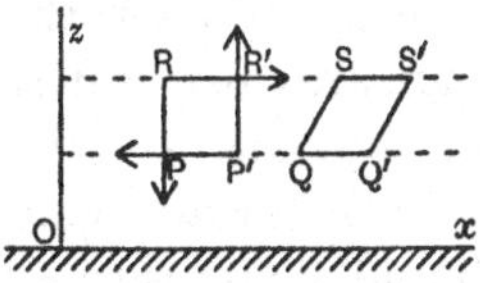

Fig. 91

The angle SQQ' is $\frac{1}{2}\pi - (dv/dz)\,dt$ and hence the rate of shearing is dv/dz.

The mass of the cube is ρl^3, where ρ is the density of the fluid, and the moment of inertia of the cube about an axis through its centre parallel to Oy is $\frac{1}{6}\rho l^5$. The moment, about the same axis, of the pair of equal and opposite forces $l^2\eta\,dv/dz$ acting on the faces RR', PP' is $l^3\eta\,dv/dz$. The ratio of this couple to the moment of inertia is proportional to $1/l^2$ and thus increases indefinitely as l is diminished. There would therefore be an indefinitely great angular acceleration of an infinitesimal cube unless tangential stresses acted on the faces which are parallel to Oyz as well as on the faces which are parallel to Oxy. It is clear that the tangential stresses on the faces RP, $R'P'$ must be T per unit area and that they must act in the directions indicated in Fig. 91.

The tangential stresses on the four faces are equivalent to a pressure T normal to the plane $R'P$, and a tension T normal to the plane RP'. The pressure and tension combined cause the fluid to expand along the direction PR' and to contract along RP', and so cause the square $PRR'P'$ to be distorted into the parallelogram $QSS'Q'$.

146. Heat produced by shearing. In Fig. 90, PQ, RS are the lines in which the plane Oxz is cut by two planes parallel to Oxy at distance dz apart. The liquid moves parallel to Ox, and the velocity of the liquid is v at P and $v + (dv/dz)\,dz$ at R. The rate of shearing of the layer of liquid between these planes is dv/dz radians per unit time. The shearing is unaltered if the planes be actual surfaces of, say, glass or metal which move with velocities v and $v + (dv/dz)\,dz$. The force which the layer of liquid between the planes exerts upon the plane RS is $\eta\,dv/dz$ per unit area pulling it back; the force exerted on PQ is $\eta\,dv/dz$ per unit area pulling it forward. Each unit area of the surface moving with velocity $v+(dv/dz)\,dz$ does work on the layer at the rate of

$$(v+dv/dz)\,\eta\,dv/dz \text{ per unit time.}$$

The layer does work on the other plane at the rate of $v\,.\,\eta\,dv/dz$ per unit area per unit time. The resultant work done per unit time on unit area of layer—of volume dz—is $\eta\,(dv/dz)^2\,dz$. Hence, if work h per unit volume per unit time be expended in shearing a liquid of viscosity η at R radians per unit time,

$$h=\eta R^2. \quad \text{.........................(3)}$$

The ordinary kinetic energy of the liquid does not increase, for the motion is "steady," and hence the work h appears as heat in the liquid.

147. Turbulent motion. In the simple case considered in §§ 144, 145, each particle of the liquid moves parallel to a straight line. In other cases, the particles though not moving parallel to a straight line yet follow definite paths called stream lines. So long as the velocity is everywhere sufficiently small, it will be multiplied by n when the driving forces are multiplied by n. But a stage is soon reached when this simple law breaks down. The liquid then ceases to flow in stream lines and is thrown into a state of turbulent motion. Thus, when a ship is gliding slowly through the water, a stream-line flow similar to that of Fig. 90 exists in the water near the ship's side. But, when the ship is going at full speed, the motion of the water near the side is turbulent.

The case of liquid flowing along a uniform tube of circular section was investigated by Osborne Reynolds.* If the radius of the tube be a cm. and a volume $\pi a^2 U$ c.c. pass along the tube per sec., then U cm. per sec. is the mean velocity for the section. When U is sufficiently small, the liquid flows in stream lines parallel to the axis of the tube. Reynolds found that, for a certain critical value, U_c, of U, the stream-line motion breaks down and is replaced by turbulent motion. Approximately

$$U_c=1000\eta/(\rho a), \quad \text{.......................(4)}$$

where ρ grm. cm.$^{-3}$ is the density and η dyne cm.$^{-2}$ per unit rate of shearing (one radian per sec.) is the viscosity of the liquid.

When the motion is turbulent, the ratio which the difference

* *Philosophical Transactions, Royal Society*, p. 935 (1883).

of pressure between the ends of the tube bears to U is larger than the constant value which holds as long as there is stream-line motion.

148. Distribution of velocity in a tube deduced from minimum heat production. Let liquid of viscosity η flow with steady stream-line motion through a tube of circular section and of radius a and let a volume Q pass through in unit time. The velocity u, which depends upon the distance r from the axis, is subject to the conditions (i) that $u=0$ when $r=a$ and (ii) that the volume passing per unit time is Q, or that

$$2\pi\int_0^a ur\,dr = Q. \quad \text{.......................(5)}$$

For any distribution of velocity, say $u=f(r)$, the rate of heat production per unit length of tube is H, where

$$H = 2\pi\eta\int_0^a (du/dr)^2 r\,dr, \quad \text{..................(6)}$$

since, by (3), the rate of heat production is ηR^2 per unit volume, R being the rate of shearing; in our case $R^2=(du/dr)^2$. The value of H depends upon the function $f(r)$. We may, however, be certain that the actual distribution of velocity is such as to make H the least possible for a given value of Q. By the aid of this principle we can determine the velocity at any point in terms of r and Q. Let v be the actual velocity at distance r from the axis when volume Q flows through the tube in unit time, and let H_0 be the rate of heat production. Then

$$2\pi\int_0^a vr\,dr = Q, \quad \text{.......................(7)}$$

and $$H_0 = 2\pi\eta\int_0^a (dv/dr)^2 r\,dr. \quad \text{..................(8)}$$

If the velocity be changed from v to $u=v+w$, where u satisfies conditions (i) and (ii), we have, by (5),

$$2\pi\int_0^a wr\,dr = 0, \quad \text{.......................(9)}$$

and, by (6), $$H = H_0 + 2\pi\eta\,(2F+G),$$

where $$F = \int_0^a \frac{dv}{dr}\cdot\frac{dw}{dr}\cdot r\,dr, \qquad G = \int_0^a \left(\frac{dw}{dr}\right)^2 r\,dr.$$

If H be a minimum when $w=0$ for all values of r, there will be no first order difference between H and H_0 when w is everywhere small compared with v, provided that dw/dr be small compared with dv/dr. The term $2\pi\eta G$ involving $(dw/dr)^2$, which is positive, is then of the second order compared with H_0.

On integration by parts,

$$F=\int_0^a \frac{dw}{dr}\,.\,r\frac{dv}{dr}\,dr=\left[w\,.\,r\frac{dv}{dr}\right]_0^a-\int_0^a w\frac{d}{dr}\left(r\frac{dv}{dr}\right)dr.$$

Since $w=0$ when $r=a$, and since $r=0$ on the axis, the first term vanishes at both limits. By (9),

$$\int_0^a wr\,dr=0,$$

and thus we can write

$$F=\int_0^a w\left\{Ar-\frac{d}{dr}\left(r\frac{dv}{dr}\right)\right\}dr=\int_0^a wJ\,dr,$$

where $J=Ar-d(r\,dv/dr)/dr$ and A is any constant.

Unless $J=0$, we shall, by a proper choice of the arbitrary velocity w, be able to give F any small value we please. Hence, if H suffer no first order change for any arbitrary small variation of the velocity distribution, we must have $J=0$ or

$$\frac{d}{dr}\left(r\frac{dv}{dr}\right)=Ar.$$

Hence $r\,dv/dr=\frac{1}{2}Ar^2+B$, and $B=0$ since $r\,dv/dr$ vanishes when $r=0$. Then $dv/dr=\frac{1}{2}Ar$, and $v=\frac{1}{4}Ar^2+C$. Since $v=0$ when $r=a$, $C=-\frac{1}{4}Aa^2$, and thus $v=-\frac{1}{4}A(a^2-r^2)$. Since v must satisfy (7) we have

$$Q=-\tfrac{1}{4}A\,.\,2\pi\int_0^a(a^2-r^2)\,r\,dr=-\tfrac{1}{8}\pi Aa^4.$$

Hence $A=-8Q/\pi a^4$, and

$$v=2Q\,(a^2-r^2)/(\pi a^4). \quad\text{.................(10)}$$

This value of v agrees with that found by another method in § 150.

When v is given by (10), $F=0$, and then $H=H_0+2\pi\eta G$, where

$$H_0=8\eta Q^2/\pi a^4. \quad\text{.....................(10}a\text{)}$$

Thus, unless $w=0$ for all values of r, $H>H_0$. Hence H_0 is the least possible value of the rate of heat production for the given flow of Q per unit time.

If the data be as in § 150, the rate at which work is done by gravity and by the pressures at the ends of the tube, of length l, is

$$Q\,(p_1 - p_2 + g\rho l \cos\theta).$$

When we equate this to $H_0 l$, the rate of heat production in the length l, we obtain equation (7) of § 150.

148 A. Flow through tube of elliptic section. The axis of the tube is Oz, and Ox, Oy are the principal axes of the transverse section. At the inner surface of the tube

$$x^2/a^2 + y^2/b^2 = 1.$$

The fluid moves parallel to Oz. At x, y, z its velocity is w and w is independent of z. At x, y, z the viscous stress across a plane parallel to Oyz is $\eta\, dw/dx$. Across a parallel plane through $x + dx$, y, z it is

$$\eta\left(\frac{dw}{dx} + \frac{d^2 w^2}{dx^2}\,dx\right).$$

Similar results hold for planes defined by y and $y + dy$. The resultant, in the positive direction, of the force exerted *on* the fluid in unit length of the space defined by the four planes is

$$\eta\,(d^2 w/dx^2 + d^2 w/dy^2)\,dx\,dy.$$

If the pressure at x, y, z be p, dp/dz is independent of z. If the tube be vertical and if the flow be downwards, the driving force on the fluid (density ρ) in the unit length is

$$(g\rho - dp/dz)\,dx\,dy.$$

If p_1, p_2 be the pressures at the top and bottom of the tube, of length l,

$$g\rho - dp/dz = (g\rho l + p_1 - p_2)/l \equiv f.$$

The fluid has no acceleration when the flow is steady, and thus

$$\eta\,(d^2 w/dx^2 + d^2 w/dy^2) + f = 0.$$

This equation is satisfied by

$$w = w_0\,(1 - x^2/a^2 - y^2/b^2),$$

if

$$w_0 \eta = \tfrac{1}{2} f a^2 b^2/(a^2 + b^2).$$

If a volume Q flow through the tube per unit time,

$$Q = w_0 \iint (1 - x^2/a^2 - y^2/b^2)\, dx\, dy,$$

the integration extending over the section. Now $\iint x^2 dx\, dy$ is the "moment of inertia" of the area πab about Oy and equals $\pi ab . a^2/4$. Similarly $\iint y^2 dx\, dy = \pi ab . b^2/4$. Hence

$$Q = w_0 . \pi ab\,(1 - \tfrac{1}{4} - \tfrac{1}{4}) = w_0 . \pi ab/2,$$

and

$$\eta = \frac{fa^2 b^2}{2\,(a^2+b^2)\, w_0} = \frac{\pi f a^3 b^3}{4Q\,(a^2+b^2)}.$$

Thus the a^4 of (6), § 150, is replaced by $2a^3b^3/(a^2+b^2)$.

149. Angular velocity of liquid between a fixed and a rotating cylinder found by minimum heat method. Liquid of viscosity η is contained between a fixed cylinder of radius a and an inner coaxial cylinder of radius b, which rotates with uniform angular velocity Ω radian sec.$^{-1}$ about its axis. The cylinders are supposed of infinite length. Let the liquid so move that, at distance r from the axis, its angular velocity about the axis is ψ, where $\psi = f(r)$; the only restrictions placed on ψ are that $\psi = 0$ when $r = a$, and $\psi = \Omega$ when $r = b$, since there is no slipping, and that ψ and $d\psi/dr$ be finite. If R be the rate of shearing, we have, by § 165, $R = r\, d\psi/dr$.

The heat produced per unit volume per unit time is h, where, by § 146, $h = \eta R^2$. Hence, if H be the rate of heat production per unit length of the cylinders,

$$H = 2\pi\eta \int_b^a r^3 \left(\frac{d\psi}{dr}\right)^2 dr.$$

Let ω be the angular velocity at distance r, and H_0 the rate of heat production, in the case of the actual motion, in which, of course, $\omega = 0$ at $r = a$ and $\omega = \Omega$ at $r = b$. Then H_0 will be the minimum value of H. Let $\psi = \omega + \beta$, where $\beta = 0$ at $r = a$ and also at $r = b$. Then

$$H = H_0 + 2\pi\eta\,(2F + G),$$

where

$$H_0 = 2\pi\eta \int_b^a r^3 \left(\frac{d\omega}{dr}\right)^2 dr, \quad F = \int_b^a r^3 \frac{d\beta}{dr}\frac{d\omega}{dr}\, dr, \quad G = \int_b^a r^3 \left(\frac{d\beta}{dr}\right)^2 dr.$$

If H be a minimum when $\beta = 0$ for all values of r, there will be no first order difference between H and H_0 when β is everywhere small compared with ω, provided that $d\beta/dr$ be small compared with $d\omega/dr$. The term $2\pi\eta G$ involving $(d\beta/dr)^2$, which is positive, is then of the second order compared with H_0.

If we put $r^3 d\omega/dr = J$, we have, on integration by parts,

$$F = \int_b^a \frac{d\beta}{dr} J\,dr = \Big[\beta J\Big]_b^a - \int_b^a \beta \frac{dJ}{dr}\,dr.$$

Since $\beta = 0$ at $r = a$ and also at $r = b$, the first term vanishes. Unless $dJ/dr = 0$ for all values of r, we shall, by a proper choice of the arbitrary β, be able to give F any small value we please, and hence, if H suffer no first order change for any arbitrary small variation of the velocity distribution, we must have $dJ/dr = 0$. Hence $J = r^3 d\omega/dr = A$, and $\omega = B - \frac{1}{2}A/r^2$. The constants A, B are found from the conditions that $\omega = 0$ at $r = a$ and $\omega = \Omega$ at $r = b$. We find

$$\omega = \frac{a^2 b^2 \Omega}{a^2 - b^2}\left(\frac{1}{r^2} - \frac{1}{a^2}\right). \quad \text{..................(11)}$$

When ω is given by (11), $F = 0$, and then $H = H_0 + 2\pi\eta G$, where

$$H_0 = 4\pi\eta\, a^2 b^2 \Omega^2/(a^2 - b^2). \quad \text{..............(12)}$$

Thus, unless $\beta = 0$ for all values of r, $H > H_0$. Hence H_0 is the least possible value of the rate of heat production for the given angular velocity Ω of the inner cylinder, when there is stream-line motion.

If a couple G be required to turn a length h of the inner cylinder with angular velocity Ω, then $G\Omega$ equals $H_0 h$, the rate of heat production in a length h. Hence

$$G = 4\pi\eta\, h a^2 b^2 \Omega/(a^2 - b^2), \quad \text{..............(13)}$$

as found in § 165.

EXPERIMENT 22. **Determination of viscosity of a liquid.**

150. Introduction. The viscosity of a liquid is most easily determined by finding the volume which passes per unit time through a narrow cylindrical tube of circular section under the action of a known difference of pressure between the ends of the tube.

Let the liquid, of density ρ grm. cm.$^{-3}$ and viscosity η dyne cm.$^{-2}$ for unit rate of shearing, flow downward through a vertical tube AB (Fig. 92) of internal radius a cm. and length l cm. Fig. 92 shows a section of the tube by a plane containing the axis. Let the flow be steady and let Q c.c. of liquid pass through the tube per second. Then, if U be the mean velocity across any section, $U = Q/(\pi a^2)$. Provided that U be less than the critical value $1000\eta/(\rho a)$, or Q be less than $1000\pi\eta a/\rho$, stream-line motion will exist and the liquid will move parallel to the axis of the tube.

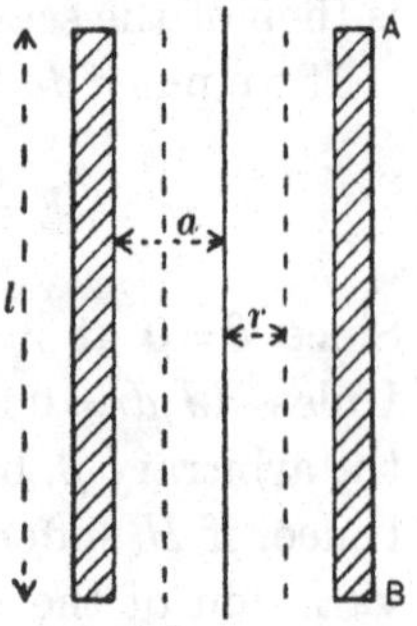

Fig. 92

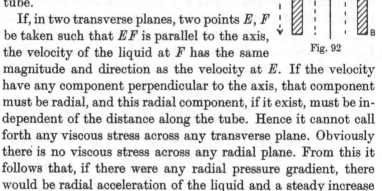

If, in two transverse planes, two points E, F be taken such that EF is parallel to the axis, the velocity of the liquid at F has the same magnitude and direction as the velocity at E. If the velocity have any component perpendicular to the axis, that component must be radial, and this radial component, if it exist, must be independent of the distance along the tube. Hence it cannot call forth any viscous stress across any transverse plane. Obviously there is no viscous stress across any radial plane. From this it follows that, if there were any radial pressure gradient, there would be radial acceleration of the liquid and a steady increase in the radial velocity. Since there is no radial velocity, there is no radial pressure gradient, and thus the pressure is the same at all points in a transverse plane.*

Let the velocity at a distance r cm. from the axis be v cm. sec.$^{-1}$. Then, at the surface of a cylinder of radius r described in the liquid, there is, by § 144, a tangential stress parallel to the axis of amount $\eta\, dv/dr$ dyne cm.$^{-2}$, the positive direction of its action on the liquid *inside* the cylinder of radius r being downward when dv/dr is positive. When the motion is steady, there is no acceleration parallel to the axis, and hence the resultant of all the forces acting on the liquid *within* the cylinder of radius r and length l is zero. If p_1 and p_2 be the pressures in the planes of the ends A and B respectively, the downward force on the upper end of the cylinder is $\pi r^2 p_1$ and the upward force on the lower end is $\pi r^2 p_2$.

* When the tube makes an angle θ with the vertical, the transverse component of gravity, viz. $g \sin\theta$, causes differences of pressure in a transverse section. These do not affect the distribution of velocity and may be disregarded.

The downward force due to gravity is $g\,.\,\pi r^2 l\rho$ and the downward force on the curved surface, of area $2\pi rl$, due to the shearing of the viscous liquid, is $2\pi rl\,.\,\eta\,dv/dr$. Equating the total downward force to zero, we have

$$2\pi rl\,.\,\eta\,dv/dr+\pi r^2(p_1-p_2)+g\,.\,\pi r^2 l\rho=0.$$

Hence $$dv/dr=-\beta r, \quad \text{......(1)}$$

where $$\beta=(p_1-p_2+g\rho l)/(2\eta l). \quad \text{......(2)}$$

On integration of (1), $$v=-\tfrac{1}{2}\beta r^2+C.$$

The constant C is determined by the condition that at the surface of the tube, where $r=a$, the velocity is zero, since the liquid does not slip at the surface. Hence $C=\frac{1}{2}\beta a^2$, and thus

$$v=\tfrac{1}{2}\beta(a^2-r^2). \quad \text{......(3)}$$

The volume of liquid which flows through the tube per second is Q, and hence

$$Q=\int_0^a v\,.\,2\pi r\,dr=\pi\beta\int_0^a(a^2-r^2)\,r\,dr=\tfrac{1}{4}\pi\beta a^4.$$

Thus $$\beta=4Q/(\pi a^4), \quad \text{......(4)}$$

and, by (3), $$v=2Q(a^2-r^2)/(\pi a^4). \quad \text{......(5)}$$

This result agrees with that given by (10), § 148, which was obtained by the minimum heat method.

Equating the values of β given by (2) and (4), we have

$$\eta=\frac{\pi a^4(p_1-p_2+g\rho l)}{8Ql}. \quad \text{......(6)}$$

If the tube, instead of being vertical, be inclined to the vertical at angle θ, the component of gravity along the tube is $g\cos\theta$, and accordingly gl must be replaced by $gl\cos\theta$, the difference of level between the centres of the two ends of the tube. We must now understand that p_1 and p_2 are the pressures at those centres. Then (6) becomes

$$\eta=\frac{\pi a^4(p_1-p_2+g\rho l\cos\theta)}{8Ql}. \quad \text{......(7)}$$

151. Method. The apparatus may be arranged as in Fig. 93. A glass tube AB, 50 to 100 cm. in length, is used; its internal section may be one to two mm.[2] A tube whose internal radius varies considerably is unsuitable. The tube is cleaned by running a little nitric acid through it, followed by water.* It is attached to a large funnel F by a short piece of rubber tube. Liquid is poured slowly into the funnel so as to keep the level of the liquid as nearly as possible to the tip of the nail N, which is carried by a piece of wood resting on the funnel. In order to remove any air bubble which may be caught in the tube of the funnel, the end B of the flow tube should be closed and a wire should be put down into the funnel to remove the bubble. If the supply of liquid be maintained, no more bubbles will be formed.

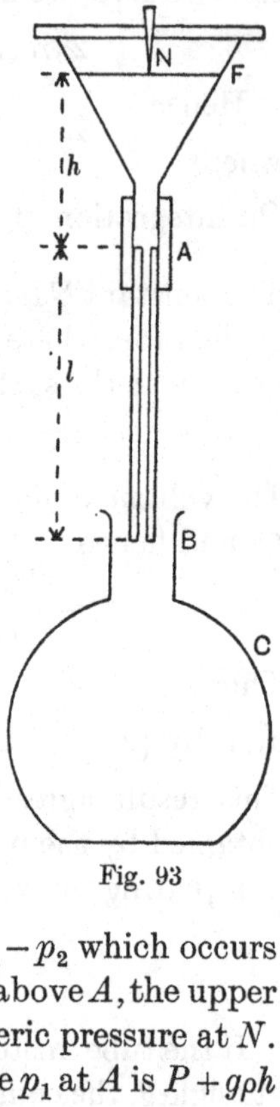

Fig. 93

The surface of the liquid is kept as nearly as possible at the standard level, and, while the liquid is running through the tube, a weighed flask C is placed below the tube to catch the liquid. After a time t seconds, the flask is removed, and is re-weighed. If M grm. be the mass of the liquid caught in time t, and if ρ be its density at the temperature prevailing in the flow tube, the volume Q c.c., which flows through per second, is given by

$$Q = M/(\rho t). \qquad \text{.............(8)}$$

We must now find the pressure difference $p_1 - p_2$ which occurs in (7). Let h be the height of the tip of the nail N above A, the upper end of the flow tube, and let P be the atmospheric pressure at N. Then, neglecting some small effects, the pressure p_1 at A is $P + g\rho h$ and at B, the lower end, the pressure p_2 is P. Hence $p_1 - p_2 = g\rho h$. Using this value in (7) we have, by (8),

$$\eta = \frac{\pi a^4 g \rho^2 (h + l\cos\theta)}{8l} \cdot \frac{t}{M}. \qquad \text{.............(9)}$$

The radius, a, of the tube is found by the aid of mercury. The tube is cleaned; it may be dried by warming it, in moderation, with a Bunsen flame and blowing dry air through it by bellows.

* If the liquid to be used be not water, the tube should be washed out with the liquid after it has been cleaned by acid and water.

The observer may *draw* air through the tube by sucking at a rubber tube attached to the glass tube. If he *blow* through the tube, his spittle will probably dribble along the bore, and, in any case, his "breath"—perhaps tobacco laden—will condense on the glass. If the tube be fouled in this way, the mercury will stick. The dried tube is filled with clean mercury. If the mass of mercury filling the tube be m grm., and if μ be the density of mercury at the temperature of the tube, the cross-section πa^2, which we have assumed to be constant along the tube, equals $m/(\mu l)$. Hence

$$\pi a^4 = m^2/(\pi \mu^2 l^2). \quad \text{......................(10)}$$

If the tube be nearly, but not quite, full of mercury, which then occupies a length s cm., we have, in place of (10),

$$\pi a^4 = m^2/(\pi \mu^2 s^2). \quad \text{......................(11)}$$

The value of πa^4 given by (10) or (11) is used in (9).

For accurate work, it is necessary to apply corrections for the buoyancy of the air. The atmospheric pressure at B is greater than that at N (Fig. 93) by $g\sigma(h + l\cos\theta)$, when the tube makes an angle θ with the vertical; the density σ of the air near the flow tube is found from its pressure and its temperature. Hence at A, $p_1 = P + g\rho h$, and at B, $p_2 = P + g\sigma(h + l\cos\theta)$, and thus

$$p_1 - p_2 + g\rho l \cos\theta = g(\rho - \sigma)(h + l\cos\theta).$$

We therefore replace ρ^2 in (9) by $\rho(\rho - \sigma)$.

If the liquid be weighed against brass weights of density β and if N grm. of brass weights balance M grm. of liquid, then

$$M(1 - \sigma/\rho) = N(1 - \sigma/\beta),$$

and thus, to the first power of σ,

$$M = N\{1 + \sigma(1/\rho - 1/\beta)\}.$$

Similarly, if the mass m grm. of mercury be balanced by n grm. of brass,

$$m = n\{1 + \sigma(1/\mu - 1/\beta)\}.$$

152. Calibration correction. Since it is impossible to obtain a glass tube of perfectly uniform bore, we must consider the case when the radius is not quite constant. Let a be the radius of the cross-section at distance x from the end A of the tube. If $|da/dx|$ be very small compared with unity, we shall make little error if we assume that the driving force on a length dx necessary to cause Q c.c. to pass per second through this portion of the tube is the same as in a *uniform* tube of radius a. If we put $l = dx$ and $p_1 - p_2 = -(dp/dx)\,dx$ in (7), we have

$$-dp/dx + g\rho\cos\theta = 8Q\eta/(\pi a^4).$$

This differential equation shows how dp/dx depends upon the radius in a non-uniform tube.

Integrating from $x=0$ to $x=l$, we have

$$p_1 - p_2 + g\rho l\cos\theta = 8Q\eta l/(\pi a_0^4),$$

where $$\frac{1}{a_0^4} = \frac{1}{l}\int_0^l \frac{dx}{a^4}. \quad \text{......................(12)}$$

When a_0^4 has been found by (12), we use it in place of a^4 in (7) or (9).

If the mass of mercury, of density μ, filling the tube be m grm.,

$$m/\mu = \pi l b^2, \quad \text{......................(13)}$$

where $$b^2 = \frac{1}{l}\int_0^l a^2 dx. \quad \text{......................(14)}$$

Hence b^2 is the mean value of the square of the radius. But b^4 is *not* equal to a_0^4, and thus we cannot find a_0^4 from (13).

If we write $$1/a_0^4 = F/b^4, \quad \text{......................(15)}$$

the correcting factor F (which is not less than unity) is only slightly greater than unity unless the tube be badly irregular in bore. The value of F is investigated in § 162.

An idea of the magnitude of the calibration correction may be gained from a simple case. We suppose that the tube is slightly conical and that $a = c - f + 2fx/l$. Then $a = c - f$ when $x = 0$, $a = c$ when $x = \frac{1}{2}l$, and $a = c + f$ when $x = l$. We easily find

$$\frac{1}{a_0^4} = \frac{1}{l}\int_0^l \frac{dx}{(c-f+2fx/l)^4} = \frac{c^2 + \frac{1}{3}f^2}{(c^2 - f^2)^3},$$

$$b^2 = \frac{1}{l}\int_0^l (c - f + 2fx/l)^2\,dx = c^2 + \tfrac{1}{3}f^2.$$

Hence $$b^4/a_0^4 = (c^2 + \tfrac{1}{3}f^2)^3\,(c^2 - f^2)^{-3}.$$

When f/c is small, $b^4/a_0^4 = 1 + 4f^2/c^2$, approximately. If $f/c = \frac{1}{10}$, so that the end radii are $0{\cdot}9c$ and $1{\cdot}1c$, $b^4/a_0^4 = 1{\cdot}041$—a serious correction. With a better tube for which $f/c = \frac{1}{100}$, $b^4/a_0^4 = 1{\cdot}0004$; this correction is negligible except in very precise work.

A detailed account of the method of calibration is given in § 162.

153. Practical example.

Messrs C. N. Wilmot-Smith and G. F. C. Searle determined the viscosity of water; they used a tube of length $l = 70{\cdot}03$ cm.

Mass of mercury filling tube $= m = 7{\cdot}5225$ grm.

Density of mercury at 19° C. $= \mu = 13{\cdot}549$ grm. cm.$^{-3}$.

If b^2 be the mean square of the radius, we have, by (13),

$$b = m^{\frac{1}{2}}/(\pi\mu l)^{\frac{1}{2}} = 5{\cdot}024 \times 10^{-2} \text{ cm.},$$

$$\pi b^4 = m^2/(\pi\mu^2 l^2) = 2{\cdot}001 \times 10^{-5} \text{ cm.}^4$$

Tube was vertical and thus $\cos\theta = 1$. Height of upper surface of water (point of nail) above top of flow tube $= h = 20{\cdot}80$ cm. Hence

$$h + l\cos\theta = 90{\cdot}83 \text{ cm.}$$

Density of water at 19° C. $= \rho = 0{\cdot}9984$. Two determinations of mass M of water passing through in time t were made. In the first, $M = 87{\cdot}67$ grm., $t = 300$ sec. and $t/M = 3{\cdot}4219$. In the second, $M = 108{\cdot}66$ grm., $t = 365$ sec. and $t/M = 3{\cdot}3591$. Mean $t/M = 3{\cdot}390$ sec. grm.$^{-1}$.

If corrections due to buoyancy of air and calibration of tube be neglected, we put $a^4 = b^4$, and then we have, by (9),

$$\eta = \frac{\pi a^4 g \rho^2 (h + l\cos\theta)}{8l} \cdot \frac{t}{M} = \frac{2{\cdot}001 \times 10^{-5} \times 981 \times 0{\cdot}9984^2 \times 90{\cdot}83}{8 \times 70{\cdot}03} \times 3{\cdot}390$$

$= 0{\cdot}01075$ dyne cm.$^{-2}$ per unit rate of shearing.

The calibration correction was found by method of § 162. Length q of mercury thread, roughly 10 cm. long, was measured when centre of thread was at approximately 5, 15, ... 55, 65 cm. from one end of tube. Results in Table give $\Sigma q = 72{\cdot}44$ and $q_0 = \frac{1}{7}\Sigma q = 10{\cdot}35$ cm. Then $d = q - q_0 = q - 10{\cdot}35$.

q	9·52	9·82	10·20	10·71	10·68	10·70	10·81	cm.
d	−0·83	−0·53	−0·15	+0·36	+0·33	+0·35	+0·46	cm.
d^2	0·689	0·281	0·022	0·130	0·109	0·122	0·212	cm.^{2}

Hence $\Sigma d^2 = 1{\cdot}565$, and, with $n = 7$,

$$F = 1 + 3\Sigma d^2/nq_0^2 = 1 + 3 \times 1{\cdot}565/(7 \times 10{\cdot}35^2) = 1{\cdot}0063.$$

Then, by (15), § 152,

$$\pi a_0^4 = \pi b^4/F = 2{\cdot}001 \times 10^{-5}/1{\cdot}0063.$$

To allow for buoyancy of air with respect to water in flow tube, we replace ρ^2 in (9) by $\rho^2(1 - \sigma/\rho)$; see § 151. Barometric height was 75·13 cm. and temperature 19° C. and $\sigma = 1{\cdot}195 \times 10^{-3}$. For water at 19° C., $\rho = 0{\cdot}9984$. Hence $1 - \sigma/\rho = 1 - 0{\cdot}00120$.

To allow for buoyancy of air with respect to water collected from flow tube, we replace M by $N\{1 + \sigma(1/\rho - 1/\beta)\}$, where N grm. of brass weights balance M grm. of water, and $\beta = 8{\cdot}5$ is density of brass weights; see § 151. Then $1/\rho - 1/\beta = 0{\cdot}884$, and

$$1 + \sigma(1/\rho - 1/\beta) = 1 + 1{\cdot}195 \times 10^{-3} \times 0{\cdot}884 = 1{\cdot}0011.$$

Buoyancy correction for mercury used in calibration is very small and is neglected.

The crude value 0·01075 found for η is corrected by multiplication by the factor

$$(1 - 0{\cdot}0012)/(1{\cdot}0063 \times 1{\cdot}0011) = 0{\cdot}9915.$$

Then

$$\eta = 0{\cdot}01075 \times 0{\cdot}9915 = 0{\cdot}01066 \text{ dyne cm.}^{-2} \text{ per unit rate of shearing.}$$

The Table of values of η in § 154 shows that 0·01066 corresponds to about 18° C.

154. Notes on the method.

The most serious obstacle in the way of an accurate determination of the viscosity of a liquid with simple apparatus is the fact that the viscosity diminishes rapidly as the temperature rises. The more viscous the liquid is, the more rapidly does the viscosity diminish with rise of temperature. The following Table derived from values given by N. E. Dorsey in the International Critical Tables shows how the viscosity, η, of water diminishes as the temperature rises:

t	η	t	η	t	η
0° C.	0·01794	20° C.	0·01009	40° C.	0·00634
5	0·01519	25	0·00895	45	0·00597
10	0·01310	30	0·00800	50	0·00549
15	0·01145	35	0·00721	55	0·00507

Glycerine at 0° C. is about 2500 times as viscous as water at 0°C. Its viscosity diminishes very rapidly with rise of temperature, and at 0°, 10°, 20°, 30° C. has the values 46, 21, 8·5, 3·5.

In the case of water at 15° C., $\eta^{-1} d\eta/dt$ is about $-0{\cdot}025$, and thus the diminution of η for a rise of one degree is about 2·5 per cent. of η. Hence we cannot expect to obtain a value of η reliable to one per cent., unless we are certain of the temperature of the water to one or two tenths of a degree. The temperature can be controlled to 0·1°, or less, by means of electric regulators and other apparatus, but these aids are not, as a rule, at the disposal of students.

It is easy to maintain a constant "head" of liquid above the top of the flow tube. But to find the mass of liquid which flows through in a given time is not easy. If the bore of the tube be so large that the liquid issues as a definite jet, the rate of efflux will be steady. The collecting vessel can be brought under the jet and be removed from under it at times which can, without any difficulty, be defined with considerable accuracy, but the rim of the vessel must cut through the jet and some splashing must occur. If the flow be so slow that the liquid collects in drops, which fall at nearly regular intervals from the end of the tube, the vessel can be brought up in the interval between the fall of one drop and the fall of the next and can be removed in a similar interval. But, of course, in this case the flow is not steady, since the pressure at the lower end of the tube is subject to small variations due to capillary action. As far, however, as the collection of the liquid is concerned, the relative importance of errors due to splashing or to periodically interrupted flow can be diminished as much as we please by sufficiently increasing the time allowed for the collection. The importance of errors due to the periodic changes of pressure can be diminished by increasing the length of the tube.

Something should be said about corrections for "kinetic energy." Whether kinetic energy is or is not involved depends upon the manner in which the liquid reaches and leaves the flow tube.

In one simple case, the flow tube AB (Fig. 94), of length l, takes liquid from a large vessel C and discharges into a large vessel D; the surfaces of the liquid in C and D are E and F. The vertical distance of F below E is z. The distance of the upper end A of the tube below E and that of B below F are large compared with the radius of the bore. The area of the upper surface of the liquid in either vessel is S. If a volume, Q, of liquid pass through the tube, and if E sink by u, per unit time, then $Su=Q$. The kinetic energy per unit volume of the liquid near E is $\frac{1}{2}\rho u^2$ and the kinetic energy in a layer of depth u is $\frac{1}{2}\rho u^2 . Su$ or $\frac{1}{2}\rho Q^3/S^2$, and this, for a given value of Q, we can make as small as we please by increasing S. Near the end A, the velocity of the liquid is appreciable; very near the centre of the opening it is of the order of $2Q/\pi a^2$, where a is the radius of the tube. Thus the liquid in C does possess some kinetic energy. But we can make the *rate of variation* of this kinetic energy, for a given value of Q, as small as we please by increasing S. Similarly, the rate of variation of the kinetic energy in D can be made as small as we please.

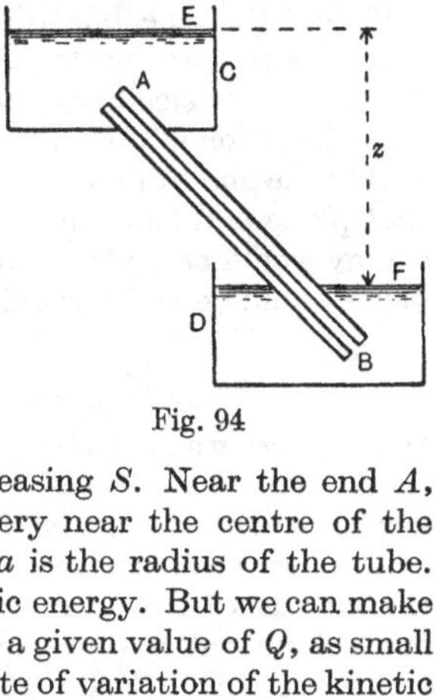

Fig. 94

When a volume Q passes from C to D, the change of potential energy is that due to the descent of Q from E to F. The rate of loss of potential energy is $g\rho zQ$, and this equals the rate of production of heat by the shearing of the viscous liquid. In the tube, except very near its ends,

$$v=2Q\,(a^2-r^2)/\pi a^4, \quad \text{.......................(16)}$$

where v is the velocity at distance r from the axis. The heat produced per unit time in the tube, of length l, is, by (8), § 148,

$$2\pi l\eta \int_0^a (dv/dr)^2\, r\,dr,$$

or $8\eta Q^2 l/\pi a^4$. If this were the whole rate of heat production, we could equate it to $g\rho zQ$, and then we should have

$$\eta = g\rho z\pi a^4/8Ql.$$

Actually, some heat is produced by the shearing of the liquid in the vessels C, D near the ends of the tube. The distribution of velocity indicated by (16) does not hold good right up to the ends of the tube, and thus the heat produced in the tube will differ somewhat from $8\eta Q^2 l/\pi a^4$. The excess of the total heat produced above this value must be proportional to ηQ^2 and may be written $8\eta Q^2(2k)/\pi a^4$, where k is a small length. Then

$$g\rho z = 8\eta Q(l+2k)/\pi a^4.$$

If we wish to consider that the liquid outside the tube is non-viscous and that the velocity within the whole tube is given by (16), we must "correct" the length of the tube by adding k to each end.

In another case, the lower vessel D is removed, and the liquid issues from B as a jet and takes kinetic energy with it at the rate J per unit time. For simplicity, we suppose that AB is horizontal. At a little distance from B, the velocity of the surface of the jet will not be zero, as it is at the inner surface of the tube, but will approximate to the velocity in the axis of the jet. Thus, near the outlet, there is a region of transition, and (16) does not represent the velocity at points in the plane of the end of the tube. Hence the calculation of J is difficult.

If in the plane of the end of the tube the velocity were perpendicular to that plane and had the value u at distance r from the axis, the kinetic energy would be $\frac{1}{2}\rho u^2$ per unit volume, and the rate at which it passes the plane would be $\frac{1}{2}\rho u^3$ per unit area. We should then have

$$J = \int_0^a \tfrac{1}{2}\rho u^3 \,.\, 2\pi r\, dr.$$

If $u = Q/\pi a^2$, we have $J = \frac{1}{2}\rho Q^3/\pi^2 a^4$. If u had the value given by (16), we should have $J = \rho Q^3/\pi^2 a^4$. We shall therefore put $J = \beta\rho Q^3/\pi^2 a^4$, where β may be expected to lie between one-half and unity.

A correction for surface tension is necessary when the liquid issues from the tube as a jet. At a short distance from B along the jet, the velocity of the liquid will be nearly uniform across the section, and thus the surface will move with velocity $\lambda Q/\pi a^2$, where λ is somewhat less than unity, since in the *tube*, the velocity decreases to zero as r increases from zero to a, and uniformity of velocity across the section of the *jet* will be established only gradually. Thus fresh surface will be formed at the rate $2\pi a \,.\, \lambda Q/\pi a^2$ or $2\lambda Q/a$, and the rate at which energy is expended is $2\lambda QT/a$, where T is the surface tension. If the liquid rise to a height f in a tube of radius a, $2T/a = g\rho f$. Hence energy is expended at the rate $g\rho\lambda fQ$. The equation of energy now becomes

$$g\rho hQ = \frac{8\eta Q^2(l+k+k')}{\pi a^4} + \frac{\beta\rho Q^3}{\pi^2 a^4} + g\rho\lambda fQ,$$

where h is the depth of B below the surface E, and k, k' are the end corrections due to heat production for the ends A, B of the tube. Hence

$$\frac{8\eta Q(l+k+k')}{\pi a^4} = g\rho(h-\lambda f) - \frac{\beta\rho Q^2}{\pi^2 a^4}.$$

The changes in the effective length of the tube due to heat production near its ends and the change of head due to surface tension are constant for a given liquid, unless the form of the jet change with Q. The effect of "kinetic energy" is to change the head by an amount proportional to Q^2.

The problem is of considerable difficulty, and it is not surprising that a variety of estimates have been given for the corrections. The relative importance of all the corrections is diminished by increasing the length of the flow tube.

EXPERIMENT 23. **Determination of viscosity of air.**

155. Introduction. In this experiment, air is pumped into a large vessel and is then allowed to escape through a capillary tube. The pressure of the air at the beginning and end of a measured interval of time is determined; it is assumed that the temperature of the air in the vessel and in the capillary tube is always equal to that of the surrounding atmosphere.

In the determination of the viscosity of a liquid, the difference of pressure forcing the liquid through the tube is kept constant, and the density of the liquid may be treated as uniform. But, in the present experiment, the density of the air is, of necessity, not uniform along the flow tube at any given time, and the difference of pressure driving the air through the tube is, on account of the method, not constant, but diminishes as the time increases. The theory must therefore take account of these two facts.

The viscosity of air and other gases depends upon the temperature but not upon the pressure, unless the pressure be very low.

156. Calculation of velocity. Consider a tube AB (Fig. 95) of radius a cm. and length l cm., through which air is passing in the direction from A to B with stream-line motion parallel to the

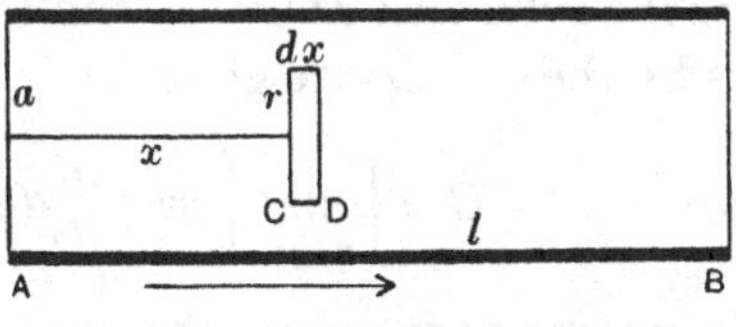

Fig. 95

axis.* Let CD be a length dx of a coaxial geometrical cylinder of radius r cm., the end C being at a distance x cm. from the end A of the tube.

* If the flow, without being turbulent, be so great that the change of pressure in a length a of the tube be not very small compared with the pressure itself, the stream lines will not be strictly parallel to the axis. We assume that the flow is so slow that no appreciable error arises from any lack of parallelism.

When the fall of pressure in a distance along the axis equal to the radius of the tube is very small compared with the pressure, the radial velocity cannot be other than exceedingly small. The argument of § 150 then shows that, to a high degree of approximation, the pressure is the same at all points in a transverse plane.*

Let the velocity of the air in the positive direction at any point defined by r and x be v cm. per sec. Then the velocity gradient at the curved surface of the cylinder CD is dv/dr sec.$^{-1}$. Hence, if the viscosity of the air be η dyne cm.$^{-2}$ per unit rate of shearing or η grm. cm.$^{-1}$ sec.$^{-1}$, the force, in the positive direction, due to viscous action on the curved surface of CD is

$$\eta\,(dv/dr)\,2\pi r\,dx \text{ dynes.}$$

If the pressure at any point on the end C of the cylinder be p dynes per sq. cm., the pressure at any point on the end D is $p+(dp/dx)\,dx$, and thus the resultant, in the positive direction, of the forces due to the pressures is

$$-\pi r^2\,(dp/dx)\,dx \text{ dynes.}$$

If the end B be lower than A by h, the component of gravity along the tube in the positive direction is gh/l, and the gravitational force in this direction on the air in the elementary cylinder CD is $(\pi r^2 g\sigma h/l)\,dx$, where σ is the density. The momentum per unit volume is σv, and thus, since a volume v passes through unit area per unit time, the rate F, at which, by the motion of the air, momentum enters the cylinder CD by the end C, and the rate G, at which it leaves by the end D, are given by

$$F=\int_0^r 2\pi r\sigma v^2\,dr, \qquad G=\int_0^r 2\pi r\left\{\sigma v^2+\frac{d}{dx}(\sigma v^2)\,dx\right\}dr.$$

The rate of gain of momentum in the positive direction is $-(G-F)$. The flow tube will maintain the absolute temperature of the stream of air at a constant value θ, and hence $\sigma=p/(R\theta)$, where R refers to unit mass of air. Since there is steady stream-

* When the tube makes an angle θ with the vertical, the transverse component of gravity, viz. $g \sin\theta$, causes very small differences of pressure in any cross-section. Except for their very small effect on the density of the air, these differences do not alter the distribution of velocity. Their effect is utterly negligible.

line motion, σv has the same value—say k—at all points along a given stream line. Thus

$$\sigma v^2 = k^2/\sigma = k^2 R\theta/p,$$

and hence $$\frac{d}{dx}(\sigma v^2) = -\frac{k^2 R\theta}{p^2}\cdot\frac{dp}{dx} = -v^2\cdot\frac{\sigma}{p}\cdot\frac{dp}{dx}.$$

If U be the velocity of sound in air at temperature θ,

$$U^2 = \gamma R\theta = \gamma p/\sigma,$$

where $\gamma\,(=1{\cdot}40\ldots)$ is the ratio of the specific heat at constant pressure to that at constant volume (see (10), (11), § 204). Thus

$$d(\sigma v^2)/dx = -\gamma(v^2/U^2)\,dp/dx,$$

and hence $$G-F = -dx\,.\,\frac{\gamma}{U^2}\frac{dp}{dx}\int_0^r 2\pi r v^2 dr.$$

For a given value of x, v has its greatest value, u, on the axis. Hence

$$|G-F| < dx\,.\,(\pi r^2\gamma u^2/U^2)\,.\,|dp/dx|\,.$$

Thus, when the greatest velocity of the air in the tube is very small compared with the velocity of sound in air at temperature θ, the rate F, at which momentum enters the cylinder CD by the end C by convection, differs from the rate G, at which it leaves CD by the end D, by an amount which involves u^2/V^2 and is negligible compared with $\pi r^2 dx\,|dp/dx|$; the latter quantity is generally large compared with $\pi r^2 g\sigma dx\,|h/l|$. Since the motion is steady, the momentum within the cylinder remains unchanged. Hence, the resultant force, in the positive direction, on the air inside the cylinder CD, which equals $-(G-F)$, is negligibly small, and thus, with sufficient accuracy,

$$\eta\,(dv/dr)\,.\,2\pi r\,dx = \pi r^2(dp/dx - g\sigma h/l)\,dx,$$

or $$\frac{dv}{dr} = \frac{r}{2\eta}(dp/dx - g\sigma h/l). \quad \text{.................(1)}$$

Since $v=0$ when $r=a$, because there is no slipping at the wall of the tube, the solution of (1) is

$$v = -\frac{1}{4\eta}(a^2-r^2)\left(\frac{dp}{dx}-\frac{g\sigma h}{l}\right). \quad \text{..............(2)}$$

If the volume of air flowing per unit time across the plane defined by x be Q,

$$Q=\int_0^a 2\pi rv\,dr=-\frac{\pi}{2\eta}\left(\frac{dp}{dx}-\frac{g\sigma h}{l}\right)\int_0^a (a^2-r^2)\,r\,dr$$

$$=-\frac{\pi a^4}{8\eta}\left(\frac{dp}{dx}-\frac{g\sigma h}{l}\right). \qquad\qquad (3)$$

The mass of air crossing this plane per unit time is σQ, where σ is the density of the air under pressure p at the temperature θ; this temperature is assumed equal to θ_0, the temperature of the surrounding atmosphere.

157. Calculation of pressure. When the motion is steady, σQ, the mass of air crossing any section of the flow tube per unit time, is the same for all sections. Hence $\sigma Q=\sigma_0 Q_0$, where σ_0 and Q_0 are the density and volume of the same mass of air at the atmospheric temperature θ_0 and at the pressure P_0 of the atmosphere at the level of the outlet end of the flow tube. Thus, since $\theta=\theta_0$, $\sigma=\sigma_0 p/P_0$ and $Q=P_0 Q_0/p$. With these values of σ and Q, (3) becomes, after multiplication by p/a^4,

$$\frac{P_0 Q_0}{a^4}=\frac{pQ}{a^4}=-\frac{\pi}{8\eta}\left\{p\frac{dp}{dx}-\beta p^2\right\}, \qquad\qquad (4)$$

where

$$\beta=gh\sigma_0/(lP_0). \qquad\qquad (5)$$

In practice, the radius of the tube will not be quite constant, but, with any but a very bad tube, it will vary so gradually from point to point along the tube that the conditions of flow for any infinitesimal length dx may be treated as if they were those which would exist there, if the whole tube had the same radius as the element dx. On these assumptions, (4) holds good for all values of x. Integrating it with respect to x from $x=0$, where $p=P$, to $x=l$, where $p=P_0$, and remembering that η is independent of the pressure, we have

$$P_0 Q_0\int_0^l \frac{dx}{a^4}=\frac{\pi}{16\eta}\left\{P^2-P_0^2+2\beta\int_0^l p^2\,dx\right\}=\frac{\pi}{16\eta}\{P^2-P_0^2+E\}.$$

Since $p<P$, except when $x=0$, $E<2gh\sigma_0 P^2/P_0$. In the experiment of § 163, E is very small compared with $P^2-P_0^2$. Of course, if the flow tube be horizontal, $E=0$, since $\beta=0$. We then have, as found by O. E. Meyer,

$$Q_0=\frac{\pi a_0^4}{16\eta l}\cdot\frac{P^2-P_0^2}{P_0}, \qquad \text{.....................(6)}$$

where

$$\frac{1}{a_0^4}=\frac{1}{l}\int_0^l\frac{dx}{a^4}. \qquad \text{.............................(7)}$$

Hence a_0^{-4} is the mean value of a^{-4}. Generally the tube is not uniform and thus calibration is necessary; the process is explained in § 162.

158. Correction for gravity. If a be constant, (4) can be integrated. Since $p=P_0$ when $x=l$, we find, on putting

$$8\eta P_0 Q_0/\pi a^4=J,$$

that

$$\frac{\beta p^2-J}{\beta P_0^2-J}=\frac{e^{2\beta x}}{e^{2\beta l}}.$$

Hence $$J(e^{-2\beta x}-e^{-2\beta l})=\beta(p^2e^{-2\beta x}-P_0^2e^{-2\beta l}). \qquad \text{.........(8)}$$

At $x=0$, $p=P$, and thus

$$J(1-e^{-2\beta l})=\beta(P^2-P_0^2e^{-2\beta l}). \qquad \text{...............(9)}$$

Now $\beta l=gh\sigma_0/P_0$, and hence $P_0e^{-\beta l}=[P_0]$, where $[P_0]$ is the atmospheric pressure at the level of the inlet end of the flow tube. If H be the "height of the homogeneous atmosphere," so that $P_0=g\sigma_0H$, we have $\beta l=h/H$. When h/H is very small, we find, on expanding $e^{-2\beta l}$, that the factor of J in (9) becomes $2\beta l(1-h/H)$. Now with $\sigma_0=1{\cdot}25\times10^{-3}$, $g=981$, $P_0=10^6$, we find $H=8{\cdot}16\times10^5$ cm., and thus we may neglect h/H in comparison with unity, unless the tube be very long. Then, with $2\beta l$ for the factor of J, we have, from (9),

$$P_0Q_0/a^4=(\pi/16\eta l)\{P^2-[P_0]^2\}. \qquad \text{...........(10)}$$

When $P-[P_0]$ is very small compared with P_0, we cannot neglect the difference between P_0 and $[P_0]$.

159. Formula for viscosity. In the experiment, the mass of the air which passes through the tube is deduced from the fall of pressure of the air in the vessel during a measured time.

Let the volume of the vessel up to the end $x=0$ of the flow tube be S, and let the mass of air in the vessel, at any time t, be M, its pressure be P and its density σ^*. Then, if, as is assumed, the temperature be θ_0, we have

$$M = S\sigma^* = S\sigma_0 P/P_0.$$

The rate at which air escapes from the vessel is $\sigma_0 Q_0$ units of mass per unit time, and this equals $-dM/dt$. Hence,

$$Q_0 = -(1/\sigma_0)\,dM/dt = -(S/P_0)\,dP/dt.$$

Using this value of Q_0 in (6), we find the differential equation

$$-\frac{S}{P_0}\frac{dP}{dt} = \frac{\pi a_0^4}{16\eta l}\cdot\frac{P^2 - P_0^2}{P_0}.$$

Putting $1/(P^2 - P_0^2)$ into partial fractions, we have

$$\frac{1}{2P_0}\left\{\frac{1}{P+P_0} - \frac{1}{P-P_0}\right\}\frac{dP}{dt} = \frac{\pi a_0^4}{16\eta l S}.$$

Integrating from $t=0$ to $t=t$, we have

$$\frac{1}{2P_0}\left[\log_e \frac{P+P_0}{P-P_0}\right]_{t=0}^{t=t} = \frac{\pi a_0^4 t}{16\eta l S}.$$

If P_1 and P_2 be the pressures in the vessel at $t=0$ and at $t=t$,

$$\pi a_0^4 P_0 t/(8\eta l S) = \lambda, \qquad \text{.................(11)}$$

where

$$\lambda = \log_e\left\{\frac{P_1 - P_0}{P_2 - P_0}\cdot\frac{P_2 + P_0}{P_1 + P_0}\right\} \qquad \text{...............(12)}$$

Hence

$$\eta = \frac{\pi a_0^4 P_0}{8lS}\cdot\frac{t}{\lambda} \qquad \text{.........................(13)}$$

From the last expression, η can be determined. The value of P_0 on the right side of (13) must be expressed in dynes per sq. cm. if the C.G.S. system be used. Thus, if the barometric height be h_0 cm. and the density of mercury at the temperature of the barometer be μ grm. per c.c.,

$$P_0 = g\mu h_0.$$

Since only the *ratios* P_1/P_0 and P_2/P_0 are involved in the formula (12) for λ, the barometric pressure P_0 and the differences $P_1 - P_0$ and $P_2 - P_0$, which occur in that formula, may be expressed in cm. of mercury.

The differences $P_1 - P_0$ and $P_2 - P_0$ are observed by means of a gauge. Then

$$P_1 + P_0 = (P_1 - P_0) + 2P_0, \qquad P_2 + P_0 = (P_2 - P_0) + 2P_0.$$

160. Conditions for stream-line motion. When the motion of the air is turbulent, the ratio of the fall of pressure along the flow tube to the flow is greater than when there is stream-line motion, and thus, if the turbulent motion be treated as stream-line motion, the value obtained for the viscosity will be too great. The mass crossing any section per unit time is $\pi a^2 \sigma U$, where U is the mean velocity over the section, and thus, assuming the tube to be uniform, σU is constant and equal to $\sigma_0 U_0$, where σ_0, U_0 refer to the outlet where the pressure is P_0 and $U_0 = Q_0/\pi a^2$. By Reynolds' result (§ 147), there will be stream-line motion, if $\sigma U < 1000\eta/a$. Since σU is constant, we have, as the criterion, $U_0 < 1000\eta/(\sigma_0 a)$. Using (6), and noting that

$$P + P_0 = P - P_0 + 2P_0,$$

we have
$$\frac{(P-P_0)(P-P_0+2P_0)}{P_0^2} < \frac{16000\eta^2 l}{\sigma_0 a^3 P_0}.$$

On the right side, P_0 must be expressed in C.G.S. units. For a barometric height of 76 cm., $P_0 = 10^6$ dyne cm.$^{-2}$ approximately. On the left, ratios of pressures occur. Let $P - P_0$ be equivalent to k cm. of mercury and P_0 to h cm. Then, with $\eta = 1{\cdot}8 \times 10^{-4}$, $\sigma_0 = 1{\cdot}25 \times 10^{-3}$, and with a and l measured in cm., we have

$$k(k+2h)/h^2 < f,$$

where $f = 4{\cdot}15 \times 10^{-7} \times l/a^3$. The equation $k^2/h^2 + 2k/h = f$ has a positive root
$$k/h = \sqrt{(1+f)} - 1.$$

For stream-line motion, we must have $k/h < \sqrt{(1+f)} - 1$. It will suffice to take the normal value 76 cm. for h; then we have, for the critical value of k,
$$k = 76\{\sqrt{(1+f)} - 1\}\text{cm.} \qquad \text{.................. (14)}$$

The Table gives, to three figures, the values of k corresponding to some values of l and a.

	l = 25 cm.	l = 50 cm.	l = 75 cm.	l = 100 cm.
a	k	k	k	k
cm.	cm.	cm.	cm.	cm.
0·01	180	278	355	419
0·02	39·2	68·1	92·1	113
0·03	13·4	25·1	35·5	45·0
0·04	5·93	11·5	16·6	21·6
0·05	3·09	6·06	8·93	11·7
0·06	1·80	3·57	5·29	6·98
0·07	1·14	2·26	3·37	4·46
0·08	0·766	1·52	2·28	3·02
0·09	0·539	1·07	1·61	2·13
0·10	0·393	0·785	1·17	1·56
0·15	0·117	0·233	0·350	0·466
0·20	0·049	0·098	0·148	0·197

In the experiment, air escapes from the vessel for a finite time t, the pressure excess falls from $P_1 - P_0$ to $P_2 - P_0$, and the flow gradually diminishes. If turbulence exist for only a small part τ of t, the value found for η will be only a little too large, but, as τ approaches t, the value will increase considerably. Hence, if $P_1 - P_0$ exceed the maximum value for which stream-line motion persists, the value of η found from the experiment will increase with $P_1 - P_0$.

161. Experimental details. A can C (Fig. 96), of about 10 litres capacity, contains the air. Into a metal disk soldered into the neck of the can is soldered a tube to which are soldered a cycle valve V and a side tube E, which is connected with the mercury gauge G by a stout rubber tube A. A gas-fitter's tap T is soldered into the can near its base, and the flow tube F is connected to T by a stout rubber tube or by a suitable gland B. The various joints must be air-tight.

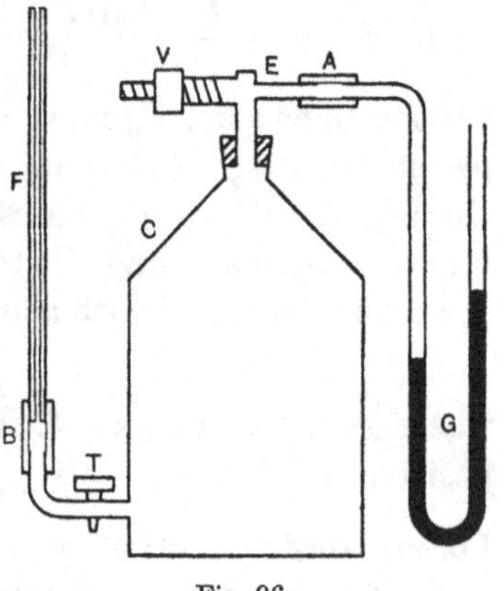

Fig. 96

A convenient gland for holding the flow tube may be made from a screw-down lubricator such as is used on motor cars. The flow tube passes through a hole bored in the cap of the lubricator and is gripped by rubber packing (a short length of thick-walled tube) which is compressed when the cap is screwed down. The volume of the can is best found from the mass of water required to fill it; small corrections are required for the various tubes connected to the can.

As the *first* step, the length l and internal radius a of the flow tube are measured in order that the length k may be found by equation (14) of § 160. If W grm. be the mass of mercury, of density μ, which fills the tube, and if b^2 be the mean value of the square of the radius, then, as in (14a), § 162, $b^2 = W/(\pi\mu l)$. It is here assumed that the tube is of uniform section and that, consequently, $a = b$. The calibration correction is found as in §162.

If the initial difference of level in the mercury gauge exceed k, turbulent motion of the air in the tube may be expected. If the flow be observed before k is known, it may easily happen that these observations are worthless because the flow was turbulent.

After k has been determined, air is pumped into the can by a cycle pump attached to the valve V, until the pressure is raised rather less than k cm. of mercury above the atmospheric pressure; the base of the can should be stout enough to endure an excess pressure of 25 cm. of mercury without serious bulging. The apparatus is then allowed to stand for some minutes in order that the rise of temperature due to the compression of the air may die away. A screen placed between the can and the observer will prevent the transfer of heat from his body to the can.

When the gauge readings have become steady, they are recorded. The tap is then opened for a time t sec.—one minute or more—and is then closed, and the new gauge readings are taken and recorded.* This process is repeated for various initial pressures. If the barometric pressure P_0 and the temperature θ_0 remain constant, we see, by (13), § 159, that the value of t/λ should have the same value for each of the experiments with a given tube.

162. Calibration correction. When the flow tube is not of uniform radius, we have to find a_0, where, by (7), § 157,

$$\frac{1}{a_0^4}=\frac{1}{l}\int_0^l \frac{dx}{a^4}.$$

The value of the integral is found as follows. Suppose that $n-1$ marks are made on the tube, dividing it into n parts, each l/n cm. in length. Let a thread of mercury of mass m grm., whose length is approximately l/n cm., be introduced into the tube. Let the thread be moved along the tube so that its centre approximately coincides with the centre of each of the n parts in turn, and let $q_1, q_2, \ldots q_n$ be its lengths in the 1st, 2nd, ... nth positions. Then, if the tube be treated as uniform over each of the n parts, we may put

$$\pi\mu a_1^2=m/q_1, \qquad \pi\mu a_2^2=m/q_2, \qquad \ldots,$$

where μ is the density of mercury.

* After the tap has been closed, the pressure in the vessel sometimes rises gradually by a few tenths of a millimetre. This indicates that, although the thermal conduction from the walls of the vessel has not kept the temperature of the expanding air quite constant, yet the fall of temperature has been very slight.

If the mass of mercury filling the whole tube be W grm.,

$$b^2 = W/(\pi\mu l), \quad \text{.....................(14}a\text{)}$$

where

$$b^2 = \frac{1}{l}\int_0^l a^2 dx.$$

Hence b^2 is the mean value of the square of the radius.

Taking l/n as an element of length instead of dx and replacing integration by summation, we have

$$W = \pi\mu \int_0^l a^2 dx = \pi\mu\, \Sigma a^2 . \frac{l}{n} = \frac{ml}{n}\, \Sigma \frac{1}{q}.$$

Hence

$$\frac{1}{m} = \frac{l}{nW}\, \Sigma \frac{1}{q} = \frac{1}{n\pi\mu b^2}\, \Sigma \frac{1}{q}. \quad \text{..................(15)}$$

Similarly we find

$$\frac{l}{a_0^{\,4}} = \int_0^l \frac{dx}{a^4} = \frac{\pi^2\mu^2 l}{m^2 n}\, \Sigma q^2 = \frac{l}{b^4}.\frac{1}{n^3}\left(\Sigma \frac{1}{q}\right)^2 \Sigma q^2 = \frac{l}{b^4}. F,$$

and thus

$$1/a_0^{\,4} = F/b^4. \quad \text{........................(16)}$$

Hence the factor by which $1/b^4$ must be multiplied to correct for the inequality of the tube is F, where

$$F = \frac{1}{n^3}\left(\Sigma \frac{1}{q}\right)^2 \Sigma q^2. \quad \text{..................(17)}$$

The factor F is only very slightly greater than unity when the tube is nearly uniform.

We may exhibit F in another form. Let q_0 be the mean value of the n quantities $q_1, q_2, \ldots$ and let $q_1 = q_0 + d_1$, $q_2 = q_0 + d_2, \ldots$. Then

$$\Sigma d = d_1 + d_2 + \ldots = 0. \quad \text{..................(18)}$$

Hence

$$\Sigma q^2 = \Sigma\,(q_0^{\,2} + 2q_0 d + d^2) = nq_0^{\,2} + \Sigma d^2. \quad \text{.........(19)}$$

With a nearly uniform tube $d_1, d_2, \ldots$ are small compared with q_0, and we may expand $1/(q_0 + d)$ in powers of d. Thus

$$\Sigma \frac{1}{q} = \Sigma \frac{1}{q_0 + d} = \Sigma \left(\frac{1}{q_0} - \frac{d}{q_0^{\,2}} + \frac{d^2}{q_0^{\,3}} - \ldots\right) = \frac{n}{q_0} + \frac{1}{q_0^{\,3}}\, \Sigma d^2 - \ldots.$$

As far as Σd^2, we have

$$F = \frac{1}{n^3}\left(\Sigma \frac{1}{q}\right)^2 \Sigma q^2 = \left(1 + \frac{2\Sigma d^2}{nq_0^{\,2}} + \ldots\right)\left(1 + \frac{\Sigma d^2}{nq_0^{\,2}}\right) = 1 + \frac{3\Sigma d^2}{nq_0^{\,2}}. \quad \text{...(20)}$$

Since $$\Sigma\frac{1}{q}=\Sigma\frac{1}{q_0+d}=\Sigma\frac{q_0-d}{q_0^2-d^2}, \quad\text{..............(21)}$$

and since q_0-d and $q_0^2-d^2$ are positive, (21) gives

$$\Sigma\frac{1}{q}>\frac{1}{q_0^2}\Sigma(q_0-d),$$

or, since $\Sigma d=0$, $$\Sigma\frac{1}{q}>\frac{n}{q_0}.$$

Hence $$F>\frac{1}{n^3}\left(\frac{n}{q_0}\right)^2(nq_0^2+\Sigma d^2),$$

or $$F>1+\Sigma d^2/nq_0^2,$$

and thus F always exceeds unity. The approximate result (20) is an example of this general inequality.

As a practical illustration, we will find the calibration correction for the flow tube No. I used in the experiment of § 163. The following Table gives the length q of a mercury thread in 11 equally spaced positions along the tube; the length of the tube was 64·82 cm.

q	q^2	$1/q$	$d=q-q_0$	d^2
cm.	cm.^{2}	cm.$^{-1}$	cm.	cm.^{2}
6·32	39·9424	0·158228	−0·15	0·0225
6·40	40·9600	0·156250	−0·07	0·0049
6·40	40·9600	0·156250	−0·07	0·0049
6·38	40·7044	0·156740	−0·09	0·0081
6·40	40·9600	0·156250	−0·07	0·0049
6·43	41·3449	0·155521	−0·04	0·0016
6·49	42·1201	0·154083	+0·02	0·0004
6·53	42·6409	0·153139	+0·06	0·0036
6·60	43·5600	0·151515	+0·13	0·0169
6·60	43·5600	0·151515	+0·13	0·0169
6·58	43·2964	0·151976	+0·11	0·0121
$q_0=6·47$	$\Sigma q^2=460·0491$	$\Sigma(1/q)=1·701467$		$\Sigma d^2=0·0968$

Hence, since $n=11$, we find, by (17),

$$F=\frac{1}{n^3}\Sigma q^2.\left(\Sigma\frac{1}{q}\right)^2=\frac{460·0491\times(1·701467)^2}{11^3}=1·00063. \quad\text{...(22)}$$

The need of tables giving q^2 and q^{-1} to several significant figures may be avoided by finding F by (20). Thus

$$F=1+\frac{3\Sigma d^2}{nq_0^2}=1+\frac{3\times0·0968}{11\times6·47^2}=1·00063. \quad\text{......(23)}$$

163. Practical example.

A flow tube of length $l = 64{\cdot}82$ cm. was used.
Volume of can and tubes (found by mass of water) $= S = 9646$ c.c.
Barometric height $= h_0 = 76{\cdot}36$ cm.
Density of mercury $= \mu = 13{\cdot}56$ grm. cm.$^{-3}$.
Barometric pressure $= P_0 = g\mu h_0 = 1{\cdot}016 \times 10^6$ dyne cm.$^{-2}$.
Mass of mercury filling flow tube $= W = 3{\cdot}671$ grm. Hence

$$b^2 = W/(\pi\mu l) = 1{\cdot}329 \times 10^{-3}\,\text{cm.}^2, \qquad b = 3{\cdot}646 \times 10^{-2}\,\text{cm.},$$
$$b^4 = 1{\cdot}767 \times 10^{-6}\,\text{cm.}^4 = Fa_0^4.$$

The calibration correction is found in § 162, and $F = 1{\cdot}00063$. Then

$$a_0^4 = 1{\cdot}767 \times 10^{-6}/1{\cdot}00063 = 1{\cdot}766 \times 10^{-6}\,\text{cm.}^4$$

By (13), $$\eta = Kt/\lambda,$$
where
$$K = \frac{\pi P_0 a_0^4}{8lS} = \frac{\pi \times 1{\cdot}016 \times 10^6 \times 1{\cdot}766 \times 10^{-6}}{8 \times 64{\cdot}82 \times 9646} = 1{\cdot}127 \times 10^{-6}\,\text{dyne cm.}^{-2}.$$

Observations made during the flow of air gave the following results. The temperature was 13·8° C.

Time	Gauge readings		$P_1 - P_0$	$P_2 - P_0$	λ	t	$\eta = Kt/\lambda$
secs.	cm.	cm.	cm.	cm.		secs.	
0	36·26	13·80	22·46	—	0·710	120	$1{\cdot}90 \times 10^{-4}$
120	30·20	19·93	—	10·27			
0	33·81	16·28	17·53	—	0·727	120	$1{\cdot}86 \times 10^{-4}$
120	29·08	21·08	—	8·00			
0	31·01	19·12	11·89	—	0·749	120	$1{\cdot}81 \times 10^{-4}$
120	27·80	22·40	—	5·40			

These observations, which were made without any previous estimation of the critical value of $P_1 - P_0$, lead to values of η which increase with $P_1 - P_0$. We therefore examine the matter in the light of § 160. With $l = 64{\cdot}82$ cm. and $a = b = 3{\cdot}646 \times 10^{-2}$ cm., we have $f = 0{\cdot}555$ and $k = 18{\cdot}8$ cm. Hence $P_1 - P_0$, measured in cm. of mercury, should not exceed 18·8. We shall discard the first two values of η; then we have, for the viscosity of air at 13·8° C.,

$$\eta_{13\cdot8} = 1{\cdot}81 \times 10^{-4}\,\text{dyne cm.}^{-2}\text{ per unit rate of shearing}$$
$$= 1{\cdot}81 \times 10^{-4}\,\text{grm. cm.}^{-1}\,\text{sec.}^{-1}.$$

E. L. Harrington (*Physical Review*, Vol. VIII, p. 749 (1916)) used a rotating cylinder method, similar in principle to that of EXPERIMENT 24. He found that, for the viscosity of air at θ° C.,

$$\eta = 1{\cdot}8226 \times 10^{-4}\{1 + 2{\cdot}705 \times 10^{-3}(\theta - 23)\}\,\text{grm. cm.}^{-1}\,\text{sec.}^{-1}.$$

This formula gives $\eta_{13\cdot8} = 1{\cdot}7772 \times 10^{-4}$.

EXPERIMENT 24. Determination of viscosity of very viscous liquid by viscometer.

164. Introduction. The viscosity of treacle is so great that it cannot be measured conveniently by allowing the liquid to flow through a tube of small bore, as is done in the case of water, for at 12° C. the viscosity of treacle is about 400 C.G.S. units, and that of Lyle's "golden syrup" is about 1400, while that of water is only 0·0124. But the great viscosity of treacle allows the use, in a simple form, of another method, in which the mechanical actions called into play are more conspicuous than when a liquid flows through a tube.

In this method, a vertical cylinder, of known radius and length, is caused to rotate uniformly within a coaxial cylinder of known radius containing the viscous liquid, and the angular velocity of the inner cylinder is measured for a known value of the couple which turns that cylinder. The length of that part of the inner cylinder which is immersed in the liquid is observed. The theory shows how the viscosity may be deduced from the measurements.

In the elementary theory, the motion of the liquid between the vertical sides of the cylinders is assumed to be the same as if the cylinders were infinite in length. In practice, the lower end of the rotating cylinder is exposed to viscous action and thus a correction for that end becomes necessary. The apparatus shown in Fig. 99 was designed, with the assistance of Mr W. G. Pye, so that this correction can be readily found.

165. Elementary theory. Let the viscosity of the liquid between the cylinders be η dynes per square cm. per unit rate of shearing (i.e. one radian per sec.). Let the inner cylinder, of radius b cm., be made to rotate about its axis, with the uniform velocity Ω radians per sec., by a couple; the internal radius of the outer (fixed) cylinder is a cm. Let the angular velocity of the liquid at a distance r cm. from the axis be ω radians per sec., and at the distance $r+dr$ let it be $\omega+d\omega$. Let PQR, STU (Fig. 97) be arcs of circles of radii $OP=r$ and $OS=r+dr$, OPS, OQT being radii. Let

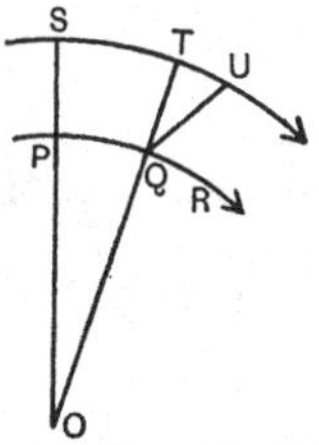

Fig. 97

$$POQ=\omega dt.$$

Then the particles which at time t are at P and S will at $t+dt$ be at Q and U, where $PQ=\omega r\,dt$ and

$$SU=(\omega+d\omega)(r+dr)\,dt=(\omega r+\omega\,dr+r\,d\omega)\,dt,$$

the product $d\omega\, dr$ being neglected. Since $ST = \omega(r + dr)\,dt$, we have $TU = SU - ST = r\,d\omega\,dt$, and thus the angle $UQT = r\,(d\omega/dr)\,dt$. At time t the line of particles along SP makes an angle $\frac{1}{2}\pi$ with the line of particles along PQ. At $t + dt$ the first set will lie along UQ and the second set along QR. The angle between the two lines has changed from $\frac{1}{2}\pi$ to $\frac{1}{2}\pi - UQT$, or to $\frac{1}{2}\pi - r\,(d\omega/dr)\,dt$. Thus, if R be the rate of shearing, $R = r\,d\omega/dr$. The tangential stress at the surface of the (geometrical) cylinder of radius r is η times the rate of shearing and so is $\eta r\,d\omega/dr$ dyne cm.$^{-2}$, and will tend to turn it round in the positive direction of ω, if $d\omega/dr$ be positive.

Since the outer metal cylinder is fixed, $d\omega/dr$ is negative, and therefore the tangential stress opposes the motion of the liquid within the cylinder of radius r.

Let the couple required to rotate a length h cm. of the inner metal cylinder, of radius b, with uniform angular velocity Ω, be G dyne cm. Since there is no angular *acceleration* of the liquid when the conditions are steady, the couple exerted by viscous action upon the liquid within a geometrical cylinder of radius r and length h is $-G$. This couple is obtained by multiplying the viscous stress by the area ($2\pi rh$) of the curved surface of that cylinder and by the radius r. Hence

$$-G = 2\pi rh\,.\,(\eta r\,d\omega/dr)\,.\,r = 2\pi\eta hr^3\,d\omega/dr, \quad \text{......(1)}$$

or

$$\frac{d\omega}{dr} = -\frac{G}{2\pi\eta h}\cdot\frac{1}{r^3}. \quad \text{....................(2)}$$

Since $\omega = 0$ when $r = a$, we have, on integration,

$$\omega = \frac{G}{4\pi\eta h}\left(\frac{1}{r^2} - \frac{1}{a^2}\right). \quad \text{....................(3)}$$

When $r = b$, the angular velocity is Ω, and so

$$\Omega = \frac{G}{4\pi\eta h}\left(\frac{1}{b^2} - \frac{1}{a^2}\right). \quad \text{....................(4)}$$

Hence

$$\eta = \frac{G\,(a^2 - b^2)}{4\pi\Omega h a^2 b^2}. \quad \text{........................(5)}$$

This result agrees with that found by the minimum heat method (see (13), § 149).

In practice, the couple G is produced by the weights of two loads, each of M grm., acting on a drum of effective diameter D cm. Hence $G = MDg$. The angular velocity is found by observing the time, nT sec., of n revolutions of the cylinder, and thus $\Omega = 2\pi/T$. The formula thus becomes

$$\eta = \frac{gD(a+b)(a-b)}{8\pi^2 a^2 b^2}\left(\frac{MT}{h}\right). \quad \text{..............(6)}$$

The method of obtaining the correction due to the finite length of the inner cylinder will be best explained after the description of the viscometer.

166. Stability of stream-line motion. In the investigation of § 165, the liquid is supposed to move steadily in circles about the axis of the cylinders. Since stream-line motion generally breaks down into turbulent motion when the rate of shearing of the liquid is sufficiently great, it would be natural to assume that turbulence will occur if the angular velocity, Ω, of the inner cylinder B relative to the fixed outer cylinder A, be increased above a critical value Ω_0, and this is found to be true in practice.

Instead of the system in which A is fixed and B rotates, a second system, in which A revolves and B is held at rest by a measured couple, has been used in determinations of viscosity. At first sight, it might seem that it is only the *relative* angular velocity Ω of one cylinder with respect to the other which concerns us, and we might expect that turbulence will occur when Ω becomes appreciably greater than the critical value Ω_0 which corresponds to the first case. There is, however, in actual fact, an important difference between the two cases. In the first case (outer cylinder at rest) there is, as G. I. Taylor has found, a definite angular velocity, Ω_0, for which the stream-line motion passes into turbulent motion. But in the second case (inner cylinder at rest) he found that the motion was non-turbulent for all the angular velocities of the outer cylinder that he was able to use. A system in which the outer cylinder revolves is therefore preferable to one in which the inner cylinder revolves. In the present experiment, however, the viscosity of the liquid is so great, and the relative angular velocity is so small, that the question of the possible instability of the stream-line motion need not be considered.

167. The viscometer. The instrument is shown diagrammatically in Fig. 98. The inner cylinder turns on pivots in bearings at top and bottom. The lower bearing is at the top of a cylindrical pillar rising from the base of the instrument, and the upper bearing is on an arm carried by a vertical rod. The axle supporting the inner cylinder also carries a drum, and the cylinder is caused to rotate by means of two masses supported by a thread wound round the drum and passing over two ball-bearing pulleys. The outer cylinder is secured to the pillar by a clamp acting on a partially split tube which makes a treacle-tight joint with the pillar. The top of the pillar carries a perforated baffle plate, loosely fitting the outer cylinder. The treacle or other liquid fills the space between the baffle plate and the bottom of the fixed cylinder and also part of the space between the two cylinders, as indicated in Fig. 98. It will be seen that the length of the inner cylinder which is immersed in the liquid will increase if the outer cylinder be made to move upwards on the pillar. The baffle plate prevents the rotation of the inner cylinder from causing any appreciable motion of the liquid in the space below the plate. For a given value of Ω, the distribution of velocity near the lower end of the inner cylinder is then practically independent of the position of the outer cylinder on the pillar, with the result that there is a *definite* "end effect."

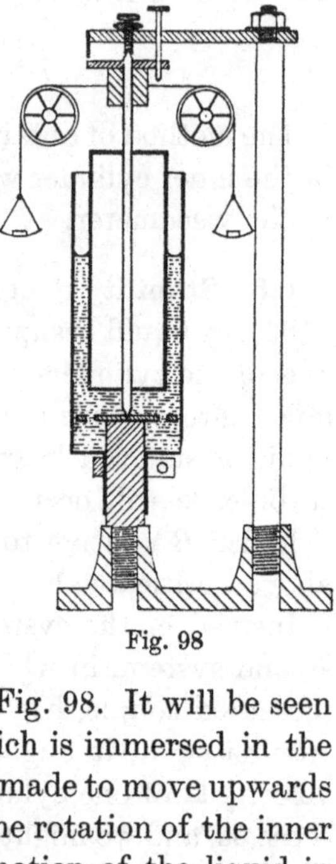

Fig. 98

The details of the apparatus are better seen in Fig. 99. A slot about 3 mm. wide is cut in the side of the fixed cylinder and is covered by a plate of glass cemented to the cylinder and secured by a brass plate. A scale of millimetres, engraved upon the inner cylinder, can be viewed through the glass plate, and thus the depth to which that cylinder is immersed in the liquid can be found. Above the drum, the axle carries a disk with a mark on its edge, by aid of which the revolutions of the cylinder are counted. In a later form, a locking pin, passing through the arm at the top of the instrument, as indicated in Fig. 98, enables the inner cylinder to be held fixed.

168. Experimental method. The outer cylinder is first clamped on the pillar in such a position that only a small length—a centimetre or so—of the inner cylinder is immersed in the liquid. If the level of the liquid have been *lowered*, it will be necessary to wait for a few minutes to allow the viscous liquid to drain off the sides of the cylinders. After the reading of the liquid on the scale engraved on the inner cylinder has been taken, that cylinder

Fig. 99

is turned round so as to wind the threads on to the drum. The cylinder is then secured by the locking pin until the observer is ready to determine the time occupied by some convenient number of revolutions. Equal masses are placed in the two pans; it is convenient if each pan be adjusted to some definite mass—say 10 grm. The time of a number of revolutions is then determined, and the time, T secs., for one revolution is deduced. If the corresponding total load hung from each thread be M grm., it will be found, on repeating the observation with various loads, that MT is constant for a given level of liquid. This result confirms the fundamental assumption that the viscous stress at each point is proportional to the rate of shearing of the liquid.

The outer cylinder is then clamped a little higher up on the pillar so as to raise the level of the liquid by about a centimetre, and the value of MT is again determined. This process is repeated with approximately equal steps in the level of the liquid so often as the construction of the instrument allows. The reading of the liquid on the inner cylinder is recorded at each stage.

The values of MT are then plotted, as in Fig. 100, against l, the length of the inner cylinder (measured from its lower end) immersed in the liquid. The diagram represents the observations recorded in § 169. It will be found that the points lie on a straight line, which cuts the axis of l at a distance from the origin on the

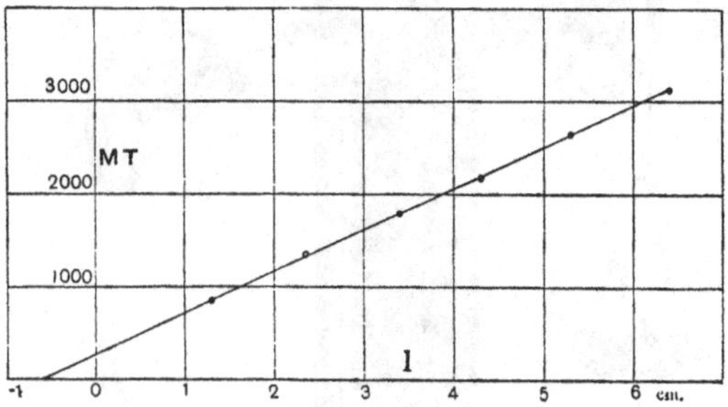

Fig. 100

negative side corresponding to a length k of something like half a centimetre. Hence, the couple required to rotate the inner cylinder at a given angular velocity is proportional to $l+k$, and thus k is the end correction for the inner cylinder. In the experiment recorded in § 169, the values of $MT/(l+k)$ were found to be nearly constant.

The other measurements are easily taken. The diameter of the inner cylinder is found by calipers. The radius of the outer cylinder is determined by finding the mass of water which fills it. For this purpose, the pillar is unscrewed from the base and is then drawn out of the outer cylinder; the hole at the bottom of the cylinder is plugged with a rubber stopper, the upper end being made flush with the floor of the cylinder. The *effective* diameter of the drum is found by adding the diameter of the thread to that of the drum itself.

The viscosity of very viscous liquids diminishes very rapidly as the temperature rises. Thus, for instance, the viscosity of treacle at 14° C. is less than half that at 7° C. Hence, care must be taken to keep the temperature of the apparatus as nearly constant as possible during the series of observations. When, however, the correction for the lower end of the inner cylinder has been found, a single observation is sufficient to determine the viscosity at the temperature of the liquid during the observation. The apparatus has been made compact so that it can be placed in a water bath for the purpose of bringing it to any desired temperature.

169. Practical example.

The viscosity of Lyle's golden syrup was determined. The temperature was 12·3° C. throughout.

Radius of outer cylinder (from mass of water contained) $= a = 2{\cdot}543$ cm.

Radius of inner cylinder $= b = 1{\cdot}865$ cm.

Effective diameter of drum $= D = 1{\cdot}88 + 0{\cdot}03 = 1{\cdot}91$ cm.

Mass of each pan $= 10$ grm.

To determine correction for lower end of inner cylinder, the following observations were made. The time for three complete revolutions was observed in each case.

Load M grm.	Time for one rev. T sec.	MT grm. sec.	Length in liquid l cm.	$l+k$ cm.	$\frac{MT}{l+k}$ grm. sec. cm.$^{-1}$
30	27·9	837	1·30	1·85	452·4
40	33·3	1332	2·35	2·90	459·3
60	29·9	1794	3·40	3·95	454·2
80	27·3	2184	4·30	4·85	450·3
90	29·5	2655	5·30	5·85	453·8
100	31·2	3120	6·40	6·95	448·9

When MT was plotted against l, as in Fig. 100, the resulting straight line indicated that k, the correction for the end, was 0·55 cm. The last column shows how nearly, with $k = 0{\cdot}55$, $MT/(l+k)$ approaches constancy. The mean value of $MT/(l+k)$ is 453·2 grm. sec. cm.$^{-1}$.

To test constancy of MT for a given depth of liquid, further observations were made with $l = 6{\cdot}40$ cm. or $l+k = 6{\cdot}95$ cm.

Load M grm.	Time for one rev. T sec.	MT grm. sec.
60	52·7	3162
80	39·2	3136
100	31·2	3120
120	26·0	3120
140	22·3	3122

The mean value of MT is 3132 grm. sec., agreeing closely with 3120, the former value for $M = 100$ grm. and $l = 6{\cdot}40$ cm. With $MT = 3132$, we find

$$MT/(l+k) = 3132/6{\cdot}95 = 450{\cdot}6 \text{ grm. sec. cm.}^{-1}.$$

Putting $l+k$ for h in (6), § 165, we find

$$\eta = \frac{gD(a+b)(a-b)}{8\pi^2 a^2 b^2}\left(\frac{MT}{l+k}\right)$$

$$= \frac{981 \times 1{\cdot}91 \times 4{\cdot}408 \times 0{\cdot}678}{8\pi^2 \times 6{\cdot}467 \times 3{\cdot}478} \times 450{\cdot}6$$

$$= 1421 \text{ dynes per square cm. per unit rate of shearing.}$$

At the same temperature of 12·3° C., the viscosity of black treacle is about 400, and that of water is about 0·0123 C.G.S. unit.

CHAPTER VIII

MATHEMATICAL DISCUSSIONS OF PROBLEMS IN CONDUCTION OF HEAT

170. Introduction. When the temperature of a body is not uniform throughout, surfaces can be drawn in the body such that the temperature is the same at all points on any one surface, different surfaces corresponding to different temperatures. Such surfaces are called isothermal surfaces. If the conditions be steady, the isothermal surfaces will be fixed in position. When the conditions change with the time, there will still be isothermal surfaces at any given instant, but their forms and positions will now depend upon the time.

Heat does not flow from a point on an isothermal surface to a neighbouring point on the same surface, and thus the lines of flow of heat are, in an isotropic body, normal to the isothermal surfaces. The rate, measured by thermal units per unit time, at which heat flows through any small area of an isothermal surface is, naturally, proportional to the area. It is also proportional to the temperature gradient, i.e. the change of temperature per unit distance along the normal to the area. The rate of flow, for a given area and a given temperature gradient, depends upon the substance and, to some extent, upon the temperature. On the C.G.S. system, rate of flow of heat is measured in (gramme) calories per second, area in square centimetres and temperature gradient in degrees centigrade per centimetre.

If we write

$$\frac{\text{Rate of flow of heat through area}}{\text{Area} \times \text{temperature gradient normal to area}} = k, \quad \ldots(1)$$

then k is called the thermal conductivity of the substance. A substance has unit thermal conductivity when the rate of flow of heat is one calorie per second per square centimetre per unit temperature gradient of one degree centigrade per centimetre.

Let θ be the temperature of the substance at any point, let n be measured along the normal to the isothermal surface, and let dh/dt

be the rate at which heat crosses unit area of the surface in the positive direction of n. Then

$$dh/dt = -k d\theta/dn. \qquad \text{......................(2)}$$

The negative sign occurs because the flow of heat is in the opposite direction to that in which the temperature increases.

The result represented by (2) can be put into a more general form. Let CD (Fig. 101), in the plane of the paper, be normal to a rectangular element "AB" of isothermal surface. Let "AB" be of length AB, and let its width at right angles to the plane ABD be dq. Let DB be normal to an element "AD" of geometrical surface of length AD and width dq. Let the angle BAD between "AB" and "AD" be ϕ. Then, since DC is perpendicular to AB, the angle CDB equals ϕ. Let the rate of flow of heat, per unit area, perpendicular to "AD" be dh'/dt. When the conditions are steady, the heat which crosses "AC" also crosses "AD," and we have

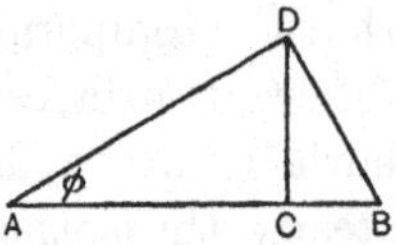

Fig. 101

$$AD \, . \, dq \, . \, dh'/dt = AC \, . \, dq \, . \, dh/dt,$$

or

$$dh'/dt = \cos\phi \, . \, dh/dt. \qquad \text{......................(3)}$$

The difference of temperature between B and D is the same as that between C and D. Hence, if $d\theta/dn$ and $d\theta/dn'$ be the temperature gradients perpendicular to AB and AD respectively,

$$DB \, . \, d\theta/dn' = CD \, . \, d\theta/dn,$$

or

$$d\theta/dn' = \cos\phi \, . \, d\theta/dn. \qquad \text{..................(4)}$$

Since $dh/dt = -k d\theta/dn$, we have, by (3) and (4),

$$dh'/dt = -k d\theta/dn'. \qquad \text{......................(5)}$$

The ratio of the rate at which the (third order) element "ADC," of volume $\frac{1}{2}AC.DC.dq$, absorbs heat, when the conditions are not steady, to the rate of flow of heat through "AC," a (second order) element of area, tends to zero as AC and dq tend to zero. Hence (5) remains true when the temperature changes with the time.

If we take rectangular axes of x, y, z, and if dh_1/dt, dh_2/dt,

dh_3/dt be the rates of flow of heat, per unit area, in the positive directions corresponding to the three axes, then, by (5),

$$\frac{dh_1}{dt} = -k\frac{d\theta}{dx}, \quad \frac{dh_2}{dt} = -k\frac{d\theta}{dy}, \quad \frac{dh_3}{dt} = -k\frac{d\theta}{dz} \quad \text{.......(6)}$$

The rate at which heat enters a rectangular volume of edges dx, dy, dz by the face for which $x=\xi$ is $-k\,dy\,dz\,.\,d\theta/dx$. Heat leaves it by the face for which $x=\xi+dx$ at the rate

$$-\left\{k\frac{d\theta}{dx} + \frac{d}{dx}\left(k\frac{d\theta}{dx}\right)dx\right\}dy\,dz.$$

Treating the other four faces in a similar way, we find that the resultant rate at which heat enters the volume is

$$\left\{\frac{d}{dx}\left(k\frac{d\theta}{dx}\right) + \frac{d}{dy}\left(k\frac{d\theta}{dy}\right) + \frac{d}{dz}\left(k\frac{d\theta}{dz}\right)\right\} dx\,dy\,dz.$$

If c be the water equivalent of unit of *volume* of the substance, the rate at which the heat within the volume increases is

$$c\,dx\,dy\,dz\,.\,d\theta/dt.$$

Hence $$\frac{d}{dx}\left(k\frac{d\theta}{dx}\right) + \frac{d}{dy}\left(k\frac{d\theta}{dy}\right) + \frac{d}{dz}\left(k\frac{d\theta}{dz}\right) = c\frac{d\theta}{dt}.$$

When the conductivity is constant,

$$\frac{d^2\theta}{dx^2} + \frac{d^2\theta}{dy^2} + \frac{d^2\theta}{dz^2} = \nabla^2\theta = \frac{c}{k}\frac{d\theta}{dt}. \quad \text{...............(7)}$$

When θ is independent of the time, $d\theta/dt=0$, and then, when k is constant,

$$\nabla^2\theta = 0. \quad \text{...........................(8)}$$

If θ_1, θ_2, ... be functions of x, y, z, such that $\nabla^2\theta_1=0$, $\nabla^2\theta_2=0$, ..., then their sum, $\theta_1+\theta_2+\ldots$, satisfies

$$\nabla^2(\theta_1+\theta_2+\ldots)=0. \quad \text{..................(8a)}$$

Thus the sum of any number of solutions of (8) is itself a solution.

171. Non-isotropic substances. In § 170 it was assumed that the substance is isotropic, i.e. has the same properties in all directions. But, in many crystals, the properties are not the same in all directions; in their case, as in that of other anistropic substances, it is not generally true that the lines of flow of heat are perpendicular to the isothermal surfaces. When heat flows out from a point into an isotropic substance, the isothermal surfaces are spheres; in a non-isotropic substance they are ellipsoids.

172. Difficulties of experimental work. The laws of the steady flow of heat are identical with those of the steady flow of electricity, in conducting substances, and the mathematical treatment is consequently the same for the two cases. But experiments in the conduction of heat are beset with difficulties from which experiments on electric conduction are practically free. When an electric current flows along a wire and the voltage between the wire and the surroundings is small, the air acts as a practically perfect insulator. Under suitable conditions, there is, it is true, a slight evaporation of metal from the surface of the wire under the influence of the electric current. Thus a small glass mirror may be silvered if it be placed in a vacuum near a silver wire carrying a current of a few amperes. No doubt the silver particles leaving the wire carry electric charges with them and so alter the distribution of current in the wire. A similar effect would be produced by coating the wire with radioactive material. Such effects can be enhanced, as in the thermionic emission of electrons from the electrodes of valves, but they do not appear unsought in experiments on electric conductivity.

It will generally be necessary to support the electrical conductor on solid bodies, but for such supports we can use insulating materials whose conductivity is extremely small compared with that of any metal.

In experiments on the conduction of heat many difficulties arise. Heat is conducted through the air surrounding a heated body, and this loss is greatly helped by the convection of the heated air. Then, again, heat passes from the body to neighbouring bodies by radiation. There are, certainly, solid substances of small thermal conductivity, such as cork, which may be termed thermal insulators, but they differ in thermal conductivity from metals far less than electrical insulators differ from metals in electrical conductivity.

Success in experiments on the conduction of heat largely depends upon a proper choice of conditions. We must, for instance, secure that the undesired, but inevitable, leakages of heat shall certainly be small compared with the heat currents which we measure.

173. Steady radial flow from axis. Let heat flow out radially from an axis so that the lines of flow are normal to the axis. Let the amount of heat flowing out be H calories sec.$^{-1}$ per cm. of the axis. At distance r from the axis, the temperature gradient is $d\theta/dr$, and hence the heat current is radial and has the value $-kd\theta/dr$ cal. sec.$^{-1}$ cm.$^{-2}$. The heat flowing through the curved surface of unit length of a coaxial cylinder of radius r is thus $-k(d\theta/dr).2\pi r$, and this equals H when the conditions are steady. Thus

$$\frac{1}{r}\frac{dr}{d\theta}=-\frac{2\pi k}{H}.$$

Hence

$$\log r=-2\pi k\theta/H+C, \qquad \text{.................(9)}$$

where C is a constant. If $\theta=\theta_a$ when $r=a$,

$$\log a=-2\pi k\theta_a/H+C. \qquad \text{..................(10)}$$

By (9) and (10),

$$\theta-\theta_a=-(H/2\pi k)\log(r/a). \qquad \text{...............(11)}$$

The logarithms are "natural" logarithms to base e.

174. Non-uniform flow in a rod. In the apparatus of § 184 the heat does not enter the bar entirely at one end but enters partly by the curved surface of the rod. The flow of heat is, therefore, not the same as that assumed in the elementary theory of § 183. It is, of course, clear that, if the length of the bar be very great compared with its diameter, the lines of flow will become more nearly parallel to the axis of the rod as the distance from the heated part increases, but one cannot, without mathematical investigation, estimate by how little the actual flow differs from the ideal uniform flow. As problems of this nature, and their analogues in electrical conduction, are of great practical importance, we give, in § 175, the solution for a special case. This solution, involving only simple mathematical functions, gives effective guidance in the design of apparatus.

175. A simple conduction problem. Let the plane $z=0$, in an infinite solid of uniform conductivity k, be divided into a set of squares by an infinite set of straight lines $x=\pm a$, $x=\pm 3a$, ... and by a set $y = \pm a$, $y = \pm 3a$, ..., the distance between consecutive lines of either set being $2a$. Let heat to the amount of J calories per cm. per sec. flow out from each of these lines. We suppose the heat to be absorbed in two planes $z = Z$ and $z = -Z$, where Z may be regarded as infinite.

The planes $x = \pm a$, ..., $y = \pm a$, ... are, by symmetry, planes of flow; thus no heat crosses them. Hence we shall not alter the flow of heat if we suppose the solid to be split up into bars of square section along these planes. Only half the heat from AB (Fig. 102), of length $2a$, goes into the bar of section $ABCD$; the other half goes into the bar of section $ABFE$. The rate at which heat enters the bar $ABCD$ is thus $\frac{1}{2} \times 8aJ$ or $4aJ$. Of this, half, viz. $2aJ$, travels along the bar in the positive direction of z and half travels in the opposite direction. Since $z = 0$ is a plane of flow, we may split the solid along that plane. If all the solid, except the bar of section $ABCD$ extending from $z = 0$ to $z = +\infty$, be removed, and if heat be supplied to this bar along $ABCD$, the circumference of the end, at the rate of $\frac{1}{4}J$ calories per cm. per sec., the temperature at any point x, y, z in this limited bar will be the same as that at x, y, z in the infinite solid, when heat is supplied to the solid at the rate of J along the two infinite sets of lines in the plane $z = 0$. The temperature θ_0, at the origin O, is, of course, supposed the same in each case.

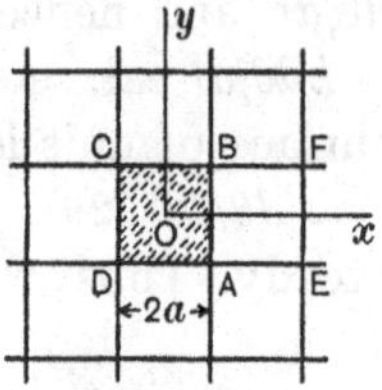

Fig. 102

From this system of heat flow, we can deduce an expression for the temperature at any point x, y, z of the solid. We assume throughout, in accordance with (8a), that, if the temperature at a point A exceed that at B by $\phi_1, \phi_2, \ldots$, when sources of heat $S_1, S_2, \ldots$ are separately in operation, the resultant excess, when the sources act simultaneously, is $\phi_1 + \phi_2 + \ldots$.

We will first find the temperature at a point P on Oz at a distance z from O. The distances of P from the lines $x = \pm a$, $x = \pm 3a$, ... in the plane $z = 0$ are

$$(a^2 + z^2)^{\frac{1}{2}}, \quad (3^2a^2 + z^2)^{\frac{1}{2}}, \quad \ldots.$$

The distances of O from these lines are a, $3a$, ... Similar results hold for the lines $y = \pm a$, ... If θ_1, θ_2 be the temperatures at distances r_1, r_2 from a straight line from which heat flows into an infinite solid at the rate J cal. cm.$^{-1}$ sec.$^{-1}$, then, by (11), § 173,

$$\theta_1-\theta_2=-\frac{J}{2\pi k}\log\left(\frac{r_1}{r_2}\right)=-\frac{J}{4\pi k}\log\frac{r_1^{\,2}}{r_2^{\,2}}.$$

Hence the difference between the temperature θ at P and that at O due to the flow J cal. cm.$^{-1}$ sec.$^{-1}$ from the single line AB, for which $x=a$, $z=0$, is given by

$$\theta-\theta_0=-\frac{J}{4\pi k}\log\frac{(a^2+z^2)}{a^2}=-\frac{J}{4\pi k}\log\left(1+\frac{z^2}{a^2}\right).$$

There are four lines, viz. $x=\pm a$ and $y=\pm a$, which are equidistant from P. Taking account of this, and adding the effects of all the other lines, we have, for the temperature on the axis Oz,

$$\theta-\theta_0=-\frac{J}{\pi k}\log\left\{\left(1+\frac{z^2}{a^2}\right)\left(1+\frac{z^2}{3^2a^2}\right)\left(1+\frac{z^2}{5^2a^2}\right)\cdots\right\}.$$

The infinite product can be expressed as a hyperbolic cosine and then

$$\theta-\theta_0=-\frac{J}{\pi k}\log\cosh\frac{\pi z}{2a}. \quad \text{..............(12)}$$

To avoid subsequent confusion we will now restrict z to positive values.

Since $\cosh\beta=\frac{1}{2}(e^{\beta}+e^{-\beta})=\frac{1}{2}e^{\beta}(1+e^{-2\beta})$, we have

$$\theta-\theta_0=-\frac{J}{\pi k}\log\{\tfrac{1}{2}e^{\pi z/2a}(1+e^{-\pi z/a})\}$$

$$=-\frac{J}{\pi k}\left\{\frac{\pi z}{2a}-\log 2+\log(1+e^{-\pi z/a})\right\}.$$

Since $|e^{-\pi z/a}|<1$, we can expand the logarithm, and then

$$\theta-\theta_0=-\frac{J}{\pi k}\left\{\frac{\pi z}{2a}-\log 2+e^{-\pi z/a}-\tfrac{1}{2}e^{-2\pi z/a}+\tfrac{1}{3}e^{-3\pi z/a}-\dots\right\}. \quad \text{......(13)}$$

For the temperature gradient on the axis Oz, we have, by (12),

$$\frac{d\theta}{dz}=-\frac{J}{2ak}\frac{\sinh\pi z/2a}{\cosh\pi z/2a}=-\frac{J}{2ak}\left\{1-\frac{2e^{-\pi z/a}}{1+e^{-\pi z/a}}\right\},$$

or, on expansion of the denominator,

$$\frac{d\theta}{dz}=-\frac{J}{2ak}\{1-2(e^{-\pi z/a}-e^{-2\pi z/a}+e^{-3\pi z/a}-\dots)\}. \quad \text{...(14)}$$

The following Table shows that $e^{-n\pi}$ diminishes very rapidly as n increases:

n	$\frac{1}{2}$	1	2	3	4
$e^{n\pi}$	4·81	23·1	535	12400	287000
$e^{-n\pi}$	$2{\cdot}08\times10^{-1}$	$4{\cdot}32\times10^{-2}$	$1{\cdot}87\times10^{-3}$	$8{\cdot}07\times10^{-5}$	$3{\cdot}49\times10^{-6}$

Since $e^{-2\pi}=1/535$, we see, by (14), that, when z is not less than $2a$, the temperature gradient does not differ from $-J/2ak$ by as much as one in 250; the discrepancy diminishes very rapidly as z increases.

The heat flowing along the limited bar of section $4a^2$ is $2aJ$. If the heat current were uniform across the section, we should have $d\theta/dz=-J/2ak$. We might have foreseen that the temperature gradient in the rod heated along $ABCD$ would approach this value as z increases.

In order to obtain the temperature at any point x, y, z of the bar, we must find $\theta-\theta_0$ as a function of x, y, z, such that θ satisfies (i) $\nabla^2\theta=0$, (ii) equation (13), if $x=y=0$, and (iii) the boundary conditions. The manner in which the heat is supplied to the solid leads us to expect that the expression for $\theta-\theta_0$, at any point x, y, z of the solid, will involve periodic trigonometrical functions of x and y. We have

$$\frac{d^2}{dx^2}\cos\frac{n\pi x}{a}=-\frac{n^2\pi^2}{a^2}\cos\frac{n\pi x}{a},\qquad \frac{d^2}{dy^2}\cos\frac{n\pi y}{a}=-\frac{n^2\pi^2}{a^2}\cos\frac{n\pi y}{a},$$

and
$$\frac{d^2}{dz^2}e^{-n\pi z/a}=\frac{n^2\pi^2}{a^2}e^{-n\pi z/a}.$$

Hence, if we write

$$\phi_n=e^{-n\pi z/a}(\cos n\pi x/a+\cos n\pi y/a),$$

we find that
$$\nabla^2\phi_n=\frac{d^2\phi_n}{dx^2}+\frac{d^2\phi_n}{dy^2}+\frac{d^2\phi_n}{dz^2}=0.$$

By comparison with (13), we see that the values of n which will occur are positive integers.

The following value of θ satisfies $\nabla^2\theta=0$, viz.

$$\theta=\theta_0-\frac{J}{\pi k}\left\{\frac{\pi z}{2a}-\log 2+\tfrac{1}{2}\left(\phi_1-\tfrac{1}{2}\phi_2+\tfrac{1}{3}\phi_3-\ldots\right)\right\}\ldots.(15)$$

On the axis of the rod, where $x=0$ and $y=0$, we have $\phi_n = 2e^{-n\pi z/a}$, and then the value of $\theta-\theta_0$ given by (15) reduces to that given by (13) for the axis Oz.

Equation (15) involves an infinite series, but can be put into a finite form as follows. We have

$$e^{-n\pi z/a}\cos n\pi x/a = \tfrac{1}{2}(e^{n\alpha}+e^{n\beta}),$$

where $\qquad \alpha=\pi(-z+ix)/a, \qquad \beta=\pi(-z-ix)/a,$

and hence, if

$$X = e^{-\pi z/a}\cos \pi x/a - \tfrac{1}{2}e^{-2\pi z/a}\cos 2\pi x/a + \ldots,$$

then $\qquad X=\tfrac{1}{2}\{e^{\alpha}-\tfrac{1}{2}e^{2\alpha}+\tfrac{1}{3}e^{3\alpha}-\ldots+e^{\beta}-\tfrac{1}{2}e^{2\beta}+\ldots\}.$

Since $\qquad |e^{n\alpha}| = e^{-n\pi z/a}|\cos n\pi x/a + i\sin n\pi x/a| = e^{-n\pi z/a},$

and since z is positive, $|e^{n\alpha}|<1$. Similarly $|e^{n\beta}|<1$. Hence the logarithmic series in e^{α} and e^{β} are absolutely convergent, and

$$X=\tfrac{1}{2}\log\{(1+e^{\alpha})(1+e^{\beta})\} = \tfrac{1}{2}\log\{2e^{-\pi z/a}(\cosh \pi z/a + \cos \pi x/a)\}.$$

Since $\cosh 2\xi = 2\cosh^2\xi - 1$ and $\cos 2\eta = 1-2\sin^2\eta$, we have

$$X = -\pi z/2a + \log 2 + \tfrac{1}{2}\log(\cosh^2 \pi z/2a - \sin^2 \pi x/2a).$$

For Y we replace x by y. Then, by (15),

$$\theta = \theta_0 - \frac{J}{\pi k}\left\{\frac{\pi z}{2a} - \log 2 + \tfrac{1}{2}(X+Y)\right\}$$

$$= \theta_0 - \frac{J}{4\pi k}\log\left\{\left(\cosh^2\frac{\pi z}{2a} - \sin^2\frac{\pi x}{2a}\right)\left(\cosh^2\frac{\pi z}{2a} - \sin^2\frac{\pi y}{2a}\right)\right\}. \qquad \ldots\ldots(16)$$

When $x=0$ and $y=0$, this value of θ reduces to that given by (12). When one, or both, of x and y equals a, the value of $\theta-\theta_0$ tends to infinity, as it should, when z tends to zero.

We know that θ must satisfy $\nabla^2\theta=0$ and that $d\theta/dx=0$ when $x=\pm a$, that $d\theta/dy=0$ when $y=\pm a$ and that $d\theta/dz=0$ when $z=0$, for these planes are planes of flow. It will be found that θ, as given by either (15) or its equivalent (16), satisfies these conditions, except, of course, when the point lies on a line of heat supply.

By (15),

$$\frac{d\theta}{dz} = -\frac{J}{2ak}\left\{1+\Sigma(-1)^n e^{-n\pi z/a}\left(\cos\frac{n\pi x}{a}+\cos\frac{n\pi y}{a}\right)\right\}. \qquad \ldots\ldots(17)$$

If the total flow of heat along the bar be F cal. sec.$^{-1}$,

$$F = -k\int_{-a}^{a}\int_{-a}^{a}\frac{d\theta}{dz}\,dx\,dy = 2aJ,$$

since the terms in $d\theta/dz$ involving cosines vanish on integration. Hence (15), or its equivalent (16), gives the correct flow along the bar.

To find the temperature gradient along one of the edges of the bar, we put $x=a$, $y=a$ in (17). Then

$$\frac{d\theta}{dz} = -\frac{J}{2ak}\{1+2(e^{-\pi z/a}+e^{-2\pi z/a}+e^{-3\pi z/a}+\ldots)\}. \quad \ldots(18)$$

This value of $d\theta/dz$ differs from the ideal $-J/2ak$ by less than one in 250 when z is not less than $2a$. The value tends to negative infinity when z tends to zero.

Along the edge, where $x=a$ and $y=a$, we have, by (15),

$$\theta_{\text{edge}} = \theta_0 - \frac{J}{\pi k}\left(\frac{\pi z}{2a} - \log 2 - e^{-\pi z/a} - \tfrac{1}{2}e^{-2\pi z/a} - \tfrac{1}{3}e^{-3\pi z/a} - \ldots\right). \quad \ldots\ldots(19)$$

Comparing (13) with (19), we find

$$\theta_{\text{edge}} - \theta_{\text{axis}} = \frac{2T}{\pi}(e^{-\pi z/a} + \tfrac{1}{3}e^{-3\pi z/a} + \ldots), \quad \ldots\ldots(20)$$

where $T = 2a\,.\,J/2ak = J/k$, so that T is the fall of temperature along a length $2a$ of the bar when z is great. When z is not less than $2a$, $\theta_{\text{edge}} - \theta_{\text{axis}}$ is less than $0{\cdot}00119T$.

When z tends to zero, the value of $\theta_{\text{edge}} - \theta_{\text{axis}}$ given by (20) tends to infinity.

CHAPTER IX

EXPERIMENTS IN HEAT

176. Temperature. Changes of temperature are measured by the changes they cause in material substances. Certain systems or arrangements of substances have very definite temperatures which are always reproduced when certain conditions are fulfilled, and these temperatures can be used as standards.

A mixture of pure ice and water in equilibrium under a pressure of one Atmosphere is such a system. The corresponding temperature, or the "ice point," is called 0° on the centigrade scale. An increase of pressure lowers the equilibrium temperature of the mixture, but only to the extent of 0·0072° C. per Atmosphere, and thus no very close specification of the pressure is necessary.

Another system is that of steam and water in equilibrium under the pressure of a Standard Atmosphere. The corresponding temperature, or the "steam point," is called 100° on the centigrade scale. Changes of pressure cause considerable changes in the steam point and a careful definition of the standard pressure is necessary. The Standard Atmosphere is defined as the pressure due to 76 cm. of mercury at 0° C. at sea level in latitude 45°. Under these conditions, the density of mercury is 13·5951 grm. cm.$^{-3}$ and $g = 980{\cdot}665$ cm. sec.$^{-2}$. Hence the Standard Atmosphere equals $76 \times 13{\cdot}5951 \times 980{\cdot}665$ or $1{\cdot}013250 \times 10^6$ dyne cm.$^{-2}$.*

When the temperature of a substance is neither 0° C. nor 100° C., the numerical value of that temperature depends upon the manner in which we connect it with the two standard temperatures of 0° C. and 100° C. defined by the ice point and the steam point. Scales of temperature are merely interpolation devices.

177. Centigrade scales of temperature. We can construct a scale of temperature in the following manner. Let F be the numerical measure of some physical quantity which is used as a

* The discovery of isotopes of hydrogen and of mercury may eventually lead to revised definitions of the Standard Atmosphere and of the Ice and the Steam Points. It will be necessary to specify the proportion of isotopes in the "mercury" and the "water" employed.

temperature indicator. Thus, F may be the resistance of a platinum wire or the pressure of a gas in a vessel of constant volume. Let F_0 and F_{100} be the values of F at 0° C. and 100° C. These temperatures of 0° C. and 100° C. depend, by definition, upon the properties of water and not upon the nature of the quantity whose measure is F. If the temperature on the centigrade scale of F be t, then t is defined by the equation

$$t = 100\,(F - F_0)/(F_{100} - F_0). \quad \dots\dots\dots\dots(1)$$

The volume of a given mass of water changes with the temperature, but we could not use that volume as a measure of temperature over a range including the point of maximum density. Near that point, there are two temperatures for the same volume, and, without further information, we could not tell which should be taken as the true temperature.

If the pressure of saturated water vapour were used as the indicator, we should have $F_0 = 4{\cdot}60$, $F_{100} = 760$ mm. of mercury, and should find $F = 36{\cdot}45$ mm. at the transition point of sodium sulphate decahydrate ($Na_2SO_4 . 10H_2O$). If the temperature at the transition point on the vapour pressure scale be $t°$,

$$t = \frac{100\,(36{\cdot}45 - 4{\cdot}60)}{760 - 4{\cdot}60} = 4{\cdot}216.$$

On the scale of a constant volume hydrogen thermometer the temperature would be 32·38°. If the resistance of a platinum wire were the indicator, the temperature would be 32·71°. The temperature on the resistance scale differs little, but that on the vapour pressure scale differs much, from that given by the gas thermometer.

We must be prepared for the disappointment of finding that (1) has only a limited range of reality. Thus a mercury in glass thermometer cannot be used when the glass is liquid or the mercury is frozen.

178. Mercury in glass thermometers. The readings of a mercury thermometer can be compared with those of a gas thermometer, and the necessary corrections can be found. This process is long and expensive and, consequently, those who use mercury thermometers have to rely generally upon a scale which depends only upon the mercury thermometer itself. Two substances are involved, viz. mercury and glass, and thus the action of the instrument depends not only upon the mercury but also upon the kind of glass employed.

The temperature on the scale of a mercury in glass thermometer is defined as follows:*

Let the whole thermometer be placed in melting ice and let a mark be made upon the stem at the level of the end of the thread of mercury. Then let the whole thermometer be placed in the steam of water boiling under a pressure of a Standard Atmosphere, and let a mark be made at the end of the thread. Let the ice point be marked 0 and the steam point 100 and let 99 other marks be made on the stem in such positions that all the spaces defined by successive marks are equal at any common temperature. Provided the glass be uniform, these marks will come on the same particles of glass whatever that common temperature may be. Then, the whole thermometer being at one temperature, that temperature is measured by the reading of the mercury upon the stem.

179. International temperature scale. The discussion in § 177 shows that the centigrade scale, as defined by the interpolation formula (1), depends upon the physical quantity employed. We saw also that the range of the scale may be limited. No system, depending upon a single physical quantity, has been found capable of practical application over the whole range of temperatures now attainable. The National Laboratories have now adopted a composite scale called the International Temperature Scale. This scale has no theoretical basis but it has been found on trial to conform closely to the ideal scale. The ideal scale is, of course, the thermodynamical scale, but it is not yet possible to make direct determinations of temperature on this scale. Except at very low temperatures, hydrogen, helium and nitrogen differ little from the "perfect gas" which would give the thermodynamical scale exactly. By experiments on these gases, it is possible to calculate the small difference between the value given by one of these gases and the value on the thermo-

* C. Chree, "Notes on thermometry," *Phil. Mag.* March 1898, p. 206.

dynamical scale of the same temperature. The accuracy of approach to the thermodynamical scale reached in this way by corrected gas thermometers is the utmost at present attainable.

The International Temperature Scale is as close to the thermodynamical scale as is at present possible. It depends upon the ice point and the steam point and upon four other temperatures which are to be realised in specified ways. The values assigned to these four temperatures have been ascertained by reference to gas thermometers, the readings of these thermometers being corrected so as to agree as nearly as possible with the thermodynamical scale. The latter is independent of the properties of any particular substance. These four temperatures are now defined, and to question their values is an act of treason against the International Temperature Scale. The highest part of the scale depends upon the value assigned to a constant in a radiation formula. The stretch of temperature from $-190°$ C. upwards to the very high temperatures of incandescent bodies is divided into four ranges. Range I is from $-190°$ C. to the ice point (0·000° C.), range II is from the ice point to 660° C., range III from 660° C. to the "gold point" (1063° C.), the temperature of equilibrium between solid and liquid gold under normal pressure, and range IV is from the gold point upwards.

For range II, a platinum resistance thermometer is used and the temperature, t, is deduced from the resistance R_t of the thermometer by the quadratic formula

$$R_t = R_0(1 + At + Bt^2). \quad \text{..................(2)}$$

The constants R_0, A, B are found by calibration at the ice point, the steam point (100·000° C.) and the "sulphur point," the temperature of equilibrium between liquid sulphur and its vapour, which, under standard pressure, is 444·60° C.

For range I, the temperature t is deduced from the resistance R_t of a platinum thermometer by the quartic formula

$$R_t = R_0\{1 + At + Bt^2 + C(t - 100)t^3\}. \quad \text{........(3)}$$

The constants R_0, A, B are determined as for range II, and the additional constant C is found by calibration at the "oxygen point," the temperature of equilibrium between liquid and gaseous oxygen, which, under standard pressure, is $-182·97°$ C.

For range III, the temperature t is deduced from the E.M.F. e of a platinum *versus* platinum-rhodium thermocouple. One junction is kept at 0° C. and the other is at the temperature t defined by the quadratic formula

$$e = a + bt + ct^2. \qquad \text{.......................(4)}$$

The constants a, b, c are determined at the freezing point of antimony, at the "silver point" (960·5° C.), the temperature of equilibrium between solid and liquid silver under normal pressure, and at the gold point. The freezing point of antimony is approximately 630·5° C., but, as it lies within range II, it is determined, for the particular sample of antimony employed in calibrating the thermocouple, by a platinum thermometer.

For range IV, the temperature t is determined in terms of the ratio of the intensity J_2 of monochromatic visible radiation, of wave-length λ cm., emitted by a black body at the temperature t, to the intensity J_1 of radiation, of the same wave-length, emitted by a black body at the temperature of the gold point. The formula to be used is

$$\log_e \frac{J_2}{J_1} = \frac{c_2}{\lambda} \left\{ \frac{1}{1336} - \frac{1}{t + 273} \right\}. \qquad \text{..............(5)}$$

The constant c_2 is taken as 1·432 cm. degrees. The formula is valid if $\lambda(t + 273)$ be less than 0·3 cm. degree. The temperature 1336 is the gold point temperature increased by 273.

For the formulae giving the temperatures of the oxygen, the steam and the sulphur points when the pressure differs from the standard value of 1,013,250 dyne cm.$^{-2}$, and for much other information, reference may be made to *The Units and Standards of Measurement employed at the National Physical Laboratory*, published (1929) by His Majesty's Stationery Office, price 1*s.* 0*d.* net.

EXPERIMENT 25. **Correction for the emergent column of a thermometer.**

180. Introduction. In the measurement of the temperature of a liquid by a mercury thermometer, it often happens that the whole thermometer is not immersed in the liquid. The bulb and a part of the stem containing mercury are in the liquid and the remainder of the stem is in the air. If part of the mercury column be above the liquid, its temperature and that of the corresponding part of the stem will, in general, differ from that of the liquid and, consequently, the reading will not be the same as when the whole thermometer is in the liquid. The necessary correction can be calculated in terms of the coefficients of expansion of mercury and of glass.

In case A (Fig. 103), let the whole thermometer be immersed in a liquid, or other medium, of temperature T°, and let the reading of the mercury on the stem be t scale divisions. In case B, let the thermometer be immersed in the liquid of temperature T° up to the mark t_1, and let the mercury read $t-x$ divisions. Let the mean temperature of the stem between the t_1 mark and the $t-x$ mark be T_2°.

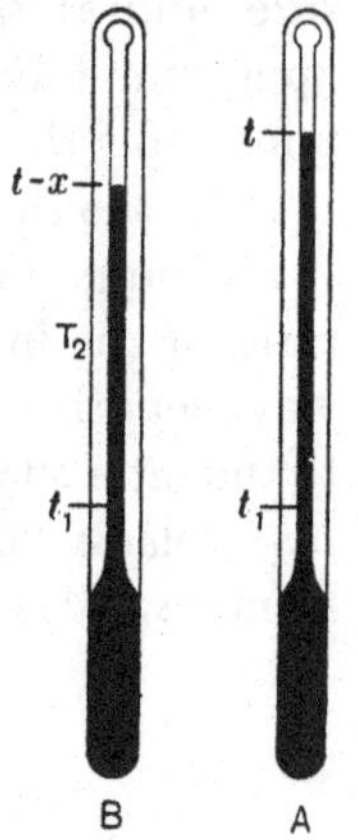

Fig. 103

Since the temperature, and therefore the volume, of the bulb and stem up to the t_1 mark are the same in both cases, the mass of mercury below this mark must also be the same. Hence, we need consider only the mercury above the t_1 mark.

Let v be the internal volume of one space of the stem, i.e. between consecutive marks, when the temperature of the stem is 0° C. Let α be the coefficient of cubical expansion of mercury and β the coefficient for glass.

In case B, the volume of a space is $v(1+\beta T_2)$, and $t-x-t_1$ spaces are filled with mercury. The volume at T_2° of the mercury above the t_1 mark is therefore

$$v(t-t_1-x)(1+\beta T_2),$$

and this equals $V(1+\alpha T_2)$, where V is the volume which the mercury above the t_1 mark would have, if cooled to 0° C. Hence

$$V=\frac{v(t-t_1-x)(1+\beta T_2)}{1+\alpha T_2} \quad \text{..................(1)}$$

In case A, the volume of a space is $v(1+\beta T)$, and $t-t_1$ spaces are filled with mercury at temperature T°. The volume of the mercury, when cooled to 0° C., is again V. Thus

$$V=\frac{v(t-t_1)(1+\beta T)}{1+\alpha T}. \quad \text{..................(2)}$$

By (1) and (2),

$$\frac{t-t_1-x}{t-t_1}=\frac{(1+\alpha T_2)(1+\beta T)}{(1+\alpha T)(1+\beta T_2)}.$$

Subtracting each side from unity, we obtain

$$\frac{x}{t-t_1}=\frac{(\alpha-\beta)(T-T_2)}{(1+\alpha T)(1+\beta T_2)}.$$

Since αT and βT_2 are small compared with unity, we may neglect them in the denominator. We then have

$$x=(\alpha-\beta)(T-T_2)(t-t_1). \qquad \text{..................(3)}$$

We can use (3) to find x when $\alpha-\beta$ is known.

In the experiment, $\alpha-\beta$ is deduced from the observed value of x by the formula

$$\alpha-\beta=\frac{x}{(T-T_2)(t-t_1)}. \qquad \text{..................(4)}$$

181. Method. In order that the correction x may be large enough to be measurable with any accuracy, t must be large compared with t_1 and T large compared with T_2. The most convenient high temperature is that of steam in a hypsometer. To hold the thermometer in position, a simple stuffing box or gland devised by Prof. E. G. Coker may be used. The lower, C (Fig.104), of two brass plates rests on the top of the hypsometer. The upper plate D can be brought nearer to C by the screws S, S. A ring of indiarubber, e.g. a short portion of a thick-walled tube, is placed between the plates, and the thermometer passes through the ring. When S, S are screwed down, the rubber packing is pressed against the thermometer and a steam-tight joint is made.

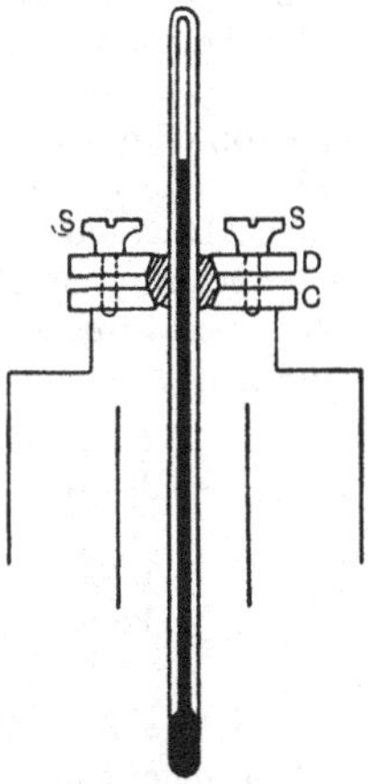

Fig. 104

The thermometer is first clamped in the gland so that as much as possible of the stem is in the air; the bulb, of course, is kept free of the gland. The mean of the stem readings of the top and bottom of the packing is taken as t_1. To find the temperature of the exposed part of the stem, two auxiliary thermometers—say G and H—are used. The bulb of G is placed near the top of the mercury column of the main thermometer and the bulb of H is placed near its stem and a few centimetres above the gland. The mean of the temperatures shown by G and H is taken as the mean temperature, T_2, of the stem. The reading of the main thermometer when conditions have become steady is $t-x$.

The thermometer is now pushed down through the packing until the top of the mercury column only just appears above the gland. The reading is now t. The hypsometer should be tall enough to ensure that the bulb of the thermometer is not splashed by the boiling water.

The difference between the coefficients of expansion is then found by (4).

For thermometers of good quality, Jena 16‴ glass is generally used. For this glass $\beta=0{\cdot}0000243$. When β is known α can be found by (4).

182. Practical example.

Readings by Mr L. Bairstow.

Reading of top of rubber packing 5·5, of bottom 0·0. Mean 2·8. Hence $t_1=2{\cdot}8$ scale divisions.

Temperature of air near stem was measured by thermometers G, H, of which G was near the 100 mark on thermometer under test and H was near top of hypsometer. Mean of temperatures T_G and T_H shown by G and H respectively is T_2. Hypsometer temperature T was assumed to be 100° C. It would have been better practice to calculate T from the barometer reading. Reading of top of thread is $t-x$ divisions.

Temperatures T_G and T_H of G and H were unsteady. Table gives some values.

$t-x$ scale divs.	T_G ° C.	T_H ° C.	T_2 ° C.
98·6	18·0	26·3	22·15
98·6	16·2	23·6	19·90
98·6	17·0	25·2	21·10
98·6	19·4	26·9	23·15

Mean $T_2=21{\cdot}6°$ C. Hence $T-T_2=100-21{\cdot}6=78{\cdot}4°$.

Reading of thread when stem was almost completely in steam in hypsometer $=t=99{\cdot}8$ divisions.

Hence $$x=t-(t-x)=99{\cdot}8-98{\cdot}6=1{\cdot}2 \text{ divisions},$$

and $$t-t_1=99{\cdot}8-2{\cdot}8=97{\cdot}0 \text{ divisions}.$$

By (4), § 180, $$\alpha-\beta=\frac{x}{(T-T_2)(t-t_1)}=\frac{1{\cdot}2}{78{\cdot}4\times 97{\cdot}0}$$

$$=1{\cdot}578\times 10^{-4} \text{ c.c. per c.c. per deg. C.}$$

For thermometer glass, $\beta = 0{\cdot}243 \times 10^{-4}$, and hence, for mercury,

$$\alpha = 1{\cdot}578 \times 10^{-4} + 0{\cdot}243 \times 10^{-4} = 1{\cdot}821 \times 10^{-4} \text{ c.c. per c.c. per deg. C.}$$

To two figures, $\alpha - \beta = 1{\cdot}6 \times 10^{-4}$ and $\alpha = 1{\cdot}8 \times 10^{-4}$.

The experiments of Callendar on the expansion of mercury show that the mean coefficient of expansion over the range 0° to 100° C. is $1{\cdot}820 \times 10^{-4}$ approximately.

Experiment 26. Determination of thermal conductivity of copper.

183. Introduction. Let a straight bar of metal of uniform cross-section A cm.^{2} be heated at one end and cooled at the other. We suppose that a steady state has been reached, and that no heat enters or leaves the bar except at the ends, where it enters and leaves in such a way that the heat current is uniform both across any transverse section and also along the length of the bar. The lines of flow of heat will then be parallel to the axis of the bar and the isothermal surfaces will be planes cutting the axis at right angles. If H calories of heat pass along the bar per sec., the heat flow is H/A calories per sec. per square cm. If k, the conductivity, be constant, and therefore independent of the temperature, it follows, from § 170, that the temperature gradient is constant. Hence, if θ_1, θ_2 be the temperatures in two transverse and isothermal planes U, V at distance l cm. apart, the temperature gradient between U and V is $(\theta_1 - \theta_2)/l$ deg. cm.$^{-1}$. The heat will flow from U to V, if $\theta_1 > \theta_2$. By (1), § 170, the conductivity is given by

$$k = \frac{H}{A(\theta_1 - \theta_2)/l} = \frac{Hl}{A(\theta_1 - \theta_2)}. \quad \ldots\ldots\ldots\ldots(1)$$

We can therefore determine k, if we know l and A and can measure H and the two temperatures θ_1 and θ_2.

184. Apparatus. The principle of continuous flow calorimetry is employed. One end of a stout bar of copper is heated by steam; the other end is kept cool by a stream of water flowing through a narrow copper tube soldered to that end of the bar. Means are provided (i) for determining the temperatures at two points on the bar between the steam jacket and the cooling tube and (ii) for measuring the temperatures of the inflowing and outflowing water. When, in addition, the mass of water passing through the tube per second is known, the conductivity of the copper can be found.

The apparatus is shown diagrammatically in Fig. 105. The copper bar XY is about 20 cm. long and about 2·54 cm. in diameter. To the end X is fixed a steam jacket in the form of a small copper cylinder, closed at the ends, and steam from a small boiler

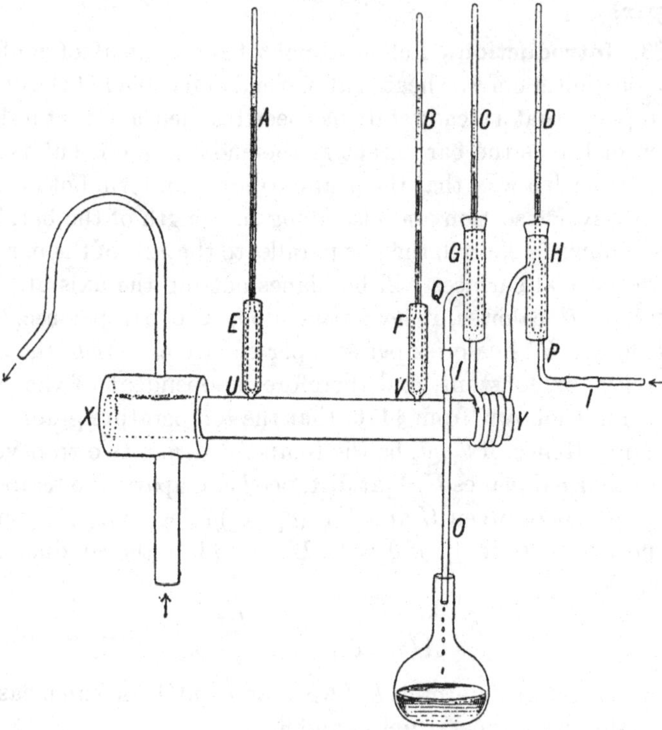

Fig. 105

passes through this cylinder. It is convenient to place the boiler below the level of the jacket and to make the steam pipe large enough to allow the water formed by condensation, about 3 or 4 grm. per min., to find its way back to the boiler; the steam pipe should be nearly vertical. The surplus steam passes out by a small escape pipe at the top of the steam jacket. A copper tube 2 mm. in internal diameter is coiled four times round the end Y and is soldered to the bar; this tube conveys the cooling water.*

The temperatures of the water before and after passing through the coiled tube are found by the thermometers D and C, which are fixed into the small copper vessels H and G by indiarubber stoppers. Water enters H by the tube P and leaves G by the tube Q; the necessary connexions are completed by the rubber joints I, I. The necessary *steady* flow of water may be obtained by aid of a small reservoir containing water, which is kept just overflowing by a stream from a tap. The inflow pipe attached to P may dip like a syphon into the reservoir or may be joined to a tube fixed to the reservoir. A Marriotte's bottle may be used as an alternative. The flow of water can be varied by changing the head, i.e. the difference of level between the surface of the water in the reservoir (or the lower end of the inflow tube of the Marriotte's bottle) and the end of the outflow tube. The head need not exceed one metre. If the head be kept constant, the flow will be constant when a steady thermal state has been reached. The viscosity of water varies rapidly with the temperature, and consequently the resistance to the flow will not be constant until the temperature of the water flowing from Q have become constant.

When the water is first drawn from the tap, the temperature of the water in the pipes near the tap may differ from that of the water in the mains, and then some time may elapse before the water flowing into the thermometer pocket H reaches a steady temperature.

To determine the temperature gradient along the part of the bar between the steam jacket and the cooling tube, we must know the temperatures at two points on the bar separated by a measured interval. These temperatures are found by means of the thermometers A, B, which fit loosely into holes bored in the stout copper rods E, F nearly to the lower ends of the rods. The lower ends of these rods are cut away so as to fit into saw cuts made in the bar at U and V, and the rods are secured to the bar by

* The bar used in the author's class is about 1·0 inch (2·54 cm.) in diameter. In the apparatus supplied by Messrs W. G. Pye and Co., the rod is 1·5 inches in diameter and the internal diameter of the cooling tube is 3·5 mm. The bar is made thicker to diminish the effects of the unavoidable errors. The greater bore of the cooling tube renders it less easily blocked.

brazing. If no heat escape from the copper rods, the thermometers in the holes bored in them will indicate the temperatures of the bar at U and V. The distance UV is about 10 cm. The isothermal surfaces passing through U and V are supposed to be plane. The theory of § 175 shows that these surfaces will differ very little from planes, if both the distance of U from the steam jacket and the distance of V from the cooling tube be not less than the diameter of the cross-section of the bar.

In the theory of the experiment, it is assumed that all the heat flowing across the section of the bar at U flows also across the

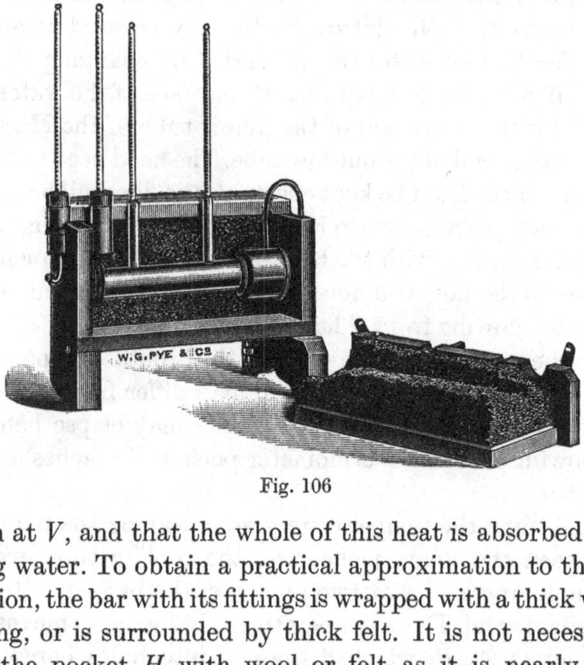

Fig. 106

section at V, and that the whole of this heat is absorbed by the cooling water. To obtain a practical approximation to this ideal condition, the bar with its fittings is wrapped with a thick woollen covering, or is surrounded by thick felt. It is not necessary to cover the pocket H with wool or felt as it is nearly at the atmospheric temperature. For convenience, the bar, when protected by wool or felt, is placed in a wooden box provided with feet. Fig. 106 shows the apparatus as supplied by W. G. Pye and Co.

On account of the high conductivity of copper and the large cross-section of the bar, the heat conducted along the bar is very large compared with that which escapes altogether from the bar and with the heat which passes from one end of the bar to the other by conduction through the protective covering.

The thermal conductivity of pure copper is very high and $k = 1{\cdot}0$ approximately. But all "copper" is not pure. Some impure copper has an electrical conductivity only one third of that of the pure metal. For impure copper, the thermal conductivity, k, may drop to 0·5 or less. To protect themselves, instrument makers should obtain, from the copper merchants, certificates of the purity of the metal.

185. Method. Let a vessel held under the exit tube O from Q (Fig. 105) catch M grm. of water in t sec. The water is, of course, allowed to flow continuously, and the time is taken from the instant when the vessel is brought up to receive the water. If a 250 c.c. flask be used, it is withdrawn when the water reaches the graduation on its neck.

Let the temperature of the water in the pocket H be θ_4 and that of the water in G be θ_3, when the conditions have become steady. The stream of water (M grm. in t sec.) abstracts heat, at the rate of H calories per sec., from the end Y of the bar. Then

$$H = Mc\,(\theta_3 - \theta_4)/t \text{ calories sec.}^{-1}, \quad \text{...........(2)}$$

where c is the specific heat of water; in the experiment, c is taken to be unity.

Let the cross-section of the bar be A cm.^{2}, and let the distance between the centres of the saw cuts at U and V be l cm. Let the temperatures at U and V, when the conditions are steady, be θ_1 and θ_2. Since no heat escapes from the bar, H calories are conducted along the bar per sec. Hence, by (1), § 183, we have, for the conductivity,

$$k = Hl/\{A\,(\theta_1 - \theta_2)\}, \quad \text{...................(3)}$$

or, by (2),

$$k = \frac{Mcl}{At} \cdot \frac{\theta_3 - \theta_4}{\theta_1 - \theta_2} \cdot \quad \text{.......................(4)}$$

Observations should be made with two or three different values of the water flow.

Four separate thermometers are used in the experiment. Each may have errors of scale as well as an error of zero and thus, apart from errors due to the "emergent columns," the difference of readings of the thermometers C and D (Fig. 105) does not necessarily give the difference of temperature between the water in the pocket G and that in H. When time allows, the observer may compare the readings of each of the four thermometers with those of a standard thermometer whose corrections have been determined at the National Physical Laboratory. The thermometers are placed in a bath of water which is gradually heated. The standard thermometer is immersed at least up to the level of the mercury in its stem. For each of the others, so much of the stem should be in the air as was the case when it was fixed in the conductivity apparatus. In this way, the correction due to the emergent column is practically avoided.

Errors of zero and those due to the emergent columns are likely to be more serious than errors of scale. It may thus suffice to compare thermometers A and B in a bath whose temperature is approximately $\frac{1}{2}(\theta_1+\theta_2)$, where θ_1 and θ_2 are the actual readings of A and B in the experiment. So much of the stem of each thermometer is exposed to the air as was the case when it was used in the apparatus. If the readings of A and B be now T_A and T_B, we may take the true excess of temperature of U over V to be $\theta_1-\theta_2-T_A+T_B$. The corrections for the emergent columns of C and D are comparatively small. We can find the relative error T_C-T_D by passing a stream of water at a constant temperature through the cooling tube *before* the bar is heated, *provided* the initial temperature of the bar equal that of the water. If this condition cannot be satisfied, C and D must be compared in a bath.

186. Practical example.

G. F. C. Searle used bar of copper about one inch in diameter. Radius of section = 1·295 cm.; area of section $=A=5{\cdot}269$ cm.^{2} Distance between axes of sockets for thermometers A, B (Fig. 105) $=l=9{\cdot}95$ cm.

Before steam was admitted to steam jacket, an attempt was made to compare C and D without removing them from their pockets. While water from the reservoir passed through the tube, pairs of readings of C, D were taken at intervals of a few minutes; these gave (C) 19·2°, 18·6°, 18·4°, (D) 19·0°, 18·3°, 18·1°, and thus C read about 0·3° higher than D.

At end of experiment, C and D were removed from their pockets and were compared in a bath at about 20°; C now read 0·05° lower than D.

The discrepancy is due to the fact that the bar was initially hotter than the water, which abstracted heat from the bar, and so was hotter on leaving than on entering the cooling tube. Flow of water was 250 grm. in about 150 sec., and rise of 0·3° corresponds to 0·5 cal./sec. Water equivalent of bar, cooling tube and steam jacket is about 100 grm. If, initially, bar be 5° hotter than water from reservoir, it can part with 500 calories and this heat will last for 1000 sec. at a steady rate of loss of 0·5 cal./sec. It is evident that, unless the initial temperature of the bar be very near to that of the water from the reservoir, no accurate comparison of *C* and *D*, while they are in their pockets, is possible.

When thermometers *A*, *B* were immersed in a bath, their readings agreed.

When water was first turned on, reading of *D* was about 19°. Reading gradually fell. From 11.0 A.M. to 11.25 it was about 17·8°; at 11.30 it was 17·2°. The changes are due to variations of temperature of water supply.

Four sets of readings of thermometers were made.

Set	Time		$A\ (\theta_1)$	$B\ (\theta_2)$	$C\ (\theta_3)$	$D\ (\theta_4)$	Time for 250 c.c.
(1)	11 hr.	35 min.	80·1°	47·3°	27·1°	17·0°	
	11	40	80·1	47·4	26·9	16·6	158·2 sec.
(2)	11	45	80·1	47·4	27·2	16·85	
	11	50	80·1	47·6	27·4	17·0	157·3
(3)	12	10	80·5	48·3	28·5	17·2	
	12	15	80·5	48·3	28·45	17·2	172·0
(4)	12	16	80·5	48·3	28·55	17·2	
	12	20	80·5	48·4	28·65	17·2	174·2

In set (1) water was collected in 250 c.c. flask in interval between 11.35 and 11.40; time occupied was 158·2 sec. In sets (1), (2), reservoir was higher and flow greater than in (3), (4).

Water at about 28° C. filling 250 c.c. flask to mark weighed 249·4 grm. against brass weights. Barometer was about 760 mm. and temperature of air about 22° C. Density of air 0·0012 grm. cm.$^{-3}$. Density of weights 8·5. At 28° C., density of water is 0·996. If m = mass of water,

$$m\,(1-0{\cdot}0012/0{\cdot}996)=249{\cdot}4\,(1-0{\cdot}0012/8{\cdot}5),$$

and $m=249{\cdot}7$ grm.

In set (1), mean readings of *A*, *B*, *C*, *D* were $\theta_1=80{\cdot}1°$, $\theta_2=47{\cdot}35°$, $\theta_3=27{\cdot}0°$, $\theta_4=16{\cdot}8°$. Allowing for difference (0·05°) between *C* and *D*, as found above, we have

$$\theta_1-\theta_2=80{\cdot}1-47{\cdot}35=32{\cdot}75°, \qquad \theta_3-\theta_4=27{\cdot}0+0{\cdot}05-16{\cdot}8=10{\cdot}25°.$$

We take the specific heat of water to be unity; thus $c=1$. Then, by (4), § 185, we have, for the conductivity,

$$k=\frac{249{\cdot}7\times 9{\cdot}95\times 10{\cdot}25}{5{\cdot}269\times 158{\cdot}2\times 32{\cdot}75}=0{\cdot}933.$$

Sets (2), (3), (4) gave 0·959, 0·964, 0·964. Mean of four values is

$k=0{\cdot}955$ calories per sec. per sq. cm. per unit temperature gradient. The unit gradient is 1° C. per cm.

Correction for emergent columns. Stems of *A*, *B* were in air, of temperature 22° C., from mark 10° upwards and those of *C*, *D* from mark −2°

upwards. These estimates are, of course, crude. If we use (7), § 186 A, to correct the value of k for set (1) for emergent column, we have $t_1 = 10$, $t_2 = -2$. Correcting factor in § 186 A is

$$1 - 1{\cdot}58 \times 10^{-4} (80{\cdot}1 + 47{\cdot}35 - 27{\cdot}0 - 16{\cdot}8 - 10 - 2) = 1 - 0{\cdot}0113.$$

Then, for set (1), $k = 0{\cdot}933 (1 - 0{\cdot}0113) = 0{\cdot}923.$

Corrected values for sets (2), (3), (4) are 0·948, 0·953, 0·953. The mean of the four is 0·944.

186 A. Correction for emergent columns.

Let the stem of each of two similar thermometers A, B, from the mark t_1 upwards, be in air at temperature T. Let the temperatures of the bulbs and stems of A, B below the mark t_1 be T_A, T_B. Let the mercury stand at t_A, t_B when the whole of A is at temperature T_A and the whole of B is at T_B. Then, by (3), § 180, the stem readings, when the stems above the mark t_1 are in air, will be θ_A, θ_B, where

$$\theta_A = t_A - x_A, \qquad \theta_B = t_B - x_B,$$

and $$x_A = \mu (T_A - T)(t_A - t_1), \qquad x_B = \mu (T_B - T)(t_B - t_1).$$

Here $\mu = \alpha - \beta$, where α, β are the coefficients of cubical expansion of mercury and glass. On the centigrade scale, $\mu = 1{\cdot}58 \times 10^{-4}$ approximately. When the thermometers are accurately graduated and are free from zero errors, $t_A = T_A$ and $t_B = T_B$. Then

$$x_A - x_B = \mu (T_A - T_B)(T_A + T_B - T - t_1),$$

and $$\theta_A - \theta_B = (T_A - T_B)\{1 - \mu (T_A + T_B - T - t_1)\}. \qquad \text{.........(5)}$$

If we have two other thermometers C, D with stems above mark t_2 in air at temperature T, with bulbs and stems below t_2 at temperatures T_C, T_D, and reading θ_C, θ_D, we have

$$\theta_C - \theta_D = (T_C - T_D)\{1 - \mu (T_C + T_D - T - t_2)\}. \qquad \text{.........(6)}$$

In the determination of the conductivity of copper, we require the ratio $(T_C - T_D)/(T_A - T_B)$. We have, by (5) and (6),

$$\frac{T_C - T_D}{T_A - T_B} = \frac{\theta_C - \theta_D}{\theta_A - \theta_B} \cdot \frac{1 - \mu (T_A + T_B - T - t_1)}{1 - \mu (T_C + T_D - T - t_2)}.$$

In the quantities multiplied by the small factor μ it will suffice to write θ_A for T_A, etc. Then, by binomial theorem, as far as the first power of μ,

$$\frac{T_C - T_D}{T_A - T_B} = \frac{\theta_C - \theta_D}{\theta_A - \theta_B}\{1 - \mu (\theta_A + \theta_B - \theta_C - \theta_D - t_1 + t_2)\}. \qquad \text{.........(7)}$$

EXPERIMENT 27. Determination of thermal conductivity of rubber.

187. Introduction. The apparatus is arranged as in Fig. 107. Steam from the small boiler A flows through the tube of rubber PP, and finally escapes into the vessel C. The tube passes through three rings of wire fixed inside the copper calorimeter B. The two screens S, S' prevent radiation from A or C from reaching B. The calorimeter contains water, which is kept in motion by a stirrer; the temperature of the water is observed by a thermometer divided to fifths of a degree. Two marks E, F are made on the tube, and the distance between them is measured while the tube is straight, i.e. before it is passed through the rings in the calorimeter. The length of tube immersed in the water can then be found by measuring the distances of E and F from the surface of the water when the tube is in place.

The calorimeter is dried and weighed. The rubber tube is then passed through the rings, care being taken that it does not touch the sides of the calorimeter. A beaker containing water, which has

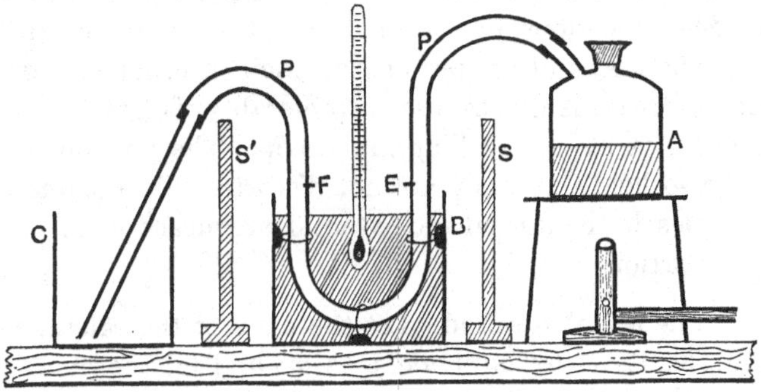

Fig. 107

been cooled five or six degrees below the temperature of the room, is weighed. Water is poured from the beaker into the calorimeter until the latter is one-half or two-thirds full, and the beaker is then weighed again; the difference gives the mass of water poured into the calorimeter.

While these operations are in progress, the water in the boiler is heated. After the cooled water has been poured into the calorimeter, the rubber tube is connected to the boiler. When the steam issues freely into the vessel C, the water in B is stirred continuously, and the thermometer is read at intervals of one minute. The observations are continued until the temperature has risen four or five degrees above that of the room.

A curve is then plotted showing how the temperature of the water in the calorimeter depends upon the time. When its temperature is equal to that of the room, the calorimeter neither loses heat to the room nor gains heat from it. Hence, if we neglect the radiation to the calorimeter from the parts of the rubber tube which are not immersed in the water, we may consider that the rate at which the calorimeter gains heat, when it is at the temperature of the room, is equal to the rate at which heat passes by conduction through the part of the tube immersed in the water. The rate of rise of temperature in degrees *per second*, when the calorimeter is at the temperature of the room, is easily found by drawing a tangent to the curve at the proper point.

The ends of the rubber tube, where it is attached to the pipe from the boiler and to the waste-pipe, soon become permanently stretched and are, therefore, useless for the purpose of measurement. Short portions, one from each end, are cut from the tube before it is used and are preserved. Their internal and external diameters are measured by a scale, preferably of glass, laid against their ends, or by a travelling microscope. The portions of tube must be so held that they are not distorted. Two diameters at right angles, both internal and external, are measured at each end of each portion.

188. Theory of method. Let the mass of the calorimeter be m grm., let its water equivalent be mc and let the mass of the water in the calorimeter be M grm.; for copper $c = 0{\cdot}095$. If the temperature of the water rise at the rate of R degrees per *second* when the calorimeter is at the temperature of the room, and if the system of calorimeter and water then gain heat by conduction through the tube at the rate H, we have

$$H = R\,(M + mc) \text{ calories per sec.} \qquad \text{...........(1)}$$

The capacity for heat of the rubber tube is small compared with that of the calorimeter and water, and the fact that the temperature of the water is not steady, but rises slowly, will not affect appreciably the rate at which heat enters the water by conduction for a given difference of temperature between the surfaces of the tube.

Since the diameter of the tube is small compared with the radius of curvature of the axis of the tube where it is immersed in the water, we may treat the tube as if it were straight. Hence, if

θ be the temperature at any point of the tube at distance r from the axis, and if k be the thermal conductivity, the radial flow of heat is $-k\,d\theta/dr$ calories per sec. per unit area of a cylinder of radius r. If the length of the tube in water be l, the curved surface of this length of cylinder is $2\pi rl$, and the rate at which heat flows outwards through it is $-2\pi rlk\,d\theta/dr$ cal. sec.$^{-1}$. Since the temperature is, to a sufficient approximation, steady, this flow is the same for all values of r and equals H. Hence

$$H = -2\pi rlk\,d\theta/dr,$$

and thus
$$(1/r)\,dr/d\theta = -2\pi lk/H.$$

Hence $\log_e r = -2\pi lk\theta/H + C$, where C is a constant. If a and b be the external and internal radii of the tube, and if θ_a and θ_b be the temperatures of the corresponding surfaces, we have

$$\log_e a = -2\pi lk\theta_a/H + C, \qquad \log_e b = -2\pi lk\theta_b/H + C,$$

and thus
$$\theta_b - \theta_a = \frac{H}{2\pi lk}\log_e\frac{a}{b}.$$

Hence, by (1),
$$k = \frac{R(M+mc)}{2\pi l(\theta_b-\theta_a)}\log_e\left(\frac{a}{b}\right). \qquad \text{.................(2)}$$

Since $a/b = (1+q)/(1-q)$, where $q = (a-b)/(a+b)$, we have

$$\tfrac{1}{2}\log_e\left(\frac{a}{b}\right) = q + \tfrac{1}{3}q^3 + \ldots = \frac{a-b}{a+b} + \frac{1}{3}\left(\frac{a-b}{a+b}\right)^3 + \ldots,$$

and thus, since $(a-b)/(a+b)$ is small, it may be sufficiently accurate to write $(a-b)/(a+b)$ for $\frac{1}{2}\log_e(a/b)$. Then

$$k = \frac{R(M+mc)(a-b)}{\pi l(\theta_b-\theta_a)(a+b)}. \qquad \text{.................(3)}$$

To this degree of accuracy, the result is the same as if the tube were replaced by a flat plate of thickness $a-b$ and area $\pi l(a+b)$.

The thermal conductivity of rubber (0·0004) is only about one-third of that of water (0·0014) and thus we may assume that θ_b is practically identical with the temperature of the steam passing through the tube and that θ_a is practically identical with the temperature shown by the thermometer in the calorimeter, when the water is well stirred.

189. Practical example.

Tube of red rubber was used.

Mass of copper calorimeter and stirrer $= m = 162{\cdot}6$ grm.; $c = 0{\cdot}095$.

Water in calorimeter $= M = 578{\cdot}5$ grm.

Water equivalent of whole $= M + mc = 578{\cdot}5 + 162{\cdot}6 \times 0{\cdot}095 = 593{\cdot}9$ grm.

Length of tube immersed $= l = 15{\cdot}6$ cm.

Two external diameters at right angles were measured at each end of each of the two portions cut from tube; similar measurements were made for internal diameters. Results were:

$2a$ ... 1·06, 1·07, 1·06, 1·07, 1·06, 1·07, 1·06, 1·08. Mean 1·066 cm.

$2b$ ... 0·62, 0·62, 0·62, 0·61, 0·61, 0·61, 0·60, 0·62. Mean 0·614 cm.

Hence $a = 0{\cdot}533$, $b = 0{\cdot}307$ cm., and $\log_e(a/b) = 0{\cdot}5517$.

While steam was passing through tube, temperature of water in calorimeter was observed. Temperature of room was 16·5° C.

Time min.	Temp. ° C.	Time min.	Temp. ° C.	Time min.	Temp. ° C.
0	12·30	5	15·80	10	19·00
1	13·00	6	16·45	11	19·55
2	13·75	7	17·05	12	20·20
3	14·40	8	17·75	13	20·80
4	15·15	9	18·35	14	21·35

Slope of temperature-time curve at 16·5° indicated that temperature of water was rising at rate of 0·654 degree per *minute*. Hence

$$R = 0{\cdot}654/60 = 0{\cdot}0109 \text{ deg. sec.}^{-1}.$$

Then, by (2),

$$k = \frac{R(M+mc)}{2\pi l(\theta_b - \theta_a)} \log_e\left(\frac{a}{b}\right) = \frac{0{\cdot}0109 \times 593{\cdot}9 \times 0{\cdot}5517}{2\pi \times 15{\cdot}6 \times (100 - 16{\cdot}5)}$$

$$= 4{\cdot}36 \times 10^{-4} \text{ cal. cm.}^{-2} \text{ sec.}^{-1} \text{ per unit temp. grad. of } 1° \text{ C. per cm.}$$

The *approximate* formula (3) gives

$$k = \frac{R(M+mc)(a-b)}{\pi l(\theta_b - \theta_a)(a+b)} = \frac{0{\cdot}0109 \times 593{\cdot}9 \times 0{\cdot}226}{\pi \times 15{\cdot}6 \times 83{\cdot}5 \times 0{\cdot}840} = 4{\cdot}26 \times 10^{-4}.$$

EXPERIMENT 28. Distribution of temperature along a bar heated at one end.

190. Introduction. When one end of a long bar is maintained at a constant temperature higher than that of the surrounding air, heat flows along the bar from the hotter towards the colder end and, at the same time, heat escapes from the surface of the bar. The distribution of temperature gradually approaches a steady state, which, theoretically, is reached only after an infinite time. When, however, a considerable, though finite, time has elapsed since the heating began, the changes of temperature with increase of time become too slow to be measurable. We may then treat the problem as if the steady state had been reached. In the steady state, the rate at which heat crosses any transverse section of the bar equals the rate at which heat escapes from the whole surface of the bar on the cooler side of the section.

The rate at which heat escapes from a given surface depends upon the excess of the temperature of the surface above that of its surroundings. When the excess is not too great, the rate of loss obeys Newton's Law of Cooling and is proportional to the excess of temperature. The escape of heat is a complex affair; it involves not only conduction into the air but also radiation through the space separating the body from neighbouring bodies. The rate of loss by conduction is much increased by convection, which continually brings cool air to the surface of the heated body and so maintains a steep gradient of temperature in the air near the surface.

The rate of loss of heat for given temperatures of the surface and of its environment depends upon the nature of the surface; it is greater for a blackened than for a bright metal surface. For a given environment, the rate of loss of heat depends upon the temperature of the heated body and increases with that temperature. For temperatures such that the body is luminous, the rate of loss of heat is largely due to radiation and increases very rapidly with rise of temperature. When, however, the temperature of the body is not far above that of the room, the emissivity, i.e. the loss of heat, measured in calories, per square cm. per sec. per degree centigrade of excess of temperature, has a definite value.

We will now investigate the distribution of temperature along a long uniform rod of metal which is heated at one end. Let the emissivity of the surface be η and the conductivity of the metal be k. Let the area of a cross-section be A and the area of the surface of unit length of the bar be S. Let O (Fig. 108), a point on the surface of the bar, be taken as origin, and let OPP' be a line on the surface parallel to the axis. Let

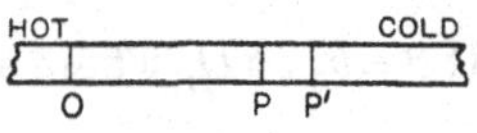

Fig. 108

$OP=x$, and let OP be positive when P is farther from the heated end than O. Let $OP'=x+dx$. Let θ be the excess of the temperature of the surface at P above that of the air in the room. When the conductivity is large and the emissivity is small, the isothermal surfaces differ little from planes transverse to the bar. We shall make the usual assumption that they are transverse planes.

The temperature gradient at P is $d\theta/dx$ and thus, by (2), § 170, heat crosses the section at P in the direction OP at the rate of $-kA\,d\theta/dx$ per sec. The rate at which heat crosses the section at P' in the same direction is

$$-A\left\{k\frac{d\theta}{dx}+\frac{d}{dx}\left(k\frac{d\theta}{dx}\right)dx\right\} \text{ per sec.}$$

Hence the resultant rate at which heat enters the volume PP' by conduction is $kA\,(d^2\theta/dx^2)\,dx$ per sec., when the conductivity is independent of the temperature.

The rate at which heat escapes from the surface of the element PP' is $\eta\theta S\,.\,dx$ per sec. When the temperature of the element has become steady, the two rates are equal and thus

$$d^2\theta/dx^2=\mu^2\theta, \quad \text{......................(1)}$$

where

$$\mu^2=\eta S/(kA). \quad \text{......................(2)}$$

If we write $\theta=Fe^{mx}$, where F is a constant,

$$d^2\theta/dx^2=m^2Fe^{mx}=m^2\theta,$$

and thus $\theta=Fe^{mx}$ is a solution of (1), if $m^2=\mu^2$. Hence m has *two* values, viz. μ and $-\mu$, and the complete solution of (1) becomes

$$\theta=Fe^{\mu x}+Ge^{-\mu x}, \quad \text{..................(3)}$$

where F and G are constants.

If the bar extend to infinity in the positive direction of x, F must be zero, for the term $Fe^{\mu x}$ would become infinite with x, if F were not zero. Hence (3) reduces to

$$\theta=Ge^{-\mu x}. \quad \text{.........................(4)}$$

If $\theta=\theta_0$ at O, where $x=0$, we have $G=\theta_0$, and then

$$\theta=\theta_0e^{-\mu x}. \quad \text{.........................(5)}$$

Taking logarithms to base e, we have

$$\log_e\theta=\log_e\theta_0-\mu x. \quad \text{....................(6)}$$

If we use logarithms to base 10,

$$\log_{10}\theta = \log_{10}\theta_0 - 0{\cdot}4343\mu x, \quad \ldots\ldots\ldots\ldots\ldots(7)$$

since $$\log_{10}\theta = \log_{10} e \,.\, \log_e \theta = 0{\cdot}4343 \log_e \theta.$$

Hence, if we plot $\log_e\theta$ or $\log_{10}\theta$ against x, we shall obtain a straight line.

When the bar is a circular cylinder of radius r cm., $S = 2\pi r$ cm., $A = \pi r^2$ cm., and $\mu^2 = \eta S/kA = 2\eta/kr$ cm.$^{-2}$. Hence

$$\eta/k = \tfrac{1}{2}\mu^2 r \text{ cm.}^{-1}. \ldots\ldots\ldots\ldots\ldots\ldots\ldots(8)$$

When k and μ are known, η can be found from (8).

191. Measurement of temperature by thermocouple. The difference of temperature between any point of the bar and the surrounding air is most conveniently measured by means of a thermocouple. A thermocouple consists of two wires AHB, AKB (Fig. 109) of different metals or alloys, such as German silver and iron, or copper and constantan, joined at A and B by soldering or brazing so as to form a closed circuit.

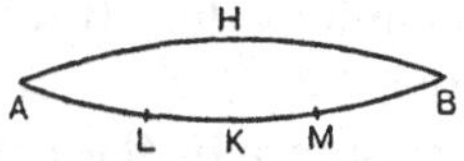

Fig. 109

When the temperatures t_1, t_2 of the junctions A, B are unequal, a current flows round the circuit and may be detected by its magnetic effects. If the resistance of the circuit be known, the E.M.F. due to the difference of temperature between A and B can be found if the current be measured. If we cut one wire—say AKB—and connect the cut ends by the wire LM of any metal, the E.M.F. in the circuit for given values of t_1 and t_2 will be the same as before, provided that the temperature of M equal that of L; this temperature need not be equal to either t_1 or t_2. The points L, M may be the terminals of a galvanometer whose wire is LM. Though the temperatures t_1, t_2 may be definite, the temperatures of intermediate points on the wires will be beyond our control, and thus the resistance of the circuit will be uncertain. If, however, the resistance of the galvanometer be large compared with that of the rest of the circuit, and if the temperature of the galvanometer be constant, the variations of resistance will be very small compared with the resistance itself and the current may then be considered to be proportional to the E.M.F.

For more accurate work, the E.M.F. is measured by a potentiometer. One cut end of the wire AKB is joined at L to a wire PQ (Fig. 110) and the other cut end is joined to the terminal M of a two-way switch FMN. A battery C sends a current through PQ and through a resistance QS. A standard cell W joins N to S, and the galvanometer G joins F to Q. By an adjustable resistance T, the current from C is adjusted so that, when the switch joins F to N, no current flows through the galvanometer. The E.M.F. between the ends of the resistance QS then equals that of the cell W. If W and QS be kept at constant temperatures, the current through QS and the wire PQ has a definite value. If, when the switch joins F to M, no current flow through G, the E.M.F. due to the couple is proportional to the resistance of the part LQ of the wire PQ. The points L, M, F, Q, N, S should be at the same temperature.

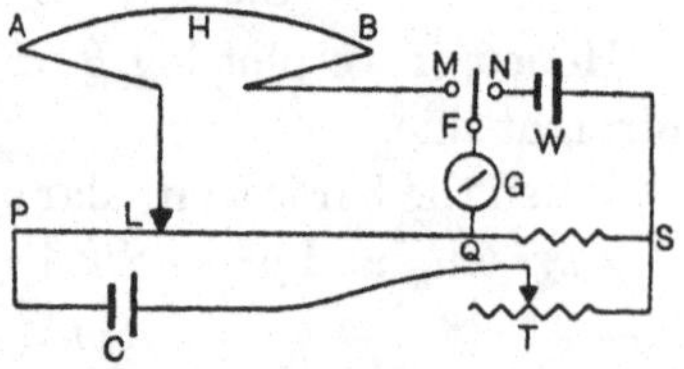

Fig. 110

If E_1 be the E.M.F. when one junction, A, is at 0° C. and the other, B, is at t_1, and if the E.M.F. be E_2 when A is at 0° and B is at t_2, the E.M.F. when B is at t_2 and A is at t_1 is $E_2 - E_1$. Hence it is sufficient to know how E depends upon t, the temperature of B, when A is kept at 0° C. There is, as a rule, no quite simple formula giving E in terms of t, which applies when the range of temperature is large. In some cases E, with A at 0°, increases to a maximum at a comparatively high value of t and then decreases with a further increase in t. When a suitable couple, such as copper-constantan, is used, the E.M.F. is *approximately* proportional to $t_2 - t_1$ over a range of 100° of t_2, when the cold junction is at a constant temperature t_1, where t_1 is such a temperature as that of the room. If one temperature be 0° C., a quadratic formula

$$E = at + bt^2$$

will represent E with considerable accuracy over a limited range of the other temperature t. If b be small, E will be nearly proportional to t.

Adams* investigated the E.M.F. of a copper-constantan couple,

* *Jour. American Chem. Soc.* **36**, p. 65 (1914). See also Ezer Griffiths, *Methods of Measuring Temperature*, p. 76 (1918).

with one junction at 0° C., and found that, when E is expressed in microvolts, the formula

$$E = 74\cdot672t - 13892\,(1 - e^{-0\cdot00261t}) \quad \text{...........(9)}$$

represented the E.M.F. to a fraction of a microvolt over the range from 0° to 350° C. of the temperature, t, of the other junction. The value of E given by (9) increases with t over this range, as does also dE/dt. The formula is not intended for use outside this range.

Some values of E yielded by (9), for the range from 0° to 350°, are, to the nearest tenth of a microvolt, as follows:

t	E by (9)	E by (10)	t	E by (9)	E by (10)
0° C.	0·0	0·0	70° C.	2907·3	2905·8
10	388·8	390·2	80	3355·9	3354·2
20	786·9	788·7	90	3812·2	3810·9
30	1193·9	1195·5	100	4275·9	4275·9
40	1609·7	1610·6	200	9285·0	9383·4
50	2034·0	2034·0	300	14858·7	15322·5
60	2466·6	2465·7	350	17815·5	18603·9

The quadratic formula

$$E = 38\cdot601t + 0\cdot04158t^2 \quad \text{.................(10)}$$

gives the same values as (9) for $t = 0$, 50, 100. Other values given by (10) are shown in the Table. From 0° to 100°, the differences are so small that, for any but the most precise work, (10) may be used over that range of t. When t much exceeds 100°, the differences become serious.

The linear formula $E = 42\cdot759t$ gives the same values as (9) for $t = 0°, 100°$. The values for $t = 20°, 40°, 60°, 80°, 200°, 300°$ are 855, 1710, 2566, 3421, 8552, 12828. The accuracy of the linear formula is thus much inferior to that of the quadratic.

If
$$E = at + bt^2, \quad \text{......................(11)}$$

when junction A is at 0° C. and B is at $t°$, then, when A is at $T°$ and B is at $(T + \theta)°$, we have

$$E = a\,(T + \theta) + b\,(T + \theta)^2 - \{aT + bT^2\},$$

or
$$E = c\theta + b\theta^2, \quad \text{......................(12)}$$

where $c = a + 2bT$. Thus, if T be kept constant, E is a quadratic function of θ.

We can, by (12), find E when T and θ are known. In the experiment, however, we observe E and wish to find θ, the temperature excess. Solving (12) for θ, we have

$$\theta = \{-c+\sqrt{(c^2+4bE)}\}/2b, \quad \text{..............(13)}$$

where the positive sign is prefixed to the square root because $\theta = 0$ when $E = 0$. If we expand the square root, we have

$$\theta = \frac{E}{c}\left\{1-\left(\frac{bE}{c^2}\right)+2\left(\frac{bE}{c^2}\right)^2-5\left(\frac{bE}{c^2}\right)^3+\ldots\right\}. \quad \text{......(14)}$$

The series in (14) converges too slowly to be of service unless θ be small compared with 100.

We must therefore find θ by the formula (13), which may be put into the more convenient form

$$\theta = \frac{2E}{c+\sqrt{(c^2+4bE)}}. \quad \text{.................(15)}$$

If, for a constant temperature, $T°$, of A, E be measured for two values of θ, the excess of the temperature of B over that of A, we can find c and b from (12), and then θ can be found by (15) for other values of E. It is, however, better to take a series of values of θ and to plot E/θ against θ. The straight line lying most evenly among the plotted points, as found graphically or by Awbery's method (NOTE I), gives c and b. This method should be followed if time allow. As an alternative, we may measure E for a series of values of θ and plot E against θ. We can then read from the curve the value of θ for any given E.

If (12) represent the relation between E and θ with little error over the range of θ employed, the straight line is to be preferred to the curve, for a straight line (i) takes a better account of the whole body of observations than does a freehand curve and (ii) can be drawn more accurately than a curve.

When the thermocouple causes a deflexion, D, of a galvanometer and the deflexion is proportional to E, we may put $D = \gamma\theta + \beta\theta^2$, and then $D/\theta = \gamma + \beta\theta$. If we plot D/θ against θ, we shall obtain a straight line from which γ and β can be found. Then

$$\theta = \frac{2D}{\gamma+\sqrt{(\gamma^2+4\beta D)}}. \quad \text{.................(16)}$$

192. Method. One end of a brass bar, about 1 cm. in diameter and 1 metre in length, is heated by a steam jacket; the rest of the bar is exposed to the air. The temperature of the cooler end will not exceed that of the surrounding air by more than $\frac{1}{10}$ degree and thus the distribution of temperature along the warmer half of the bar will differ very little from that which would exist if the bar extended to infinity.

The apparatus is arranged as in Fig. 111. The bar AB rests on the wooden supports U, V; the end A is surrounded by the metal jacket D through which steam from a boiler passes by the pipes F, H. The jacket may be covered with a wrapping of flannel to diminish the loss of heat from the jacket to the air.* A screen S prevents heat from D reaching the cooler parts of the bar by

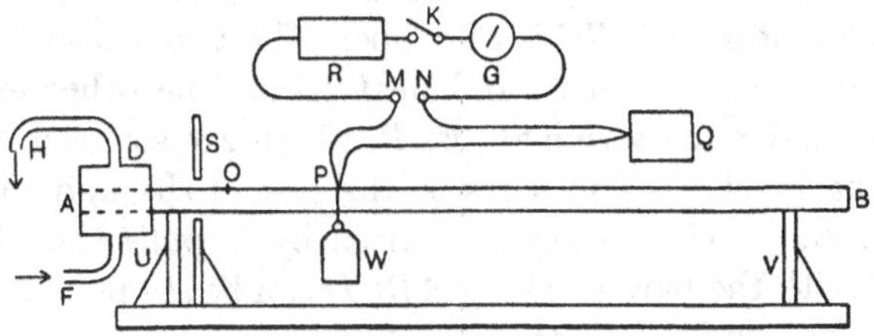

Fig. 111

radiation. The excess of the temperature of any point of the bar above that of the air is found by a thermoelectric couple. One of the junctions is pressed upon the bar at P by the weight W. The other junction is soldered to a small plate of metal Q, which is exposed to the air and is placed so far from A that its temperature is not sensibly raised, above that of the air in the room, by radiation from A. The constantan wire of the couple connects P to Q. From P and Q the copper wires of the couple lead to M and N, where they are joined to the wires leading to the resistance box R and the galvanometer G. The circuit can be completed at will by the key K. The junctions M, N should be close together so as to be at

* A covering of "heat insulator" does not necessarily diminish the loss of heat from a hot body to the air. If the emissivity of the surface of the covering be high compared with that of the surface of the body itself, and if the covering be not very thick, the loss may be *increased* by the covering. The recommendation in the text is an example of unreasoning acceptance of convention—like submission to vaccination, only less dangerous.

the same temperature.* The resistance of the circuit is adjusted so that, for the greatest excess of the temperature of P above that of Q, the galvanometer deflexion is not inconveniently large. A point O on the bar is chosen as origin; it should be as near A as the screen S allows, so that the temperature of O may be high.

When the thermocouple is to be calibrated, the junction P and the plate Q are placed in two vessels containing water. The water in the Q-vessel is as nearly as may be at the temperature of the air. The temperatures are measured by a mercury thermometer. The deflexion, D, of the galvanometer is observed for a number of temperatures of P. A curve is plotted showing how D depends upon $t_P - t_Q$; from this curve we can read the value of $t_P - t_Q$ for any given D. Care must be taken to keep t_Q constant.

The thermocouple may be formed as in Fig. 112. The wire AB is of constantan and BC is of copper. The two are soldered together at B with a small and neat joint. The other ends are clamped under the terminals E, F, which are screwed into the insulating block D. The wires to the rest of the apparatus are fixed to E, F. The weight W carried by D pulls the wire into contact with the bar. The loop $ABCD$ can be slipped off the bar at the cool end when the couple is to be calibrated. The terminals E, F must be at the same temperature. The wire touches the bar over the arc MBN (Fig. 112), but this does not affect the E.M.F. in the circuit, provided, of course, that MBN be all at one temperature.

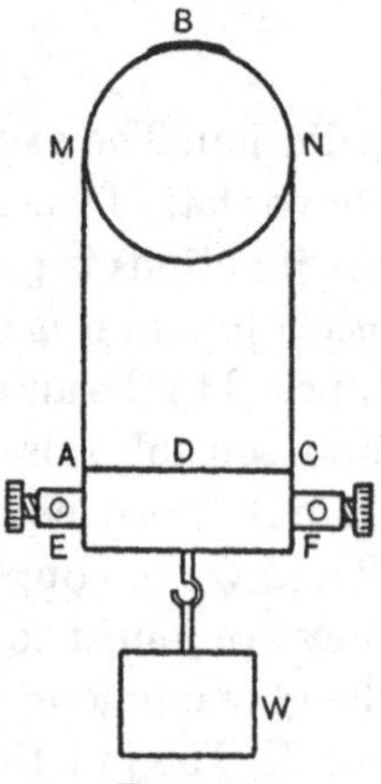

Fig. 112

A series of small transverse grooves filed in the bar at intervals of 1 cm., starting from O, facilitates the placing of the junction B at definite distances from O.

Steam should be allowed to pass through the steam jacket for at least ten minutes before any observations are made with the thermocouple; in this way we shall ensure that the temperature of each point of the bar has attained very

* Equality of temperature of the junctions M and N (Fig. 111) is essential if a German silver-iron couple be used. In that case, the wires MP, NQ are of German silver but the wires leading from M to R and N to G are of copper.

nearly to a steady value which will not alter in the course of the observations. When the junction has been moved to a new position, a minute or two will elapse before it attains a definite temperature, as shown by a steady deflexion of the galvanometer.

193. Practical example.

Mr W. H. Hadley used a copper-constantan couple. Radius of section of bar, of brass, was $r = 0{\cdot}56$ cm.

Calibration of thermocouple. Cooler junction was nearly at the temperature, $T°$, of the air; this was 16° C. The temperature excess, $\theta°$, of the warmer junction was measured by mercury thermometer. Deflexion on galvanometer scale $= D$ cm.

θ	D observed	D/θ	D calculated
	cm.	cm./deg.	cm.
10·1°	3·59	0·3554	3·607
21·3	7·80	0·3662	7·777
33·1	12·31	0·3719	12·363
47·0	18·04	0·3838	18·018
54·1	20·95	0·3872	21·013
59·4	23·35	0·3931	23·295

When D/θ was plotted against θ, best straight line through points gave $D/\theta = 0{\cdot}35 + 0{\cdot}00071\theta$. Values in fourth column are calculated from

$$D = 0{\cdot}35\theta + 0{\cdot}00071\theta^2;$$

they agree closely with observed values.

Distribution of temperature along bar. The excess, θ, of temperature of bar above that of room was deduced from galvanometer deflexion by formula (16), § 191, with $\gamma = 0{\cdot}35$ and $\beta = 0{\cdot}00071$. Thus

$$\theta = \frac{2D}{0{\cdot}35 + \sqrt{(0{\cdot}35^2 + 4 \times 0{\cdot}00071D)}}.$$

x	D	$\log_{10}\theta$ (obs.)	θ (obs.)	$\log_{10}\theta$ (cal.)	θ (cal.)
0 cm.	22·43	1·7589	57·40	1·7576	57·23
4	16·53	1·6376	43·41	1·6374	43·39
6	14·16	1·5751	37·59	1·5772	37·78
10	10·62	1·4575	28·67	1·4570	28·64
16	6·89	1·2778	18·96	1·2766	18·90
20	5·12	1·1528	14·22	1·1563	14·33
30	2·56	0·8579	7·21	0·8557	7·17

In accordance with (7), § 190, it was assumed that $\log_{10}\theta = mx + c$. Awbery's method (NOTE I) of finding m and c, when applied to numbers in first and third columns, gave

$$\log_{10}\theta = -0{\cdot}030065x + 1{\cdot}75761. \quad \text{.................(17)}$$

Values of $\log_{10}\theta$ and of θ given by (17) are in fifth and sixth columns. Observed and calculated values of θ agree closely.

By (17) and (7), we find $0{\cdot}4343\mu = 0{\cdot}030065$; then

$$\mu = 0{\cdot}06923 \text{ cm.}^{-1}, \qquad \mu^2 = 0{\cdot}004792 \text{ cm.}^{-2}.$$

Since $r = 0{\cdot}56$ cm., we have, by (8),

$$\eta/k = \tfrac{1}{2}\mu^2 r = 0{\cdot}28 \times 0{\cdot}004792 = 1{\cdot}342 \times 10^{-3} \text{ cm.}^{-1}.$$

For brass $k = 0{\cdot}26$ cal. per sq. cm. per sec. per (deg. per cm.), and hence $\eta = 0{\cdot}26 \times 1{\cdot}342 \times 10^{-3} = 3{\cdot}49 \times 10^{-4}$ cal. per sq. cm. per sec. per degree.

EXPERIMENT 29. Determination of mechanical equivalent of heat.

194. Introduction. When forces are applied to a body in such a way as to distort it, as when a wire is bent or hammered or a liquid is stirred in a calorimeter, the temperature of the body is, as a rule, changed. The temperature can be restored to its original value by applying water to the body or to the calorimeter, and the heat received by the cooling water can be measured. If the energy of the body return to its original value when the original temperature is restored, the heat received by the water is the equivalent of the work done by the forces. But, in many cases, we cannot tell whether the internal energy of the body has or has not returned to its original value. In such cases, we are not entitled to say that the heat received by the cooling water is the equivalent of the work spent upon the body. If a mixture of ice and water be churned in a calorimeter, the ice will be melted, but, until all the ice have melted, the temperature will not rise. If the churning stop before all the ice has melted, no heat will be received by ice-cold water applied to the outside of the calorimeter, in spite of the fact that much work was spent in churning the mixture. In this case, all the work has been spent in changing the internal energy of the ice.

In other cases, however, the circumstances are such that we can say that the final value of the internal energy, after the original temperature has been restored, is identical with its initial value. Such a case is that of Joule's experiment in which water is churned in a calorimeter. If means were provided for removing heat continuously from the calorimeter, the churning might go on indefinitely without producing any change in the churned water. Joule did not actually employ this method of continuous cooling, for he allowed the heat produced to remain in the water. But, if the temperature of the churned water were reduced to its original value by cooling water applied to the calorimeter, the internal energy of the churned water would be restored to its original value. In this case, the work spent in churning the water in the calorimeter equals the heat which would be given to the cooling water by the calorimeter and the churned water in falling to their initial temperature, and this heat can be calculated when we know the masses and the specific heats of the various parts and also the rise of temperature due to the churning.

In the present experiment, one cone turns in another. They are lubricated with a little oil, but at times they touch, as is shown by slight signs of wear. The amount of metal abraded is very small, and hence the difference between the final value of the internal energy, when the initial temperature has been restored, and the initial value is entirely negligible.

195. Apparatus. The machine supplied by W. G. Pye and Co. is a modified form of that originally made by the author for use at the Cavendish Laboratory. A vertical spindle *A* (Fig. 113) carries at its upper end a metal cup *B*. At the top of the cup is fixed an ebonite ring *C* and at the bottom an ebonite block *D*; ebonite is a poor conductor of heat. The truncated cone *E* is supported by *C* and is kept in position by two pins which project from *D* and enter holes at the bottom of the cone. These pins compel the cone to turn with the cup. A second cone *F*, provided with two pins *P*, *P*,

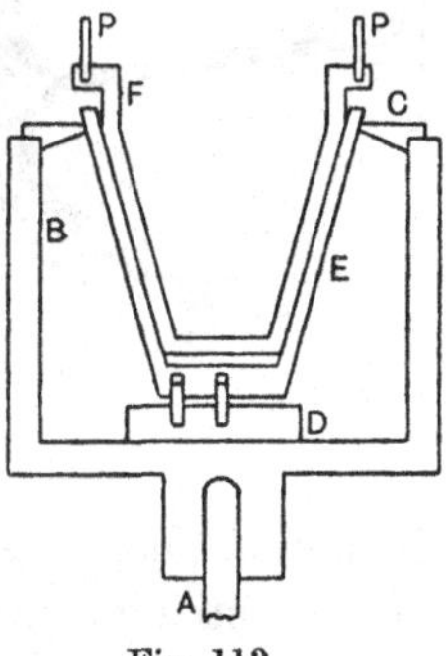

Fig. 113

fits into the cone *E*. In the experiment, a wooden disk rests on the cone *F*; motion of the disk relative to the cone is prevented by the pins *P*, *P* which enter holes in the disk. These holes must be accurately located, so that the disk may be coaxial with the spindle. If the disk be prevented from turning while the cup revolves, the outer cone will revolve and the inner cone will remain at rest. The friction between the cones converts into heat the mechanical energy supplied. An iron ring, resting on the disk and held in position by two pins, serves to give a suitable loading to the inner cone.

The complete instrument is shown in Fig. 114. A grooved pulley is fixed to the spindle; by a belt passing round the pulley and round a hand-wheel, motion is imparted to the spindle. A revolution counter, actuated by a pair of bevelled cog-wheels, registers the number of revolutions made by the spindle. A bent steel arm attached to the frame carries a cradle in which runs a small guide pulley; the top of this pulley is on the same level as the groove in the wooden disk. The cradle turns freely about a vertical axis. A *plaited* silk string (fishing line) attached to the disk and passing along the groove in its edge, and over the guide pulley, supports a load of 200 or 300 grm. The bottom of the groove is cylindrical and should be coaxial with the spindle. After a

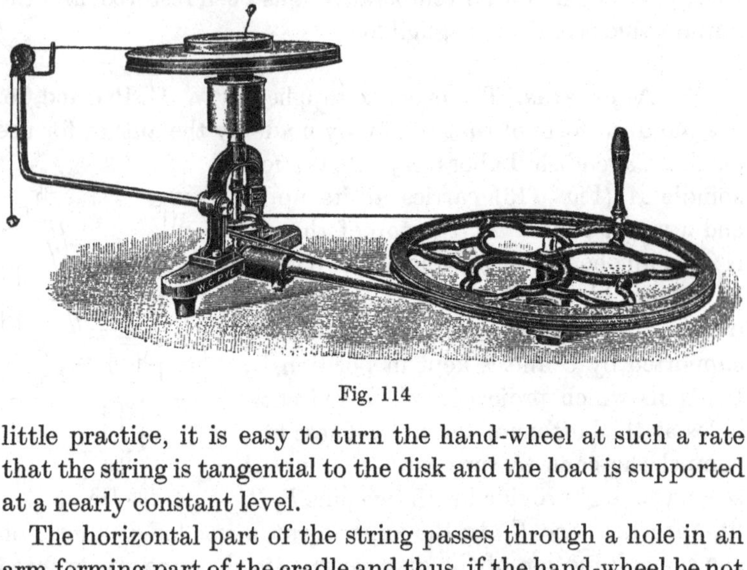

Fig. 114

little practice, it is easy to turn the hand-wheel at such a rate that the string is tangential to the disk and the load is supported at a nearly constant level.

The horizontal part of the string passes through a hole in an arm forming part of the cradle and thus, if the hand-wheel be not turned fast enough to keep the string tangential to the disk, the cradle will be turned round by the string and the string will not run off the pulley. The vertical part of the string passes through an eye fixed to the steel rod. These devices make it impossible, even with moderately unskilful driving of the hand-wheel, either to throw the string off the pulley or to wind the load over the pulley.

The frictional machine, or "mill" as it may be called, is clamped to a table. Care must be taken that the driving belt is sufficiently tight and that it runs properly without risk of slipping off the hand-wheel. The string should be of such length that the load cannot quite reach the floor. Before the cones are put together, the rubbing surfaces are cleaned, and then one or two drops of oil are put between them. If too much oil be used, it will be difficult to turn the hand-wheel fast enough to keep the string tangential to the disk. If too little be used, the motion will be slow and the temperature will rise slowly. More oil is needed when the room is cold than when it is warm.

A thermometer, preferably one graduated to fifths of a degree, is hung from a support so that it passes through the central aperture in the wooden disk and almost touches the bottom of the inner cone. The thermometer passes through a central hole in the stirrer, which is a perforated brass disk fitted with a handle (Fig. 114).

196. Method. The cones, cleaned and oiled, are weighed together with the stirrer. The inner cone is then filled, up to about a centimetre from its top, with water two or three degrees below the temperature of the room, and the system is again weighed. The cones, the disk and the thermometer are then put into position. One observer X takes his place at the hand-wheel and a second observer Y at the mill. By working the mill, the water is warmed until its temperature is as nearly as possible equal to that of the room. After the counter has been read and the temperature, θ_1, of the water has been *carefully* observed, the observer X starts to turn the wheel and observes the time of starting by a watch. He gives this time to Y, who records it, and continues to turn the wheel at such a speed that the load is raised so far that the string is tangential to the disk. If the string be not tangential, the moment of its tension about the axis of revolution is seriously diminished. The observer X gives a signal at 1, 2, 3, ... minutes after the start, and Y, who stirs the water continuously, reads the thermometer at those instants, and records these temperatures and the corresponding times. *Very* accurate readings of these temperatures are difficult to make and are not necessary. It will be found that the time of 100 revolutions diminishes as the temperature rises, an effect due to the diminution, with rise of temperature, of the viscosity of the oil between the cones. With a load of 200 grm., the temperature will rise about 1° C. for 100 revolutions.

When the spindle has made about 1000 revolutions, the motion is stopped at a definite minute and the highest temperature, θ_2, is *carefully* read. The exact number of revolutions, n, made by the spindle is found from the initial and final numbers registered by the counter.

Observations to determine the correction for cooling are then made. Without disturbing the apparatus, the temperature is raised about two degrees above θ_2 by working the mill. The motion is then stopped, the water is allowed to cool, with *frequent* stirring, and the thermometer is read at intervals of about one minute until the temperature of the water be only two or three degrees above that of the room.

197. Calculation of correction for cooling. If no heat were lost while the mill is in action, the difference, $\theta_2 - \theta_1$, between the final and initial temperatures would need no correction. In the actual case a correction is required.

From the observations taken with the mill in action, the temperatures at 0, 1, 2, ... minutes from the start are known.

From the observations of the cooling of the water, a curve is drawn with time as abscissa and temperature as ordinate. The value of F, the rate of cooling in degrees per minute, is found, for any temperature θ, from the slope of the tangent to the curve. It is best not to *draw* the tangent. If a triangular "set square" ABC be adjusted so that one side AB touches the curve at the desired point, and if one of the other sides, say AC, be made to slide along a straight-edge, the side AB can be moved parallel to itself until it passes through an intersection of the lines ruled on the squared paper. It is now easy to read off, along the side AB of the "set square," the number of degrees corresponding to 10 minutes. Dividing this number by 10, the value of F, in *degrees per minute*, is obtained for the temperature θ at which the tangent was taken.

The rate of loss of heat at time t, when the temperature is θ, during the working of the mill, is EF calories/min., where E grm.

is the equivalent of the cones, stirrer, thermometer bulb and water, and F deg./min. is the rate of fall of temperature at temperature θ, as found by the cooling experiment.* If the mill be in action from $t=0$ to $t=T$, the total loss of heat is $E\phi$ calories, where

$$\phi = \int_0^T F\,dt. \quad \ldots\ldots\ldots\ldots\ldots\ldots\ldots(1)$$

To determine ϕ, a second curve is drawn. Against each of the times at which the temperature was read during the heating is set the value of F corresponding to that temperature; the times are abscissae and values of F are ordinates. The origin is a point on this curve, since at $t=0$ the water was at the temperature of the room and there was no cooling. The area included between the axis of t, the curve and the ordinate corresponding to the time T, when the mill was stopped, represents ϕ. If one inch (or cm.) on the paper along the t-axis correspond to p minutes and one inch (or cm.) along the F-axis correspond to q degrees per minute, then each square inch (or square cm.) represents pq degrees. If the area be A square inches (or square cm.), then $\phi = Apq$. The area may be found by the trapezoidal rule; the rule is described in NOTE IX of *Experimental Elasticity*.

A more speedy, but less accurate, method of finding ϕ may be used when time presses. For small differences of temperature between the cones and their surroundings, Newton's law of cooling may be applied, and we may write $F = k(\theta - \theta_1)$, provided the initial temperature, θ_1, of the water have been made equal to the temperature of the air by working the mill, as described in § 196. The rate of rise of the temperature decreases as the temperature rises, but, for a rough estimate, we may neglect the variation and write $\theta - \theta_1 = lt$. Hence

$$\phi = \int_0 F\,dt = \tfrac{1}{2}klT^2 = \tfrac{1}{2}F_T \, . \, T, \quad \ldots\ldots\ldots\ldots(2)$$

where F_T is the rate of cooling (at time T) at temperature θ_2, when the mill is not in action.

198. Heat produced. Let the mass of the water be W and that of the cones and stirrer be w grm. The specific heat of the metal

* For a more accurate estimate of the correction for cooling, see § 202.

may be taken as 0·095; the water equivalent* of the cones and the stirrer is thus $0{\cdot}095w$.

For accuracy, account should be taken of the water equivalent of the bulb and the part of the stem of the thermometer which is immersed in the water in the cone. A cylindrical calorimeter of thin copper, about 1 cm. in diameter and 7 cm. long, is partly filled with water and the temperature is taken with the thermometer. The thermometer is then heated in water in another vessel to as high a temperature as its construction allows with safety. It is taken out of the hot water, is wiped dry and is plunged to the proper depth into the water in the calorimeter. Let the mass of water in the calorimeter be a grm. and the equivalent of the calorimeter be b grm. Let θ' be the temperature of the thermometer just before it is plunged into the water, and let θ'' and θ''' be the initial and final temperatures of the water. Then, if the equivalent of the part of the thermometer immersed be x grm.,

$$x=(a+b)(\theta'''-\theta'')/(\theta'-\theta'''). \quad \text{...............(3)}$$

The equivalent may be calculated† approximately in terms of the volume, v c.c., of the immersed part. The specific heat of mercury is 0·033 and of glass is about 0·20; the densities are 13·6 and (about) 2·6. The equivalent of 1 c.c. of mercury is $0{\cdot}033 \times 13{\cdot}6$ or 0·45 grm. and of 1 c.c. of glass is 0·52 grm. The two values do not differ much, and, for a rough estimate, we may take

$$x=0{\cdot}45v. \quad \text{..........................(4)}$$

The volume v may be found by measurement or by immersing the thermometer to the proper depth in water in a graduated burette.

The equivalent of water, cones, stirrer and thermometer is E, and

$$E=W+0{\cdot}095w+x. \quad \text{....................(5)}$$

Let the total heat generated by friction be H calories. Part of this heat, viz. $E(\theta_2-\theta_1)$, has gone to raise the temperature

* To assume the value of the water equivalent is, of course, bad practice. The equivalent should be found by the method of mixtures.

† T. G. Bedford, *Practical Physics*, p. 133.

of the water, cones, etc. The remainder, viz. $E\phi$, has passed to other bodies by conduction and radiation. Hence

$$H = E(\theta_2 - \theta_1 + \phi) = (W + 0{\cdot}095w + x)(\theta_2 - \theta_1 + \phi). \quad \ldots(6)$$

199. Work done. Let the radius of the disk be r, and let the load carried by the string be M. Let the couple applied to the outer cone by the spindle be G, and let P be the tension of the string supporting the load M. For simplicity, we suppose that initially M rests upon the floor and that the string is tangential to the disk. If the mill be worked with sufficient skill, the load will be lifted just clear of the floor and will remain in that position so long as the mill is working. When the mill stops, the load will return to the floor with zero velocity. Except for very brief periods at starting and stopping, the load and the disk are at rest, and hence $P = Mg$. Initially, at $t = 0$, and finally, at $t = T$, the spindle is at rest. Let the angle turned through by the spindle be α at time t. On account of the diminution of the viscosity of the oil between the cones, $d\alpha/dt$ will generally increase during the working of the mill, except for the brief period when the spindle stops and $d\alpha/dt$ sinks to zero. Since, both initially and finally, $d\alpha/dt = 0$, and since the disk, the inner cone and the load are always at rest, they gain no kinetic energy. Since the load returns to the floor, there is no gain of potential energy. Hence the work done by the couple G is all expended in producing heat.

Except for the brief periods of irregularity at starting and stopping, neither the disk nor the outer cone has any angular *acceleration*, and thus G equals the moment of the tension of the string. Hence

$$G = Pr = Mgr.$$

If U be the work done while the spindle makes n revolutions,

$$U = G \times 2\pi n = 2\pi n M g r, \quad \ldots\ldots\ldots\ldots\ldots\ldots(7)$$

and this is spent in producing heat.

200. Calculation of mechanical equivalent of heat. If J denote the number of ergs of work which must be spent to raise the temperature of one gramme of water by one degree,

$$J = \frac{\text{Work spent}}{\text{Heat produced}} = \frac{U}{H} = \frac{2\pi n M g r}{(W + 0{\cdot}095w + x)(\theta_2 - \theta_1 + \phi)}.$$

The specific heat of water is not quite constant, but the accuracy of the present experiment is not so great as to require us to take account of the variations.

201. Practical example.

Messrs T. B. Rymer and J. R. Colliss found:
Mass of cones and stirrer $= w = 192{\cdot}30$ grm.; $0{\cdot}095w = 18{\cdot}27$ grm.
Mass of water $= W = 19{\cdot}25$ grm.
Water equivalent of thermometer was found by calorimeter method of § 198, with equation (3).

a	b	$a+b$	θ'	θ''	θ'''	x
grm.	grm.	grm.	° C.	° C.	° C.	grm.
2·80	0·30	3·10	40·0	17·0	20·1	0·48
2·85	0·30	3·15	40·0	20·3	22·8	0·46
2·80	0·30	3·10	39·6	19·6	22·2	0·46

Mean $x = 0{\cdot}47$ grm.

Diameter of stem of thermometer $= 0{\cdot}62$ cm.; length immersed $= 3{\cdot}5$ cm.

By (4) $\quad x = 0{\cdot}45 \times \pi \times 0{\cdot}31^2 \times 3{\cdot}5 = 0{\cdot}48$ grm.

Equivalent of water, cones and thermometer is E, and

$$E = 19{\cdot}25 + 18{\cdot}27 + 0{\cdot}48 = 38{\cdot}00 \text{ grm.}$$

The temperature was observed every minute while mill was in action. Readings of these times and temperatures are recorded in Table I, which gives also values of F, the rate of cooling, as found for these temperatures from curve drawn from numbers in Table II. At 17·60° C., the temperature of the room, $F = 0$. Values marked * were estimated. Table I also gives R, the rate of rise of temperature of the water as found from a smoothed curve. The values of $\theta + \frac{1}{3}(R+F)$ and of F' are used in § 202. Here F' is the rate of cooling when the thermometer reading is $\theta + \frac{1}{3}(R+F)$.

TABLE I.

Time	Temp. θ	R	F	$\theta + \frac{1}{3}(R+F)$	F'
min.	° C.	deg./min.	deg./min.	° C.	deg./min.
0	17·60	1·50	0	18·10	0·0078
1	18·95	1·35	0·021*	19·41	0·0291
2	20·20	1·25	0·043*	20·63	0·0503
3	21·55	1·10	0·066	21·94	0·0752
4	22·65	1·00	0·092	23·01	0·1028
5	23·45	0·85	0·116	23·77	0·1288
6	24·10	0·75	0·142	24·40	0·1574
7	24·80	0·60	0·178	25·06	0·1896
8	25·45	0·55	0·207	25·70	0·2182

Hence $\theta_1 = 17{\cdot}60$, $\theta_2 = 25{\cdot}45°$ C.

Temperature was raised to about 28° C. and the thermometer was read every minute during the cooling. Table II gives some readings.

TABLE II.

Time	Temp.	Time	Temp.	Time	Temp.
0 min.	26·80° C.	14 min.	24·10° C.	36 min.	21·85° C.
3	26·10	19	23·45	43	21·40
6	25·50	24	22·90	51	20·95
10	24·80	30	22·35	61	20·50

Since times in Table I are equally spaced, we can use the trapezoidal rule, and then

$\phi = (\frac{1}{2}$ first $+ \frac{1}{2}$ last $+$ sum of intermediate values of $F) \times$ one minute
$= 0·76°$.

Hence $\theta_2 - \theta_1 + \phi = 25·45 - 17·60 + 0·76 = 8·61°$.

Heat produced $= H = E \times 8·61 = 38·00 \times 8·61 = 327·2$ calories.

Load on string $= M = 198·14$ grm. Radius of disk $= r = 12·01$ cm.

Initial reading of counter $= 33152$; final 34130. Hence $n = 978$.

Work done $= U = 2\pi nMgr = 2\pi \times 978 \times 198·14 \times 981 \times 12·01$
$= 1·4345 \times 10^{10}$ ergs.

Hence $J = U/H = 1·4345 \times 10^{10}/327·2 = 4·38 \times 10^7$ ergs/calorie.

202. Further discussion of correction for cooling.

The estimate made in § 197 of the heat lost by cooling needs a correction which was brought to the author's attention by Dr J. K. Roberts. While heat is being generated by the friction between the cones, heat passes from the inner cone, which has high thermal conductivity, into the water, which is of low conductivity. If heat flow normally across the interface between one substance and another, and if a steady state have been reached, the temperature gradients in the substances are inversely proportional to their conductivities. The conductivity of brass is about 0·2 and of water about 0·0014 cal. per sq. cm. per sec. per unit temp. gradient of 1° C. per cm. In the experiment recorded in § 201, the temperature of the water rose on the average about 1·0° per min. or 0·017° per sec. The water equivalent of the water, cones, etc. was 38·00 grm., and hence the rate of generation of heat was rather greater than 38 × 0·017 or 0·64 cal. per sec., since some heat escaped. The water equivalent of the cones, etc. was about 19 grm., and hence about half the heat entered the water. Thus 0·32 cal. per sec. passed from the inner cone to the water. The surface of contact between the cones is about 15 cm.^{2} When the curvature of the surfaces is neglected, the flow of heat is 0·32/15 or 0·021 cal. per sec. per. sq. cm. The rate of rise of temperature is small and thus, for a rough approximation, we may treat the conditions at any moment as if they were steady. Hence the temperature gradient in the brass is 0·021/0·2 or 0·105 deg. cm.$^{-1}$. The whole thickness of the brass does not exceed 0·5 cm., and hence the difference of temperature between the inner surface of the inner cone and the outer surface of the outer cone does not exceed 0·05°. Thus the temperature differences in the cones themselves are negligible.

In the water the gradient is 0·021/0·0014 or 15 deg. cm.$^{-1}$. Thus there is a fall of temperature of about 1·5 degrees in a layer of water one millimetre thick next to the cone. Hence, unless the water be very well stirred, the temperature of the water, as shown by the thermometer whose bulb is on the axis of the cones, will be considerably less than that of the cones.

During the cooling experiment the water will be hotter than the cones.

If we take the rate of loss of heat for any given temperature θ of the water (as shown by the thermometer) during the heating to be that deduced from the rate of fall of temperature found in the cooling experiment when the thermometer reads θ, we shall considerably under-estimate the correction to be made for the heat lost while the mill is in action.

No exact account is possible, but a rough estimate may be made, if we know how the difference between the temperature θ shown by a thermometer in the water and ω, the temperature of the outer surface of the outer cone, is related to the rate of change of θ during the cooling.

Messrs T. B. Rymer and J. R. Colliss used a pair of cones similar to those used in the experiment of § 201. The difference $\theta-\omega$ was measured by a copper-constantan thermocouple. One junction was soldered to the outer surface of the outer cone. The other junction did not dip directly into the water, but was surrounded by a fine tube of thin glass, containing a little paraffin oil.* The glass tube was necessary to avoid an electrolytic E.M.F. In one set of readings, the water was constantly stirred; in the other set the water had been left unstirred for half a minute when the reading for $\theta-\omega$ was taken. The temperature of the air was 14·5° C. By the nature of the case, very accurate results are impossible. The following values of $\theta-\omega$ and of $-d\theta/dt$ are from smoothed curves lying evenly among the plotted points.

TABLE III.

θ	$-d\theta/dt$ deg./min.	$\theta-\omega$, water stirred	$\theta-\omega$, water not stirred
26·5° C.	0·35	0·125°	0·300°
23·5	0·25	0·080	0·205
20·5	0·16	0·045	0·120
18·5	0·08	0·027	0·075

The excess of temperature, $\theta-\omega$, when the water is not stirred is about 2·5 times the excess when it is stirred—a result which makes it clear that constant stirring is essential.

When the water is stirred, $\theta-\omega$ is about $\frac{1}{3}F$, where F is the rate of fall of θ in degrees per minute. We may conclude that, during the heating, $\omega-\theta$ is about $\frac{1}{3}R$, where R is the rate of rise of θ.

The rate of loss of heat depends upon ω, and thus the rate of loss during heating when the thermometer reads $\omega-\frac{1}{3}R$ equals the rate of loss during cooling when the thermometer reads $\omega+\frac{1}{3}F$. Roughly, we may say that the rate of loss of heat when the thermometer reads θ during the heating equals the rate of loss when the thermometer reads $\theta+\frac{1}{3}(R+F)$ during the cooling.

* For use in the fine tube, water would have been better than paraffin oil, because the thermal conductivity of water is three times that of paraffin oil.

Table I, § 201, gives $\theta+\frac{1}{3}(R+F)$ and also F', where F' is the rate of cooling at $\theta+\frac{1}{3}(R+F)$. The value of F' is found by interpolation. Thus when $\theta=21{\cdot}55°$, $\frac{1}{3}(R+F)=0{\cdot}39°$. At $\theta=21{\cdot}55°$, $F=0{\cdot}066$, and at $\theta=26{\cdot}65°$, $F=0{\cdot}092$, and so F increases by 0·026 when θ increases by 1·10°. Hence, when θ increases by 0·39°, F increases by $0{\cdot}39\times 0{\cdot}026/1{\cdot}10$ or by 0·0092, and thus at $\theta=21{\cdot}94°$, $F'=0{\cdot}066+0{\cdot}0092=0{\cdot}0752$.

If ϕ' be the corrected value of ϕ, we have

$$\phi'=(\tfrac{1}{2}\text{ first}+\tfrac{1}{2}\text{ last}+\text{sum of intermediate values of } F')\times\text{one minute}$$
$$=0{\cdot}85°.$$

Then $\theta_2-\theta_1+\phi'=8{\cdot}70°$, instead of 8·61° as found in § 201. Hence

$$H=38{\cdot}00\times 8{\cdot}70=330{\cdot}6 \text{ calories,}$$

and $$J=1{\cdot}4345\times 10^{10}/330{\cdot}6=4{\cdot}34\times 10^{7} \text{ ergs/calorie.}$$

Table III shows how greatly $\theta-\omega$ is affected by stirring and makes it clear that even "continuous" stirring cannot be entirely effective. The stirring during cooling—which is a long process—was probably, through human frailty, less continuous than that during the heating. Thus the value found for F is likely to be too small, and hence the cooling correction of 0·85° is probably too small.

Some further slight corrections are needed. Since the stem of the thermometer is at the temperature of the air, a correction for the emergent column of mercury is required. The correction to be applied to the higher temperature θ_2 is greater than that for the lower θ_1, and hence the observed value of $\theta_2-\theta_1$ is too small by two or three hundredths of a degree.

When the mill is stopped, heat continues to flow into the water for a short time, but during this time heat has been escaping from the cones. Hence the highest temperature θ_2 is slightly too small.

This discussion has been given to draw the attention of students to the difficulties attending experiments in which heat flows from a metal into water.

The use of the thermocouple for finding the difference of temperature between the water and the outside of the cones forms a useful exercise.

CHAPTER X

MATHEMATICAL DISCUSSIONS OF PROBLEMS IN SOUND

203. Introduction. The phenomenon of resonance in a straight tube and the corresponding frequency of vibration of the air in the tube are connected in a simple and obvious way with the velocity of sound in air. But, in the case of such a resonator as an ordinary bottle, the connexion between the frequency and the velocity of sound is not apparent at first sight. In some experiments, in which the form of the resonator is altered and the resulting changes of frequency are observed, we need consider only the relation of the frequency to the geometrical form, so long as the temperature of the air remains unchanged. We cannot, however, calculate the frequency in terms of the geometrical data alone, for it depends also upon ρ, the density of the air, and upon E_ϕ, the adiabatic elasticity of the air.

It will be found that in resonator problems E_ϕ and ρ always appear linked together in the form E_ϕ/ρ, which is exactly how they are combined in the formula for the velocity of sound. If, therefore, we can discover how E_ϕ/ρ is connected with the velocity of sound, we shall, if we have sufficient mathematical skill, be able, with the help of the geometrical data, to find the frequency of a given resonator at a given temperature, when we know the velocity of sound at that temperature. The calculation of the velocity in terms of E_ϕ and ρ is thus an essential step in the mathematical treatment of the vibrations of air in resonators.

204. Calculation of velocity of plane waves of sound. When plane waves of sound advance into previously undisturbed gas, there is, by the nature of waves, a space ahead of them which they have not yet reached. The gas in this region we shall consider to be at rest; we shall estimate all velocities relative to it. In Fig. 115 the direction of propagation is indicated by the arrows marked U, which are normals to the wave fronts. Let B be a mathematical plane, in the disturbed part of the gas, perpendicular to the direction of propagation and therefore parallel to the wave fronts. Let both the disturbance at B and the plane B move forward with velocity U relative to the undisturbed gas, as already explained. Let a second mathematical plane A, parallel to B, move forward in the *undisturbed* part of the gas with the same velocity.

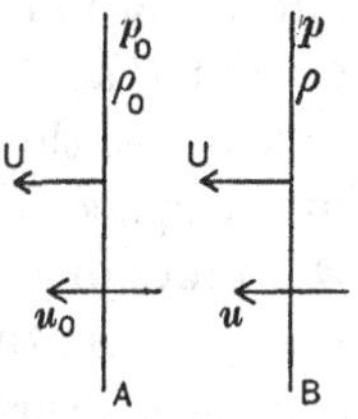

Fig. 115

If the disturbance be propagated with the definite velocity U, the state of the gas at any point, which lies on A or B, or between them, and moves forward with that velocity, remains unchanged. Hence (I) the mass of the gas between A and B and (II) its momentum remain unchanged as these mathematical planes advance.

If Q be any point in the disturbed region and if Q be fixed relative to the undisturbed part of the gas, the density, pressure and forward velocity of the gas at Q depend upon the time, but, when it happens that the plane B passes through Q, the density, pressure and velocity have definite values ρ, p and u, which are characteristic of B and occur at any other point Q' when B reaches that point. We must, of course, distinguish between the velocity u of the gas at B and the velocity U of the mathematical plane B.

In the same way, the density, pressure and velocity at A will always be ρ_0, p_0 and u_0, the values at any point in the undisturbed gas. Since, however, the gas is at *rest* at A, $u_0 = 0$.

The gas through which A moves is undisturbed, and hence the rate, measured in grm. per sec. per sq. cm., at which gas enters the moving space AB through A is $\rho_0 U$. The velocity of the plane B relative to the gas at B is $U - u$ in the forward direction and hence the rate at which gas is left behind by B, and, therefore, the rate at which it leaves the moving space AB, is $\rho(U-u)$. By condition I, the mass between A and B remains unchanged, and hence the two rates must be equal. Thus

$$\rho_0 U = \rho(U - u). \quad \text{.......................(1)}$$

It is convenient to re-write (1) in the forms

$$\rho u = (\rho - \rho_0) U, \quad \text{.......................(2)}$$

and

$$U - u = \rho_0 U / \rho. \quad \text{.......................(3)}$$

The forward momentum of the gas at B is ρu per unit volume, and $U-u$ units of volume are left behind per unit time per unit area of B as that plane advances. Hence, on account of the passage of gas through B, forward momentum $\rho u(U-u)$ passes out of the space AB per unit time per unit area. The moving space AB takes in gas through A, but, as that gas is at rest, it does not bring momentum into AB. Thus, so far as convection of momentum is concerned, AB loses forward momentum at the rate $\rho u(U-u)$. By condition II, the momentum in AB remains constant and, therefore, forward momentum must be produced by some agency. That agency is the excess, $p-p_0$, of the pressure p at B above the pressure p_0 at A, and this excess of pressure generates forward momentum at the rate of $p-p_0$ units per unit time per unit area.

Equating the rate of gain of momentum to the rate of loss, we have

$$p-p_0=\rho u(U-u). \quad \text{......................(4)}$$

Hence, by (2) and (3),

$$p-p_0=(\rho-\rho_0)\rho_0 U^2/\rho.$$

Thus $$U^2=\frac{p-p_0}{\rho-\rho_0}\cdot\frac{\rho}{\rho_0}=\frac{(p-p_0)/v}{(1/v-1/v_0)\rho_0}=-\frac{1}{\rho_0}\cdot\frac{v_0(p-p_0)}{v-v_0}, \quad \text{...(5)}$$

where $v=1/\rho$ and $v_0=1/\rho_0$ are the volumes of unit mass of gas at B and A.

By (5), we see that, although U depends upon p and v, it would be constant if p and v were so connected that $(p-p_0)/(v-v_0)$ had the same value at all points of the medium. This would require that, under the conditions prevailing in the propagation of the waves, p should be a linear function of v—say $p=m-nv$, where m and n are constants. But the relation between p and v is not linear. In the propagation of sound, the changes of pressure are so rapid that the heat developed by the compression of the gas does not have time to spread appreciably by radiation or by conduction before the compression is changed to rarefaction. Hence the conditions are adiabatic, and the actual relation between p and v in the wave system is

$$pv^\gamma=p_0 v_0^\gamma, \quad \text{........................(6)}$$

where p_0, v_0 refer to the undisturbed state of the gas, and γ is the ratio of the specific heat of the gas at constant pressure to its specific heat at constant volume. Since p and v are connected by (6), the ratio of $p-p_0$ to $v-v_0$ is not constant, and we conclude that, in strictness, no definite value can be assigned to the velocity of waves of sound. Another way of expressing the same fact is to say that, in strictness, waves cannot be propagated without change of form.

If p/p_0 differ very little from unity, we shall make little error if for $(p-p_0)/(v-v_0)$ we substitute the differential coefficient $[dp/dv_\phi]_0$, found for the value p_0 of p, where the subscript ϕ means that dp/dv has to be found from the adiabatic relation (6). In this case, (5) becomes

$$U^2 = -\frac{1}{\rho_0} v_0 \left[\frac{dp}{dv_\phi}\right]_0,$$

and this gives U^2 when $p=p_0$ and $v=v_0$. For the general case, we have

$$U^2 = -\frac{1}{\rho} v \frac{dp}{dv_\phi}. \quad \text{.......................(7)}$$

Differentiating (6) with respect to v, we have

$$-dp/dv_\phi = \gamma p/v. \quad \text{....................(8)}$$

If E_ϕ be the adiabatic elasticity, we have, by definition,

$$E_\phi = -v\,dp/dv_\phi.$$

Thus, by (8),

$$E_\phi = -v\,dp/dv_\phi = \gamma p. \quad \text{..................(9)}$$

By (7), (8), (9),

$$U^2 = \gamma p/\rho = E_\phi/\rho = \gamma pv. \quad \text{..............(10)}$$

By the gas equation, $pv = R\theta$, where θ is the absolute temperature and R is the constant corresponding to unit mass of gas. Hence

$$U = \sqrt{(\gamma R\theta)}, \quad \text{......................(11)}$$

and thus the velocity is independent of the pressure and is proportional to the square root of the absolute temperature of the gas.

It is sometimes convenient to use a in place of U for the velocity. Then

$$a^2 = \gamma p/\rho = E_\phi/\rho = \gamma R\theta. \quad \text{.................(12)}$$

If a_0 be the velocity of sound in air at 0° C., $a_0 = 3{\cdot}313 \times 10^4$ cm./sec. approximately. At the centigrade temperature t, when $\theta = 273 + t$,

$$a = 3{\cdot}313 \times 10^4 \{(273+t)/273\}^{\frac{1}{2}} \text{ cm./sec.} \quad \text{......(13)}$$

When there are finite changes of pressure, there is no definite "velocity of sound," for the waves change in form as they advance. Conspicuous effects occur with the very intense disturbances due to the firing of cannon. But in ordinary speech, and in such sounds as those produced by tuning forks, the *changes* of pressure which occur are very small compared with the atmospheric pressure. We may legitimately assume, and it is verified by experiment, that, for a given temperature, sounds of very small amplitude have a definite velocity.

If, over a small range of values of $p-p_0$, defined by $|p-p_0|<q$, where q is a small pressure, the quantity denoted by U in (5) have a very small range, we may conclude that, if the range of pressure in a train of sound waves do not exceed $2q$, the waves will not suffer any appreciable change of form as they advance.

By (6),

$$v/v_0 = p^{-1/\gamma}/p_0{}^{-1/\gamma}.$$

If we write $p-p_0=p_0x$ or $p=p_0(1+x)$, we have, when x is small,

$$(v-v_0)/v_0=(1+x)^{-1/\gamma}-1=-x/\gamma+\tfrac{1}{2}(\gamma+1)x^2/\gamma^2-\ldots.$$

Then, since $1/\rho_0=v_0$, the square root U of the quantity U^2 given by (5) becomes

$$U=(\gamma p_0v_0)^{\frac{1}{2}}\{1-\tfrac{1}{2}(\gamma+1)x/\gamma\}^{-\frac{1}{2}}=(\gamma p_0v_0)^{\frac{1}{2}}\{1+\tfrac{1}{4}(\gamma+1)x/\gamma\}.$$

If p lie between $p_0(1+x)$ and $p_0(1-x)$, U will lie between

$$(\gamma p_0v_0)^{\frac{1}{2}}\{1+\tfrac{1}{4}(\gamma+1)x/\gamma\} \quad \text{and} \quad (\gamma p_0v_0)^{\frac{1}{2}}\{1-\tfrac{1}{4}(\gamma+1)x/\gamma\}.$$

Lord Rayleigh found that sounds were still audible when x was as small as 10^{-8}. In this case, since $\gamma=1{\cdot}4$ for air, U lies between

$$(\gamma p_0v_0)^{\frac{1}{2}}(1+0{\cdot}43\times10^{-8}) \quad \text{and} \quad (\gamma p_0v_0)^{\frac{1}{2}}(1-0{\cdot}43\times10^{-8}).$$

In such cases, we make little error if we assert that the sounds have a definite velocity U, where

$$U=\sqrt{(\gamma p_0v_0)}=\sqrt{(\gamma R\theta)}.$$

SPHERICAL WAVES AND RADIATION OF ENERGY

205. Spherical waves. In order that the sound due to any vibrating body may reach the ear or a telephone receiver, it must spread out from the body. The waves of sound will travel out in all directions and will carry energy away with them. To gain some idea of this process, and to form an estimate of the rate at which the waves carry off energy, we shall consider the case in which a sphere expands and shrinks radially according to the harmonic law of vibration, the amplitude being very small compared with the mean radius of the sphere.

We suppose that the undisturbed atmosphere has absolute temperature θ_0, pressure p_0 and density ρ_0. In the waves, the density ρ will vary and we write

$$\rho=\rho_0(1+s). \quad \text{.......................(1)}$$

The number s is called the "condensation." Since ρ cannot be negative, $s>-1$. By compression pumps we can make s exceed 2000. In the case of sounds of small intensity, s is very small—say 10^{-6}. To avoid grave mathematical difficulties, we are driven, at times, to treat s as very small.

Since the vibrations advance into and disturb, adiabatically, air of pressure p_0 and density ρ_0, the relation between p and ρ in the waves is

$$p/\rho^\gamma=p_0/\rho_0{}^\gamma=k, \quad \text{....................(2)}$$

where γ is the ratio of the specific heat of air at constant pressure to that at constant volume, and k is constant for the particular adiabatic described. Then, for adiabatic changes,

$$dp/d\rho_\phi=\gamma k\rho^{\gamma-1}=\gamma p/\rho=\gamma R\theta=a^2, \quad \text{........(3)}$$

where a is a velocity. We have seen in § 204 that, as the amplitude of plane waves tends to zero, the velocity of their propagation tends to a definite limit a, where $a^2=\gamma R\theta$, and θ is the absolute temperature of the air in which they travel.

In terms of the condensation,

$$p=p_0(1+s)^\gamma, \quad dp/ds=\gamma p_0(1+s)^{\gamma-1}.$$

When $|s|$ is very small compared with unity, we may write

$$p=p_0+\gamma p_0 s=p_0+a^2\rho_0 s, \quad \text{..............(4)}$$

and

$$dp/ds=\gamma p_0=a^2\rho_0, \quad \text{....................(5)}$$

where $a^2=p_0\gamma/\rho_0$, and thus a is the velocity of sound at temperature θ_0. Strictly, we should put $p_0\gamma/\rho_0=a_0{}^2$, but, when we use (4) or (5), we disregard the (very small) difference between a and a_0.

We will now obtain the equation of continuity appropriate to the case when the air moves radially with respect to a centre O, and the velocity and density depend only upon the radius and the time. The mass of air contained in the shell between the spheres of radii r and $r+dr$, centered at O, depends upon the time. Air enters the shell through the smaller sphere at the rate of $\rho u \,.\, 4\pi r^2$ units of mass per unit time, if the positive direction of u be outwards from O. Air leaves the shell through the larger sphere at the rate

$$4\pi \{\rho u r^2 + dr \,.\, d(\rho u r^2)/dr\}.$$

If the density of the air in the shell increase at the rate $d\rho/dt$, the mass of air in the shell increases at the rate $4\pi r^2 dr \,.\, d\rho/dt$. Equating this to the resultant rate of *increase* of mass due to the passage of air through the surfaces bounding the shell, we have, as the equation of continuity,

$$r^2 d\rho/dt = -d(\rho u r^2)/dr. \quad \text{.................(6)}$$

Since r is, of course, independent of t, we have

$$\frac{d^2(r\rho)}{dt^2} = -\frac{1}{r}\frac{d}{dt}\frac{d(\rho u r^2)}{dr} = -\frac{1}{r}\frac{d}{dr}r^2\frac{d(\rho u)}{dt}. \quad \text{......(7)}$$

We next consider the dynamics of the problem. It will be necessary, later on, to make approximations which are justifiable when the changes of density are very small. It is best, however, to begin with the exact equations, for the differentiation of approximate results is apt to be hazardous.

Let ACB, $A'C'B'$ (Fig. 115a) represent two hemispheres of radii r and $r+dr$, centered at O, with a common base; let OCC' be normal to the base. If p be the pressure at distance r from O, the resultant force, in the direction OC, on the inner surface of ACB equals the force on the base AB due to a *constant* pressure p, and thus is $\pi r^2 p$. The resultant force, in the direction OC, on the outer surface of $A'C'B'$ is

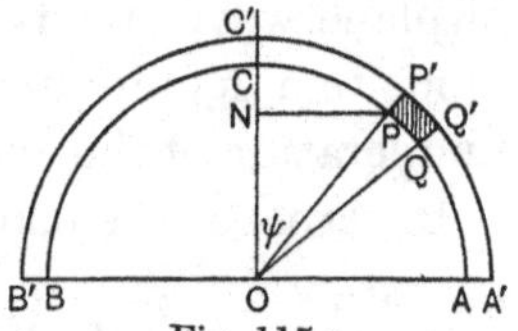

Fig. 115 a

$$-\pi\{r^2 p + dr \,.\, d(r^2 p)/dr\}.$$

The force in the direction OC due to pressure on the annulus of area $2\pi r\, dr$, corresponding to AA', is $2\pi r p\, dr$. Hence, if the total force, in the direction OC, which the pressure, due to the surrounding air, exerts on the air within the hemispherical shell, be $F dr$, we have

$$F = 2\pi r p - \pi d(r^2 p)/dr = -\pi r^2 dp/dr.$$

By (3), we have, for the adiabatic changes concerned,

$$F = -\pi r^2 \frac{dp}{d\rho}\frac{d\rho}{dr} = -\pi\gamma R\theta r^2 \frac{d\rho}{dr}. \quad \text{..............(8)}$$

The force $F dr$ equals the rate at which momentum, in the direction OC, increases within the hemispherical shell, together with the rate at which momentum, in the direction OC, escapes from the shell by convection.

If $OP = r$, if $POC = \psi$ and if PN be perpendicular to OC, $PN = r \sin\psi$. The volume traced out when the element $PP'Q'Q$, of area $dr \,.\, r d\psi$, revolves about OC is $2\pi r^2 \sin\psi \, dr \, d\psi$. The momentum of the air per unit volume is ρu along the radius; the component parallel to OC is $\rho u \cos\psi$. Hence the component, in the direction OC, of the momentum in this ring element is

$$\rho u \cos\psi \,.\, 2\pi r^2 \sin\psi \, dr \, d\psi.$$

The momentum of the air in the shell is

$$dr \,.\, \pi\rho u r^2 \int_0^{\frac{1}{2}\pi} 2 \sin\psi \cos\psi \, d\psi = dr \,.\, \pi\rho u r^2.$$

If the rate of increase of this momentum be $G dr$, we have

$$G = \pi r^2 d(\rho u)/dt. \quad \text{....................(9)}$$

The air which passes through the surface ACB carries momentum into the shell at the rate

$$\int_0^{\frac{1}{2}\pi} \rho u \cos\psi \,.\, u \,.\, 2\pi r^2 \sin\psi \, d\psi = \pi\rho u^2 r^2.$$

Momentum is carried out from the shell through the surface $A'C'B'$ at the rate

$$\pi \{\rho u^2 r^2 + dr \,.\, d(\rho u^2 r^2)/dr\}.$$

If the resultant rate at which momentum in the direction OC leaves the shell be $H dr$, we have

$$H = \pi d(\rho u^2 r^2)/dr. \quad \text{....................(10)}$$

Since the velocity is radial, no momentum enters or leaves the shell by the annulus corresponding to AA'.

Newton's law, connecting force with rate of increase of momentum, requires that

$$F = G + H.$$

Hence $G=F-H$, and thus, by (8), (9) and (10),

$$r^2\frac{d(\rho u)}{dt}=-\gamma R\theta r^2\frac{d\rho}{dr}-\frac{d(\rho u^2r^2)}{dr}. \quad\text{............(11)}$$

If we use this value of $r^2 d(\rho u)/dt$ in (7), we obtain

$$\frac{d^2(r\rho)}{dt^2}=\frac{1}{r}\frac{d}{dr}\left\{\gamma R\theta r^2\frac{d\rho}{dr}+\frac{d(\rho u^2r^2)}{dr}\right\}. \quad\text{.........(12)}$$

But

$$\frac{1}{r}\frac{d}{dr}r^2\frac{d\rho}{dr}=2\frac{d\rho}{dr}+r\frac{d^2\rho}{dr^2}=\frac{d^2(r\rho)}{dr^2},$$

and hence (12) becomes

$$\frac{d^2(r\rho)}{dt^2}=\gamma R\theta\frac{d^2(r\rho)}{dr^2}+\gamma Rr\frac{d\rho}{dr}\frac{d\theta}{dr}+\frac{1}{r}\frac{d^2(\rho u^2r^2)}{dr^2}. \quad\text{...(13)}$$

This exact equation* is too complicated for further use, and we must confine ourselves to the case where waves of very small amplitude are propagated through air which, before it is disturbed, has uniform temperature θ_0 and uniform density ρ_0. It will be convenient to consider the "condensation" s rather than the density ρ. By (1) we have

$$d\rho/dt=\rho_0 ds/dt, \qquad d\rho/dr=\rho_0 ds/dr,$$

and hence (13) becomes

$$A=B+C+D+E,$$

where

$$A=\frac{d^2(rs)}{dt^2}, \qquad B=\gamma R\theta\frac{d^2(rs)}{dr^2}, \qquad C=\gamma Rr\frac{ds}{dr}\frac{d\theta}{dr},$$

$$D=\frac{1}{r}\frac{d^2}{dr^2}(u^2r^2), \qquad E=\frac{1}{r}\frac{d^2}{dr^2}(su^2r^2).$$

Let the amplitude of the pulsating sphere which produces the waves be h, so that its radius varies harmonically between R_0-h and R_0+h. If h be small, A and B will, for a given frequency and a given value of r, be proportional to h, C and D will be proportional to h^2 and E to h^3. We cannot, of course, say that C, D and E are *always* very small compared with the first order terms A and B, for these will periodically vanish. What is true is that, at any given point, the amplitudes of C, D and E can be made as small as we please compared with the amplitude of A or B, by

* Equation (13) is found in a different way in Note II.

making h small enough. When h is very small, the variations of θ will be very small and $|\theta-\theta_0|$ will tend to zero, as h tends to zero. We conclude that the error in the equation

$$\frac{d^2(rs)}{dt^2}=\gamma R\theta_0\frac{d^2(rs)}{dr^2}=a_0{}^2\frac{d^2(rs)}{dr^2}$$

tends to zero as h tends to zero. Under these conditions, we may drop the subscript 0 from $a_0{}^2$ and call a "the" velocity of sound. Thus, finally, we have

$$\frac{d^2(rs)}{dt^2}=a^2\frac{d^2(rs)}{dr^2}. \quad \text{..................(14)}$$

In terms of s, (11) becomes

$$r^2\rho_0\frac{du}{dt}+r^2\rho_0\frac{d(su)}{dt}=-\gamma R\theta r^2\rho_0\frac{ds}{dr}-\rho_0\frac{d(u^2r^2)}{dr}-\rho_0\frac{d(su^2r^2)}{dr}.$$

When we neglect terms of the second and third orders, we have

$$\frac{du}{dt}=-\gamma R\theta\frac{ds}{dr}=-a^2\frac{ds}{dr}. \quad \text{..............(15)}$$

The differential equation, (14), for the product rs, is satisfied by

$$rs=f(r-at)+F(r+at), \quad \text{..............(16)}$$

where $f(r-at)$ is any function of $r-at$ and similarly for F. The first term $f(r-at)$ represents a wave travelling outwards from O with velocity a. The term $F(r+at)$ represents a wave travelling inwards with velocity a, but with this we are not here concerned.

We suppose that the sphere pulsates with simple harmonic motion in the periodic time T_0; we put $2\pi/aT_0=m$.

A periodic solution of (16), with periodic time T_0, is

$$s=(C/r)\cos m(r-at), \quad \text{..................(17)}$$

where C is a constant. If λ be the wave-length, $\lambda=aT_0=2\pi/m$.

By (4) and (17),

$$p=p_0+(a^2\rho_0C/r)\cos m(r-at). \quad \text{............(18)}$$

By (15),

$$du/dt=-a^2ds/dr. \quad \text{.....................(19)}$$

Then, by (17),

$$\frac{du}{dt}=\frac{a^2C}{r}\left\{m\sin m(r-at)+\frac{1}{r}\cos m(r-at)\right\}.$$

Integrating with respect to t and noting that, since it is periodic, u can contain no constant term, we find that, at distance r from O and at time t, the velocity, u, of the air is given by

$$u = aC/(mr^2) . \{mr \cos m(r-at) - \sin m(r-at)\}. \quad \ldots(20)$$

The volume of air which, per unit time, passes outwards through a sphere of radius r is $4\pi r^2 u$. We note that, by (20), the amplitude of $4\pi r^2 u$ depends upon r. By (17), the amplitude of rs is independent of r. By (20),

$$4\pi r^2 u = q \cos\{m(r-at)+\phi\}, \quad \ldots\ldots\ldots\ldots(21)$$

where $\tan\phi = 1/(mr)$ and

$$q = 4\pi C(a/m)\sqrt{(1+m^2r^2)} = 4\pi C(a/m)\sqrt{(1+4\pi^2r^2/\lambda^2)},$$

or

$$C = mq\{4\pi a\sqrt{(1+4\pi^2r^2/\lambda^2)}\}^{-1}. \quad \ldots\ldots\ldots\ldots(22)$$

The last equation gives C in terms of the amplitude q of the flow of air through a sphere of radius r. When mr or $2\pi r/\lambda$ is small, ϕ is nearly $\frac{1}{2}\pi$, and q may be taken to be independent of r and to equal $4\pi Ca/m$ or $2Ca^2T_0$.

206. Radiation of energy. Air passes outwards through the spherical surface of radius r at the rate of $4\pi r^2 u$ units of volume per unit time, and hence the rate at which the air inside the sphere does work upon the air outside the sphere is $4\pi r^2 up$. If W be the mean rate of working, we have

$$W = \frac{1}{T_0}\int_0^{T_0} 4\pi r^2 up\, dt. \quad \ldots\ldots\ldots\ldots\ldots\ldots(23)$$

Now the integral of $\cos m(r-at)$ with respect to the time over a complete period T_0 vanishes, and the same is true of the integrals of $\sin m(r-at)$ and $\cos m(r-at)\sin m(r-at)$. Hence, by (18) and (20),

$$W = \frac{4\pi a^3C^2\rho_0}{T_0}\int_0^{T_0}\cos^2 m(r-at)\,dt$$

$$= \frac{2\pi a^3C^2\rho_0}{T_0}\int_0^{T_0}\{1+\cos 2m(r-at)\}\,dt.$$

In the second integral the cosine term vanishes on integration, and thus

$$W = 2\pi a^3C^2\rho_0. \quad \ldots\ldots\ldots\ldots\ldots\ldots\ldots\ldots(24)$$

By (24), we see that the mean rate at which energy is transmitted through the surface of any sphere centered at the origin O is independent of the radius of that sphere. This result might have been foreseen, since the energy within any sphere cannot increase indefinitely as the time increases.

Using (22) and writing $\lambda = aT_0$, $m = 2\pi/aT_0$, we find, by (24), in terms of the amplitude q of the flow through a sphere of radius r,

$$W = \frac{\rho_0 a q^2 m^2}{8\pi(1 + 4\pi^2 r^2/\lambda^2)} = \frac{\pi\rho_0 a q^2}{2(a^2 T_0^2 + 4\pi^2 r^2)}. \quad \text{......(25)}$$

If $2\pi r$ be small compared with λ, we may treat q as constant with the value $4\pi Ca/m$, and then

$$W = \pi\rho_0 q^2/(2aT_0^2), \quad \text{..................(26)}$$

where q is, to this approximation, the amplitude of the flow through the surface of any small sphere centered at O.

207. Radiation from a pulsating sphere. To give definite ideas as to the magnitudes involved, we will calculate the energy radiated from a pulsating sphere. We must first find how a sphere centered at O pulsates when it produces the train of spherical waves corresponding to (21). Let its radius R be given at time t by

$$R = R_0 - h\sin\{m(R_0 - at) + \phi_0\}, \quad \text{...........(27)}$$

where the positive quantity h is small compared with R_0, and $\tan\phi_0 = 1/(mR_0)$. Then $R_0 = R + h\eta$, where $|\eta| \leqslant 1$. By (27),

$$\begin{aligned} dR/dt &= mah\cos\{m(R_0 - at) + \phi_0\} \quad \text{.....................(28)} \\ &= mah\cos\{m(R - at) + \phi + mh\eta + \phi_0 - \phi\}, \end{aligned}$$

where we suppose that $\tan\phi = 1/(mR)$.

Since $m = 2\pi/\lambda = 2\pi/(aT_0)$, $mh|\eta|$ does not exceed $2\pi h/\lambda$. Since

$$\tan(\phi - \phi_0) = \frac{m(R_0 - R)}{1 + m^2 R R_0} = \frac{\eta \,.\, 2\pi h/\lambda}{1 + 4\pi^2 R R_0/\lambda^2},$$

$|\tan(\phi - \phi_0)| < 2\pi h/\lambda$. Hence, when h/λ is small, we have

$$|mh\eta + \phi - \phi_0| \leqslant 4\pi h/\lambda.$$

Thus, if we write

$$dR/dt = mah\cos\{m(R - at) + \phi\}, \quad \text{...........(29)}$$

in place of (28), the amplitude of dR/dt will be unchanged, and only a small error of phase less than $4\pi h/\lambda$ will be introduced.

Comparing (29) with (21), we see that, when we neglect the difference between R/R_0 and unity, dR/dt will be identical with u as given by (21), if

$$q = 8\pi^2 h R_0^2 / T_0. \quad \text{.....................(30)}$$

The pulsating sphere defined by (27) will then generate the waves corresponding to (20). When we use this value of q in (25) we find that W, the rate at which energy is carried away by sound waves, is given by

$$W = \frac{32\pi^5 \rho_0 a h^2 R_0^4}{T_0^2 (a^2 T_0^2 + 4\pi^2 R_0^2)}. \quad \text{..............(31)}$$

When $a^2 T_0^2$ or λ^2 is large compared with $4\pi^2 R_0^2$, we have

$$W = \frac{32\pi^5 \rho_0 h^2 R_0^4}{a T_0^4}. \quad \text{.....................(32)}$$

If the amplitude of the pressure variation be β, we find, by (18), that, at the surface of the pulsating sphere, $\beta = a^2 \rho_0 C / R_0$, where, approximately, $C = mq/4\pi a$. Hence, by (30),

$$\beta = 4\pi^2 \rho_0 R_0 h / T_0^2. \quad \text{.....................(33)}$$

For air at 0° C. and 760 mm. pressure, $\rho_0 = 1{\cdot}3 \times 10^{-3}$ grm. cm.$^{-3}$, $a = 3{\cdot}3 \times 10^4$ cm. sec.$^{-1}$. For water $\rho_0 = 1{\cdot}0$ grm. cm.$^{-3}$ and $a = 1{\cdot}4 \times 10^5$ cm. sec.$^{-1}$. The following Table gives λ and W for some values of T_0 for both air and water, for the case in which the pulsating sphere has the mean radius $R_0 = 1$ cm. and the amplitude of the pulsation is $h = 10^{-4}$ cm. It has been assumed, in effect, that β is small compared with the mean pressure. At a frequency

	Air		Water	
T_0	λ	W	λ	W
sec.	cm.	ergs/sec.	cm.	ergs/sec.
1	$3{\cdot}3 \times 10^4$	$3{\cdot}858 \times 10^{-12}$	$1{\cdot}4 \times 10^5$	$7{\cdot}000 \times 10^{-10}$
10^{-1}	$3{\cdot}3 \times 10^3$	$3{\cdot}858 \times 10^{-8}$	$1{\cdot}4 \times 10^4$	$7{\cdot}000 \times 10^{-6}$
10^{-2}	$3{\cdot}3 \times 10^2$	$3{\cdot}856 \times 10^{-4}$	$1{\cdot}4 \times 10^3$	$7{\cdot}000 \times 10^{-2}$
10^{-3}	$3{\cdot}3 \times 10$	$3{\cdot}723$	$1{\cdot}4 \times 10^2$	$6{\cdot}981 \times 10^2$
10^{-4}	$3{\cdot}3$	$8{\cdot}341 \times 10^3$	$1{\cdot}4 \times 10$	$5{\cdot}822 \times 10^6$
10^{-5}	$3{\cdot}3 \times 10^{-1}$	$1{\cdot}061 \times 10^6$	$1{\cdot}4$	$3{\cdot}308 \times 10^9$
10^{-6}	$3{\cdot}3 \times 10^{-2}$	$1{\cdot}064 \times 10^8$	$1{\cdot}4 \times 10^{-1}$	$3{\cdot}471 \times 10^{11}$

of 10^6 vibrations per sec., when $T_0 = 10^{-6}$ sec., the flow of energy is, in the case of air, $1{\cdot}064 \times 10^8$ ergs per sec. or 10·64 watts; for water it is $3{\cdot}471 \times 10^4$ watts or 46·5 horse-power.

The Table shows what W would be if the approximations were valid. But (33) shows that, when $T_0 = 10^{-6}$ sec., the amplitude, β, of the pressure variation at the surface of the sphere would be $5{\cdot}1 \times 10^6$ dyne cm.$^{-2}$ for air, which is *not* small compared with the atmospheric pressure of 10^6 dyne cm.$^{-2}$. If h were 10^{-8} cm., β would be $5{\cdot}1 \times 10^2$ dyne cm.$^{-2}$, and W would then be 1·064 ergs/sec., and this, doubtless, is a good approximation.

208. Radiation in a conical space. Since the flow is radial from O, it will be unchanged if we divide the space surrounding O into two parts by a conical surface with O as its vertex. Let the solid angle of the cone be Ω. If, for a given amplitude of total flow from O, the rate of loss of energy in the cone be W, and from the whole space surrounding O be W_s, then $W = W_s \Omega/4\pi$. If the amplitude of flow in the cone at distance r from O be q, that of the total flow from O at distance r is $q \,.\, 4\pi/\Omega$. Hence, by (25), we have for the cone

$$W = \frac{\Omega}{4\pi} \cdot \frac{\pi\rho_0 a\,(4\pi q/\Omega)^2}{2\,(a^2 T_0{}^2 + 4\pi^2 r^2)} = \frac{4\pi}{\Omega} \cdot \frac{\pi \rho_0 a q^2}{2\,(a^2 T_0{}^2 + 4\pi^2 r^2)} \quad \text{....(34)}$$

When the cone becomes a plane, $\Omega = 2\pi$.

In obtaining (34) we supposed the air to be free from viscosity. Actually, on account of viscosity, the radial velocity at the surface of the cone will vanish. Near the surface, the flow will be modified in a manner sufficiently indicated by the results of § 226.

When $2\pi r$ is small compared with $\lambda\,(=aT_0)$, (34) becomes

$$W = 2\pi^2 \rho_0 q^2/(a T_0{}^2 \Omega), \quad \text{..................(35)}$$

and we may treat q as independent of r (see § 205).

Theory of Resonators

209. Kinetic energy of radial stream of liquid. Consider a hemisphere of radius R placed with its base on an infinite plane, and suppose that, by some means, an incompressible liquid, of density ρ, is made to flow radially outwards from the hemisphere, the velocity being the same at all points of the hemisphere. Let a volume v pass outwards per unit time. Then, if u be the velocity of the liquid at distance r, greater than R, from O, the centre of the hemisphere, $u = v/(2\pi r^2)$, and the kinetic energy of the liquid per unit volume is $\frac{1}{2}\rho u^2$ or $\rho v^2/(8\pi^2 r^4)$. The volume contained between two hemispheres of radii r and $r+dr$, with their centres at O, is $2\pi r^2 dr$, and the kinetic energy of the liquid in that volume is

$$\frac{\rho v^2}{8\pi^2 r^4} \cdot 2\pi r^2 dr = \frac{\rho v^2 dr}{4\pi r^2}.$$

If T_r be the kinetic energy outside the hemisphere of radius r,

$$T_r = \frac{\rho v^2}{4\pi} \int_r^\infty \frac{dr}{r^2} = \frac{\rho v^2}{4\pi} \cdot \frac{1}{r}.$$

If T_R be the energy outside the hemisphere of radius R,

$$T_R = \rho v^2/(4\pi R).$$

The energy between the hemispheres is $T_R - T_r$, and

$$T_R - T_r = \frac{\rho v^2}{4\pi}\left(\frac{1}{R} - \frac{1}{r}\right).$$

If R be very small compared with r, the kinetic energy, T_r, of the liquid outside the hemisphere of radius r, is negligible compared with $T_R - T_r$, the energy of the liquid between the hemispheres of radii R and r. In other words, so long as r is large compared with R, the kinetic energy between the two hemispheres differs by a negligible amount from the kinetic energy of the whole of the liquid outside the hemisphere of radius R.

If the liquid flow out from a complete sphere instead of from a hemisphere, $u = v/(4\pi r^2)$. The energy between the spheres of radii r and $r+dr$ is now $\rho v^2 dr/(8\pi r^2)$, and the energy outside a sphere of radius r becomes $\rho v^2/(8\pi r)$.

When the liquid flows out through an opening of any form in an infinite plate, the flow, though not radial near the opening, becomes more and more nearly radial as the distance from the opening increases. Hence the energy of the liquid outside a hemisphere of radius r with its centre somewhere in the area of the opening—say at the centre of gravity of the area—tends to the value $\rho v^2/(4\pi r)$ as r increases, and thus tends to be negligible compared with the energy within the hemisphere.

Similarly, if the liquid flow out from the end of a tube of any section, the energy outside a sphere of radius r, with its centre at the end of the tube, tends to $\rho v^2/(8\pi r)$ and tends to be negligible compared with the energy within the sphere.

210. Resonators. A resonator may be defined as a vessel which is closed except for an opening of any form through which the air within the vessel is in communication with the atmosphere. We define the "volume" S of the vessel as the volume enclosed within the walls of the vessel and a plane or curved (geometrical) surface A (Fig. 116) situated somewhere in the opening. If the resonator have a wide-spreading mouth, as in Fig. 117, we must clearly take A somewhere in the waist or narrow part of the opening. The pressure at any point between the waist and G can differ very little from that of the atmosphere, and thus the greater part of the volume between the waist and G does not really belong to the resonator. Let the mouth of the vessel be closed by some obstacle and let air be forced into the vessel so as

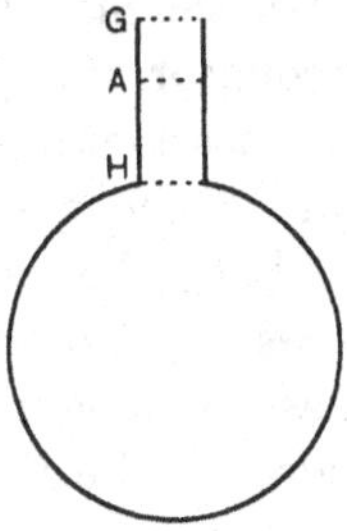

Fig. 116

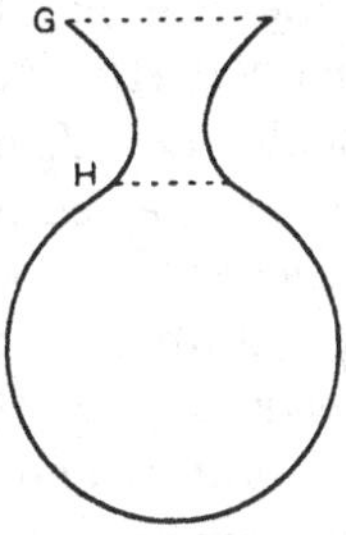

Fig. 117

to raise the pressure above P, the atmospheric pressure. If, now, the obstacle be suddenly annihilated, air will stream out of the vessel and the mass of air in S will diminish. The outward flow will not cease when the mass in S has fallen to its normal value, for the momentum acquired by the air, particularly near the neck, will cause it to continue to flow outwards for a time. A stage will, however, be reached when the mass enclosed by S has fallen so far below the normal value that the outward motion of the air through A ceases. A reverse inward flow then sets in, and in this way an oscillatory motion of the air is set up; the amplitude gradually diminishes, because energy is (1) carried off by waves to distant parts of the atmosphere and (2) is converted into heat through the action of viscosity. The "periodic time" of the oscillatory motion, i.e. the interval between successive maxima of outward flow, is somewhat greater than it would be if there were no radiation of energy.

A tuning fork, or other source of sound, in action near the resonator will cause waves to impinge on the opening of the resonator. If the periodic time of the fork nearly agree with that of the resonator, forced oscillations of considerable amplitude will be set up. If the free oscillations of the resonator be strongly damped, the resonator will respond to the fork over a considerable range of frequency of the fork, but the response will never be vigorous. Such a resonator will, of course, be of little value as a means of identifying the frequency of a sound. In order that sharper tuning may be possible, it is necessary that the damping of the free vibrations of the resonator should be comparatively small.

211. Frequency of resonator. Let the mass of air contained within the volume S (§210) of the resonator, at any instant during the vibration, be M. Let M_0 be the mass contained when the air is at rest and is at the temperature θ and the pressure P of the atmosphere near the resonator. Let ρ_0 be the density of the air at this temperature and this pressure, and let $M - M_0 = \rho_0 v$, so that M exceeds M_0 by the mass of a volume v of air at atmospheric temperature and pressure. As the air streams to and fro through the opening, waves of wave-length λ travel out to the distant parts of the atmosphere and, consequently, the motion of the air is not everywhere in the same phase—a type of motion which is, of course, impossible with an incompressible liquid. Although differences of phase occur in the case of the vibrating air, they will be small, if we keep within a distance of $\frac{1}{4}\lambda$ from the opening. It is, of course, necessary that λ should be large compared with the greatest width of the opening. Hence, if the distribution of velocity over the surface A (Fig. 116) be the same for the flow of air as for the flow of an incompressible liquid, the velocity of the air at any point near the opening (within $\frac{1}{4}\lambda$) will not differ appreciably from that of the liquid.

If T be the kinetic energy of the air in the neighbourhood of the opening, and if c be an appropriate constant, T will be of the form

$$T = \tfrac{1}{2}\frac{\rho_0}{c}\left(\frac{dv}{dt}\right)^2, \quad \ldots\ldots\ldots\ldots\ldots\ldots\ldots\ldots (1)$$

for T must be proportional to ρ_0 and also to $(dv/dt)^2$. It is sometimes convenient to write

$$dv/dt = f, \qquad \text{.........................(2)}$$

and then

$$T = \tfrac{1}{2}(\rho_0/c)f^2. \qquad \text{.......................(3)}$$

The constant c depends upon the form and size of the opening. The "dimensions" of kinetic energy are $\mathbf{ML^2T^{-2}}$, of ρ_0 are $\mathbf{ML^{-3}}$ and of $(dv/dt)^2$ or f^2 are $\mathbf{L^6T^{-2}}$. Hence c has the "dimensions" of length, and thus, on the C.G.S. system, c will be measured in centimetres. For a circular opening in a thin infinite plane plate, c equals the diameter of the opening (see § 218).

For geometrically similar openings in plane plates, such as rectangular openings with a fixed ratio between the sides, c is necessarily proportional to the square root of the area A of the opening. For a circle of radius r, $c = 2r$, and thus $c = 2\sqrt{(A/\pi)}$. Hence we can write

$$c = N \, . \, 2\sqrt{(A/\pi)},$$

where the numerical factor N depends upon the geometrical form of A, and not upon the magnitude of A. It appears from § 220 that N cannot be less than unity, its value for a circle. The less compact the area is, the greater is N.

We will now find the potential energy of the air when a small volume v has entered the resonator through the surface A (Fig. 116). The air which, when undisturbed, occupied the volume S now occupies $S - v$, if we neglect the very small change of volume of the small volume v. If the pressure be thereby raised from the atmospheric pressure P to $P + p$,

$$p = E_\phi v/S, \qquad \text{.........................(4)}$$

where E_ϕ is the adiabatic elasticity; by (9), § 204, $E_\phi = \gamma P$.

The potential energy of the system, made up of (1) the atmosphere and (2) the air within the volume S, equals the work which must be spent by some agent in forcing a piston into the vessel so as to diminish the volume of the contained air by v. At any stage, the piston is exposed to the atmospheric pressure P on the outside and to the pressure $P + p$ on the inside, and consequently it is only the *excess* pressure p which we have to consider. Now, for the small change considered, p is proportional to v, and thus the work spent is half the product of the final change of volume and the final excess pressure. Hence, if V be the potential energy,

$$V = \tfrac{1}{2}pv = \tfrac{1}{2}E_\phi v^2/S, \quad \text{.....................(5)}$$

or, by (4),

$$V = \tfrac{1}{2}p^2 S/E_\phi. \quad \text{.......................(6)}$$

Energy is carried off by waves to distant parts of the atmosphere and is also dissipated into heat by viscous action in the neck of the resonator. If the energy so "lost" during one cycle be small compared with $T+V$, the error in the periodic time, calculated on the assumption that the system is a "conservative" dynamical system in which the total energy $T+V$ is constant, will be very small. If $T+V$ be constant,

$$dT/dt + dV/dt = 0.$$

By (1) and (5) we have, on differentiation with regard to t,

$$\frac{\rho_0}{c}\left(\frac{dv}{dt}\frac{d^2v}{dt^2}\right) + \frac{E_\phi}{S}\left(v\frac{dv}{dt}\right) = 0. \quad \text{...................(7)}$$

By (10), § 204, $E_\phi/\rho_0 = U^2$, where U is the velocity of sound at the temperature θ of the atmosphere. Thus (7) becomes

$$\frac{d^2v}{dt^2} = -\frac{cU^2}{S}v.$$

Hence the motion is harmonic, and, if T_0 be the periodic time,*

$$T_0 = \frac{2\pi}{U}\sqrt{\frac{S}{c}}. \quad \text{.......................(8)}$$

If the frequency be n, so that $n = 1/T_0$, we have

$$n = \frac{U}{2\pi}\sqrt{\frac{c}{S}}. \quad \text{.......................(9)}$$

The velocity of sound is often denoted by a, and then

$$1/T_0 = n = (a/2\pi)\sqrt{(c/S)}. \quad \text{..............(10)}$$

If n_0 be the frequency at absolute temperature θ_0, we have, by (11), § 204,

$$n/n_0 = \sqrt{(\theta/\theta_0)}. \quad \text{....................(11)}$$

* In the earlier parts of this section, the symbol T has been used for kinetic energy in accordance with the custom of books on dynamics. It is unfortunate that a quantity which is not a time should have T for its symbol. Distinguishing marks can be used when there is risk of confusion.

212. Criticisms of theory of resonator. In the course of § 211, many assumptions and approximations were made. We assumed that the pressure within the volume S at any time is definite, i.e. is the same for all points within S, and this implies that the volume is so small and so compact that the pressure may be considered to have the same phase throughout S. This will be secured if the greatest distance from the surface A to any point of S be small compared with λ. The pressure outside A has been assumed to be constant and equal to the atmospheric pressure P. This implies that the amplitude of the waves generated is small and that the wave-length is large. The energy of the system has been treated as constant. This is, of course, true when we take account of the whole atmosphere and of any distant bodies which absorb sound, and allow for the effect of the viscosity of the air in dissipating energy into heat. But, when we consider only a small part of the atmosphere near the mouth of the resonator, energy is "lost," and the amplitude of the vibrations diminishes. If, however, the energy carried away or dissipated during one vibration be small compared with the potential energy of the air in the resonator at its greatest compression during the vibration, the periodic time, as calculated, will differ little from the interval between two successive states of maximum pressure.

213. Correction of volume due to potential energy of air in neck of resonator. We now examine the case when the volume of the neck GH (Fig. 116) is not very small compared with the whole volume of the resonator. If P be the atmospheric pressure and $P+p$ the pressure in the main volume of the resonator, $|p/P|$ is usually very small, perhaps not exceeding 10^{-6}. The changes of density of the air in the neck may, therefore, be neglected, and our estimate of the kinetic energy in terms of the geometrical quantity c holds good.

When the pressure is not constant throughout the resonator, we find, by (6), § 211, that the potential energy is given by

$$V = \tfrac{1}{2}(E_\phi)^{-1}\int p_1^2\, dS, \quad \ldots\ldots\ldots\ldots\ldots\ldots(12)$$

where $P+p_1$ is the pressure at the element dS of volume.

We may divide the whole volume S_0 of a resonator, such as that of Fig. 116, into the neck of volume W, bounded by the planes G and H, and the body of volume $S_0 - W$. Throughout $S_0 - W$ the pressure is approximately $P + p$, but in W its value, $P + p_1$, depends upon the position of the element, which we will call dW. Hence

$$V = \tfrac{1}{2}(E_\phi)^{-1}\{p^2(S_0 - W) + \int p_1^2 dW\}. \quad \text{.........(13)}$$

If the neck be of uniform section A, the pressure difference p_1 will be proportional to z, the distance from the mouth, for in the neck the air moves as an incompressible fluid, since the differences of phase are negligible when the length l of the neck is small compared with the wave-length. Thus $p_1 = \rho\beta z$, where β is the acceleration and ρ the density of the air. Since $dW = A\,dz$, and $p = \rho\beta l$,

$$\int p_1^2 dW = \rho^2\beta^2 A \int_0^l z^2 dz = \tfrac{1}{3}\rho^2\beta^2 A l^3 = \tfrac{1}{3}p^2 W.$$

Thus, by (13), $\quad V = \tfrac{1}{2}(E_\phi)^{-1} p^2 (S_0 - \tfrac{2}{3}W).$

Hence the potential energy is less than that calculated on the supposition of uniform pressure throughout the *whole* volume S_0 in the ratio of $S_0 - \tfrac{2}{3}W$ to S_0. Thus, if a be the velocity of sound, we see, by § 211, that the corrected frequency will be

$$n = \frac{a}{2\pi}\sqrt{\frac{c}{S_0 - \tfrac{2}{3}W}}.$$

Whatever the form of the neck, V must be greater than $\tfrac{1}{2}(E_\phi)^{-1} p^2 (S_0 - W)$ and less than $\tfrac{1}{2}(E_\phi)^{-1} p^2 S_0$. We shall not be far wrong in any case if we take $S_0 - \tfrac{1}{2}W$ as the "effective" volume.

The space W has, of course, no really definite boundary, not even when the neck is cylindrical, and thus its volume is somewhat uncertain. In many cases, however, W/S_0 is small, and then an approximate estimate of W suffices.

214. Resonator with multiple and equal openings. Let the mouth of a vessel J (Fig. 118), such as a deflagrating jar, be covered by a flat metal plate L, which is pierced by a number of equal and similar holes. By a plate or plates of glass laid on L, one or more of the holes can be closed. The volume S of the air in the resonator can be so adjusted, by varying the amount of water in J, that there is resonance when a fork of frequency n is held

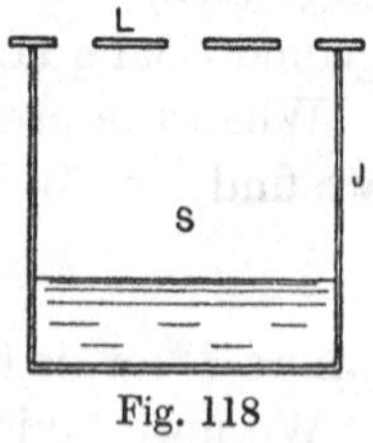

Fig. 118

near the holes. We assume that the dimensions of the space S are small compared with the wave-length.

Let S contain, at time t, in excess of the air it contains when the air is at rest, a volume v of air measured at the temperature θ and the pressure P of the atmosphere, when the density is ρ_0. Then, if T_1 be the kinetic energy of the air when one hole is open, we have, by (1), § 211,

$$T_1 = \tfrac{1}{2}\frac{\rho_0}{c}\left(\frac{dv}{dt}\right)^2,$$

where c is the length appropriate to the hole.

When r holes are open, the flow dv/dt is distributed among the holes. If the least distance between any two holes be large compared with the diameter of a hole, the lines of flow near one hole are not appreciably affected by the flow through other holes. The flow is then equally distributed between the holes and is $(dv/dt)/r$ through each hole. The kinetic energy for each hole, being proportional to the square of the flow, is

$$\tfrac{1}{2}(dv/dt)^2(\rho_0/cr^2),$$

and, since there are r holes, the kinetic energy is r times that just found for one hole. Thus

$$T = \tfrac{1}{2}\frac{\rho_0}{rc}\left(\frac{dv}{dt}\right)^2 \quad \text{.........................(14)}$$

By (5), § 211, the potential energy is V, where

$$V = \tfrac{1}{2}E_\phi v^2/S. \quad \text{.........................(15)}$$

If S be such that the resonator, with r holes open, responds to a fork of frequency n, we have, from (14) and (15), as in § 211,

$$n = \frac{a}{2\pi}\left(\frac{rc}{S}\right)^{\frac{1}{2}}, \quad \text{.........................(16)}$$

where a is the velocity of sound at the temperature θ. Thus, if n be kept constant, S is proportional to r. If S_0 be the volume corresponding to a single hole,

$$S = rS_0. \quad \text{.........................(17)}$$

215. General case of multiple openings. In § 214 the openings were supposed to be all of the same form and size. But the theory is easily extended to the case in which the openings have any forms, as in Fig. 119, provided they be sufficiently far apart.

Let the volume S of the resonator be adjusted so that it responds to a fork of frequency n. Let S be supposed divided into partial spaces S_1, S_2, ..., as in Fig. 119. Then $S_1 + S_2 + \ldots = S$. We have to arrange S_1, S_2, ... so that the air which passes in and out of the first opening does not invade the partial spaces S_2, S_3, ..., and similarly for each of the other openings. Since the pressure is the same at all points within the resonator, these partial spaces are proportional to the excess volumes v_1, v_2, ... of air which have entered through the separate holes. The geometrical surfaces separating the spaces might be replaced by thin rigid walls without affecting the frequency.

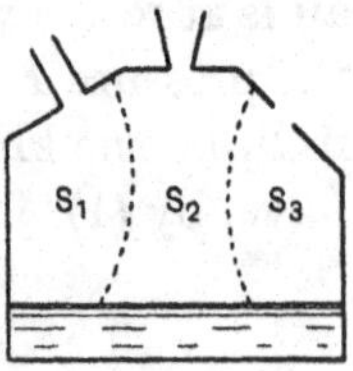

Fig. 119

Each of the spaces S_1, S_2, ..., with the corresponding opening, forms a resonator, and each of these component resonators has the same frequency as the whole resonator.

Let $c_1, c_2, \ldots, c_r$ be the constants for the r individual openings, and k the constant for the complete resonator, i.e. when all the openings are in use. Then, by (10), § 211,

$$\frac{a^2}{4\pi^2 n^2} = \frac{S_1}{c_1} = \frac{S_2}{c_2} = \ldots = \frac{S_1 + S_2 + \ldots}{c_1 + c_2 + \ldots} = \frac{S}{c_1 + c_2 + \ldots}, \quad \ldots(18)$$

and also
$$a^2/4\pi^2 n^2 = S/k.$$

Hence
$$k = c_1 + c_2 + \ldots + c_r. \quad \ldots\ldots\ldots\ldots\ldots\ldots(18a)$$

It follows from (18) that, if an extra hole be opened in a resonator of given volume S, the frequency will be raised.

The volumes S_1, S_2, ... can be found by opening each hole in turn and observing the volume which gives resonance to a fork of frequency n for that hole; the remaining $r-1$ openings are, of course, closed. The adjustment may be made by water. If S_1 be found in this way, the shape of the space will differ from that suggested in Fig. 119, but it is only the volume, and not the form, of the space which is involved. It is, of course, essential to the theory that the dimensions of the vessel should be small compared with the wave-length.

When the openings are equal, the partial spaces S_1, S_2, ... are all equal to S_0, the space for a single opening. Then, since $S_1+S_2+...+S_r=S$, we have, as in § 214,

$$S=rS_0.$$

216. "Conductivity" of opening of resonator. Let AB (Fig. 120) represent the neck of the resonator and let O be a point in the neck. If non-viscous air stream through the neck, there will be lines of flow, as indicated by the dotted lines. Let F, G be two geometrical surfaces which cut at right angles the lines of flow

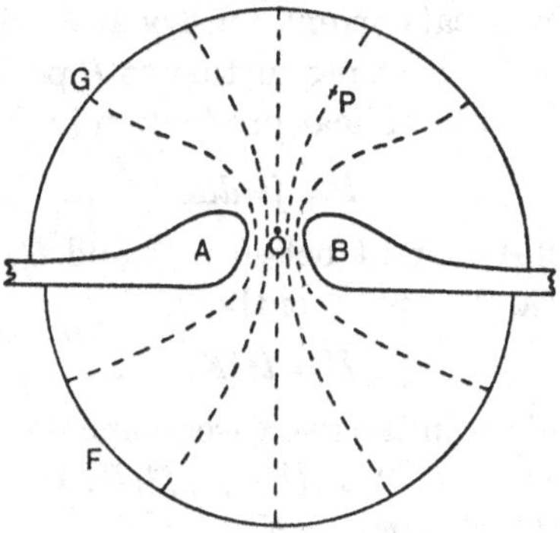

Fig. 120

and are at distances from O large compared with the dimensions of the neck.

Let T be the kinetic energy of the air in the space defined by $ABFG$ when the flow through the opening, i.e. the volume passing through per unit time, is f. Then we have, by definition of c (§ 211),

$$T=\tfrac{1}{2}\rho_0 f^2/c, \quad \text{.......................(19)}$$

where ρ_0 is the density. Here f denotes dv/dt.

If u be the velocity of the air at any point P, the kinetic energy per unit volume at P is $\tfrac{1}{2}\rho_0 u^2$, and thus

$$T=\tfrac{1}{2}\rho_0 \int u^2 d\omega, \quad \text{.......................(20)}$$

where $d\omega$ is an element of volume, and the integration extends throughout the space $ABFG$.

Now let the space $ABFG$ be filled with a substance of unit electrical resistivity or, what is the same thing, of unit specific resistance. Let the walls of the resonator be non-conducting, and let F, G be perfectly conducting electrodes. When a difference of potential is maintained between F and G, a current flows from F to G. The lines of flow will cut F and G at right angles and will be identical with those of the streaming air, since, in the neighbourhood of the neck, the air behaves like an incompressible fluid. Let K be the electrical conductance, i.e. the reciprocal of the resistance, between the electrodes F and G. Then, as we shall see, the constant c equals K.

Let i be the electrical current density at P. Since there is unit resistivity, the rate of heat production at P per unit volume is i^2. Hence, if H be the rate of heat production in the space $ABFG$,

$$H = \int i^2 d\omega. \quad \text{.......................(21)}$$

If I be the total current between F and G, the difference of potential is I/K, and, consequently,

$$H = I^2/K. \quad \text{.......................(22)}$$

Since the lines of flow in the two cases coincide, we have $u/i = f/I$, and hence, by (21) and (22), $T/H = \frac{1}{2}\rho_0 f^2/I^2$. But, by (19) and (22), $T/H = \frac{1}{2}\rho_0 f^2 K/(cI^2)$. Hence $c = K$.

217. Opening in a thin plane plate. When the "neck" takes the form of an opening in a thin plane plate, the constant c can be expressed in another way. By symmetry, the lines of flow of the air are normal to the plane of the opening, and on either side of the opening they have the same forms as the lines of electric force due to a charge upon the conducting lamina which would just fill the opening. The lines of electric force between the lamina AB (Fig. 121) and a large equipotential surface FG are indicated by dotted lines. Let Q be the charge on AB and let E be the electric force at P; then the electric displacement is $E/4\pi$, if the medium have unit inductivity, as air has in the electrostatic system.

Since the lines of electric force have the same forms as the lines of flow of the air, we have

$$\frac{u}{E/4\pi} = \frac{f}{\frac{1}{2}Q},$$

or

$$u/E = f/2\pi Q, \quad \text{.......................(23)}$$

since the flux of electric displacement from each side of the lamina is $\frac{1}{2}Q$.

If the electrical capacity of the lamina, when surrounded by the equipotential surface FG, be C, the potential of AB, when FG is at zero potential, is Q/C, and the electrical energy is W, where

$$W = \tfrac{1}{2}Q^2/C. \quad \text{.......................(24)}$$

But also

$$W = \int (E^2/8\pi)\, d\omega, \quad \text{....................(25)}$$

where $d\omega$ is an element of volume, and the integration extends throughout the space FG.

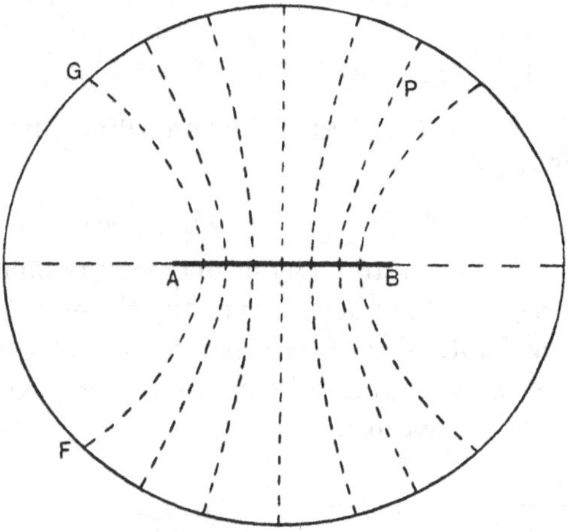

Fig. 121

Since $T = \frac{1}{2}\rho_0 \int u^2 d\omega$, we have, by (23) and (25),

$$T/W = \rho_0 f^2/\pi Q^2. \quad \text{.....................(26)}$$

Since $T = \frac{1}{2}\rho_0 f^2/c$, we have, by (24),

$$T/W = \rho_0 f^2 C/Q^2 c. \quad \text{....................(27)}$$

Hence, by (26) and (27), $c = \pi C$.

In the case of a resonator, the surfaces F, G may be taken at distances from the opening which are considerable compared with the dimensions of the opening, and this without going outside the region where the air flows (practically) as if incompressible. Hence C differs little from the capacity of the lamina in free space.

218. Circular opening. When the opening is circular and of radius r, the lamina is a disk of radius r, and then $C=2r/\pi$. Hence $c=2r$.

219. Elliptical opening. In the case of an elliptical opening with major and minor axes $2r_1$, $2r_2$ and area $A=\pi r_1 r_2$, it can be shown that

$$c=2N\sqrt{(A/\pi)}, \quad \text{..................(28)}$$

where

$$N=\frac{\pi}{2F(e).(1-e^2)^{\frac{1}{4}}}. \quad \text{..................(29)}$$

Here e is the eccentricity, so that $e^2=(r_1{}^2-r_2{}^2)/r_1{}^2$, and

$$F(e)=\int_0^{\frac{1}{2}\pi}\frac{d\theta}{\sqrt{(1-e^2\sin^2\theta)}}=\frac{\pi}{2}\left(1+\frac{1^2}{2^2}e^2+\frac{1^2.3^2}{2^2.4^2}e^4+\ldots\right).$$

The function $F(e)$ is the First Complete Elliptic Integral. When e is small, (29) gives

$$N=1+e^4/64+e^6/64+235e^8/16384+\ldots,$$

and hence, for a given area (A) of opening, c differs very little from $2\sqrt{(A/\pi)}$, its value when e is zero and the opening is circular. The following Table, from Rayleigh's *Theory of Sound*, Vol. II, § 306, shows that N differs little from unity even when the eccentricity is considerable.

e	r_2/r_1	N	$\sqrt{N}$
0·0000	1·0000	1·0000	1·0000
0·5000	0·8660	1·0013	1·0006
0·8660	0·5000	1·0301	1·0149
0·9848	0·1736	1·1954	1·0933
1·0000	0·0000	∞	∞

For a given volume, the frequency n of the resonator is proportional to $c^{\frac{1}{2}}$ or to $N^{\frac{1}{2}}A^{\frac{1}{4}}$, and thus, when r_2/r_1 is not less than 0·5, n does not differ from its value for a circular opening of equal area by as much as 1·5 per cent.

220. Opening of any form. If we start with a plane conducting lamina of any form and, keeping it plane, change its shape, but not its area, in such a way as to make it less compact, we shall increase its capacity C. We shall increase the quantity c in the formula $T=\frac{1}{2}(\rho_0/c)f^2$, of § 211, since $c=\pi C$. The surface density is greatest at those parts which project most, and thus at some central point, such as the centre of gravity of the lamina, and therefore at all points of the lamina, the potential for a given charge will be diminished by the change of form, and the capacity will be increased. Thus, the capacity of a rectangular lamina of sides nh and h/n, where h is given, increases with n from the minimum value it has when $n=1$.

It follows, therefore, that the capacity of a circular lamina is less than that of any non-circular lamina of equal area.

A more violent change is made if we separate the lamina into two or more parts in the same plane. We may suppose the parts connected by very fine wire so as to be at the same potential. The capacity of the divided lamina increases as the distances between the parts increase and tends to a limit as the distances tend to infinity. The limiting capacity is the sum of the separate capacities of the separate parts.

The capacity of a circular disk of radius r and area πr^2 is $2r/\pi$, and of a disk of radius $r/\sqrt{2}$ is $(r\sqrt{2})/\pi$. The capacity of two disks, each of radius $r/\sqrt{2}$ and of area $\frac{1}{2}\pi r^2$, tends to $2\times(r\sqrt{2})/\pi$ or $(2r/\pi)\sqrt{2}$, as the distance between them tends to infinity. For any finite distance between the disks of radius $r/\sqrt{2}$, the capacity lies between $2r/\pi$ and $(2r/\pi)\sqrt{2}$. Hence, for a pair of circular openings of radius $r/\sqrt{2}$, the acoustical quantity c, which is $\pi\times$ capacity, lies between $2r$ and $2r\sqrt{2}$.

The capacity can be calculated when the lamina is circular or elliptic but in no other cases.

The capacity of a square of side h is greater than that of a circle of equal area. The radius of this circle is $h/\sqrt{\pi}$, and the capacity is $(2/\pi)h/\sqrt{\pi}$ or $h\times 0{\cdot}3592$. Hence the capacity of the square is greater than $h\times 0{\cdot}3592$. According to an approximate numerical estimation by Maxwell,* the capacity is $h\times 0{\cdot}3607$. Henry Cavendish, about 1770, compared by experiment the capacity of a square plate with that of a circular plate and with that of a sphere. His results are roughly in agreement with Maxwell's estimate.

In the kinetic energy formula $T=\frac{1}{2}(\rho_0/c)f^2$, of § 211, c is a length. When the channel through which the air flows is an opening in an infinitely thin plane lamina, c must be proportional to the square

* *The Scientific Papers of Hon. Henry Cavendish*, edited by J. C. Maxwell, revised by J. Larmor, Vol. I, p. 412 (1921).

root of the area, for openings which are geometrically similar. If the sides h, k of a rectangle of area A be $h=(1+\alpha)A^{\frac{1}{2}}$ and $k=(1+\alpha)^{-1}A^{\frac{1}{2}}$, c will be unchanged if h become $(1+\alpha)^{-1}A^{\frac{1}{2}}$ and k become $(1+\alpha)A^{\frac{1}{2}}$. When α is small, we may replace $(1+\alpha)^{-1}$ by $1-\alpha$, and thus c has the same value for $h=(1-\alpha)A^{\frac{1}{2}}$ as for $h=(1+\alpha)A^{\frac{1}{2}}$. Hence c is stationary for variations of h, when $h=A^{\frac{1}{2}}$, and thus, rectangles of area A, which are nearly squares, have values of c which differ little from the minimum value pertaining to the square of area A. The result for an ellipse leads us to expect that, unless the rectangle be elongated, c will not be greatly increased by the change of shape.

221. Resonator with thin-plate opening. If n be the frequency of a resonator of volume S, we have, by (10), § 211,

$$n=\tfrac{1}{2}(a/\pi)\sqrt{(c/S)},$$

where a is the velocity of sound, and c is a linear quantity depending upon the opening. When this is a circular opening of area A and radius r in a thin plane plate, we have, by § 218,

$$c=2r=2\sqrt{(A/\pi)}.$$

Hence
$$n=\frac{aA^{\frac{1}{4}}}{2^{\frac{1}{2}}\pi^{\frac{5}{4}}S^{\frac{1}{2}}}. \qquad \qquad (30)$$

The discussions of §§ 219, 220 lead us to expect that formula (30) will serve for any form of opening of area A which is nowhere very narrow in comparison with its length.

222. Correction for thickness of plate. In some cases, e.g. in EXPERIMENTS 32, 34, the plate has a finite thickness not very small compared with the width of the opening. By § 216, c is the conductivity measured between two large equipotential surfaces, one on each side of the opening, the walls of the mouth being non-conducting. If we keep the area of the opening unchanged in size and shape, but give the plate a finite thickness t, we diminish c, because we displace conducting by non-conducting material. Hence, if n be constant, the area of the opening, for a given volume S, will be somewhat greater for the thick plate than for the ideal "thin" plate.

223. Decay by radiation of vibrations in a resonator. Since energy is carried away from a sounding resonator by waves of sound, the amplitude of the vibration must diminish. A complete calculation of the rate of decay due to radiation would be far beyond our limits, but an approximate estimate is easily obtained. In § 208, it is proved that, when the flow is limited to a conical space of solid angle Ω and the amplitude of the flow at distance r from the apex is q, the energy WT_0 carried off during one period T_0 is independent of r, and is given by

$$WT_0 = \frac{4\pi}{\Omega} \cdot \frac{\pi \rho_0 a q^2 T_0}{2(a^2 T_0^2 + 4\pi^2 r^2)}.$$

If $2\pi r$ be small compared with aT_0, the wave-length, we may write

$$WT_0 = 2\pi^2 \rho_0 q^2/(aT_0\Omega), \quad \text{..............(31)}$$

and may treat q as independent of the radius r. To fix ideas, we will suppose that, at the beginning of the period, the air at the surface of the sphere is flowing outwards with its maximum velocity.

In the case of a resonator, air flows to and fro through an opening which, for simplicity, we suppose to be circular. At distances from O, the centre of the opening, which are considerable compared with the radius of the opening but yet small compared with the wave-length, the direction of the velocity at a point P will be nearly radial from O and the velocity will be nearly independent of the direction of OP. Hence the energy lost per period will depend only upon q and T_0, and not upon the actual size of the opening, and will be given approximately by (31). The result (31) will apply approximately when the resonator has a tubular neck, as in Fig. 116, provided the length of the neck be large compared with its diameter and that the wall of the neck be thin.

Equation (31) was obtained on the assumption that the vibrations of the air were maintained by some agency, and is not strictly applicable to the vibrations associated with a resonator, for these vibrations, after their initial production, gradually die down because they are not maintained. If, however, the loss of energy in one period, when the vibrations are damped, be small compared with the energy at the beginning of the period, this loss will differ but little from that which would occur in one period if the vibrations were maintained, provided, of course, that the energy at the beginning of the period be the same in both cases. The beginning of the period has, for definiteness, been taken when the flow has its maximum value and when, in consequence, the energy of the air associated with the resonator is entirely kinetic; the air within the resonator has no potential energy of compression at that moment. Now, by (3), § 211, when the flow is f, the kinetic energy is $\frac{1}{2}\rho_0 f^2/c$, where c is a length depending upon the neck of the resonator. At the beginning of the period, the flow has its maximum value; to our approximation we may count this maximum value as the amplitude q of the flow when the vibrations do not decay. Hence we may replace f by q and then, if E be the energy at the beginning of the period,

$$E = \tfrac{1}{2}\rho_0 q^2/c.$$

If E_1 be the energy at the end of the period, $E_1 = \frac{1}{2}\rho_0 q_1{}^2/c$, where q_1 equals the flow f_1 at the end of the period. Hence

$$E - E_1 = \tfrac{1}{2}\rho_0 (q^2 - q_1{}^2)/c = \tfrac{1}{2}\rho_0 (q + q_1)(q - q_1)/c.$$

When the change is small, we may write

$$E - E_1 = \rho_0 q (q - q_1)/c.$$

By (35), § 208, the energy lost per period, when the amplitude of the flow is q, is WT_0 or $2\pi^2\rho_0 q^2/(aT_0\Omega)$, and this we are to put equal to $E - E_1$. Then, dividing by $\rho_0 q^2$, we have

$$(q - q_1)/q = 2\pi^2 c/(aT_0\Omega). \quad \text{...............(32)}$$

If μ be the damping coefficient,

$$q_1 = qe^{-\mu T_0} = q(1 - \mu T_0 + \tfrac{1}{2}\mu^2 T_0{}^2 - \ldots),$$

and thus, when μT_0 is small,

$$(q - q_1)/q = \mu T_0. \quad \text{.....................(33)}$$

Hence, by (32), $$\mu T_0 = 2\pi^2 c/(aT_0\Omega). \quad \text{.................(34)}$$

For the periodic time, T_0, we have, by (10), § 211,

$$aT_0 = 2\pi\sqrt{(S/c)}. \quad \text{.................(35)}$$

By (34), $$\mu T_0 = \pi c^{\frac{3}{2}}/(\Omega S^{\frac{1}{2}}). \quad \text{...................(35}a\text{)}$$

Now, by (33),

$$(E-E_1)/E=(q^2-q_1^2)/q^2=\mu T_0(q+q_1)/q,$$

and thus, when $(q-q_1)/q$ is small, and we may, in consequence, put $q+q_1=2q$, we have

$$(E-E_1)/E=2\mu T_0. \quad \text{.................(36)}$$

224. Application to Helmholtz resonator. A Helmholtz resonator is a spherical vessel of radius B having a relatively small opening of radius b. Treating the sphere in the neighbourhood of the opening as a plane, we have $\Omega=2\pi$ and, by § 218, $c=2b$. Hence, by (34), $\mu T_0=2\pi b/aT_0$. The volume of the resonator is $S=\frac{4}{3}\pi B^3$. With these values we have, by (35a),

$$\mu T_0=\left(\frac{3b^3}{2\pi B^3}\right)^{\frac{1}{2}}. \quad \text{...................(37)}$$

By an elaborate analysis, Mr F. P. White* obtained, in our notation,

$$\mu T_0=\left(\frac{3b^3}{2\pi B^3}\right)^{\frac{1}{2}}\left(1-\frac{39b}{40\pi B}\right). \quad \text{..............(38)}$$

Thus, when b/B is small, the error in our approximate result (37) is small.

For a Helmholtz resonator, in which $B/b=10$, we find, by (38),

$$\mu T_0=0{\cdot}02185\,(1-0{\cdot}0310)=0{\cdot}02117. \quad \text{.........(39)}$$

Hence, and by (33), the amplitude of flow diminishes by 0·02117 of itself in one period and, by (36), the energy diminishes by 0·04234 of itself.

If the globe of a Helmholtz resonator have radius $B=10$ cm. and if the opening have radius $b=1$ cm., we have

$$S=\tfrac{4}{3}\pi\times1000=4189\,\text{cm.}^3 \quad \text{and} \quad c=2\,\text{cm.}$$

If the air be at normal temperature and pressure,

$$\rho=1{\cdot}293\times10^{-3}\,\text{grm./cm.}^3 \quad \text{and} \quad a=3{\cdot}313\times10^4\,\text{cm./sec.}$$

Then (10), § 211, gives for the periodic time $T_0=8{\cdot}68\times10^{-3}$ sec.; the frequency is $n=115{\cdot}2$ sec.$^{-1}$. The wave-length is 287·6 cm. If the maximum internal pressure exceed the atmospheric pres-

* *Proc. Roy. Soc.* Vol. XCII, A, p. 555.

sure by $p = 10^3$ dyne cm.$^{-2}$, we find, by (6), § 211, and (12), § 204, that, if V be the maximum potential energy,

$$V = \tfrac{1}{2}p^2 S/E_\phi = \tfrac{1}{2}p^2 S/(\rho a^2) = 1476 \text{ ergs}.$$

By (36) and (39), the energy diminishes by 0·04234 of itself, i.e. by 62·5 ergs, in one period of $8{\cdot}68 \times 10^{-3}$ sec. The average rate of loss is $62{\cdot}5 \times 115{\cdot}2 = 7200$ ergs/sec.

In the author's "Grandfather" clock, the driving mass of 6000 grm. descends one metre in a week and thus works at the rate of 973 ergs/sec.

If the *amplitude* of the flow be q cm.^{3}/sec., the flow f has the maximum value q, and thus, by (3), § 211, the maximum kinetic energy is $T = \tfrac{1}{2}\rho q^2/c$. In the absence of damping, the maximum of T equals the maximum of V. Hence $T_{\text{max}} = 1476$ ergs and $q = 2137$ cm.^{3}/sec. If the maximum velocity at the centre of the opening be v_0, it follows from (46), § 227, that $v_0 = q/2\pi b^2 = 340$ cm./sec. The air moves harmonically, and thus the amplitude at the centre of the opening is

$$340\,T_0/2\pi = 0{\cdot}470 \text{ cm}.$$

If β be the amplitude of the pressure change at distance r from the centre of the opening, we have, by (18), § 205, $\beta = a^2\rho C/r$. By (22), § 205, $C = \tfrac{1}{2}q/(a^2 T_0)$, approximately, and thus $\beta = \tfrac{1}{2}\rho q/rT_0$. If $r = 10$ cm., $\beta = 15{\cdot}9$ dyne cm.$^{-2}$. The maximum radial velocity is $q/2\pi r^2$; for $r = 10$, this is 3·40 cm./sec.

If $d\theta$ be the increase of temperature due to adiabatic increase of pressure dp, we have

$$d\theta = (\gamma - 1)\,\theta\, dp/\gamma p,$$

where θ is the absolute temperature and γ is the ratio of the specific heat at constant pressure to that at constant volume. Since $\gamma = 1{\cdot}4$, $p = 10^6$ dyne cm.$^{-2}$ and, at 0° C., $\theta = 273$, we have

$$d\theta = 7{\cdot}8 \times 10^{-5} \times dp.$$

At a distance 10 cm. from the centre of the opening, $\beta = 15{\cdot}9$ dyne cm.$^{-2}$, and thus, with $dp = \beta$, the amplitude of the temperature change is $1{\cdot}24 \times 10^{-3}$ deg.

225. Radiation from resonator with tubular neck. If the neck be of length l and internal radius b, then $c = \pi b^2/l$. If a correction be applied for the open ends (see §§ 238, 239), c will be somewhat less than $\pi b^2/l$. Approximately, $\Omega = 4\pi$, and thus, by (34),

$$\mu T_0 = \frac{2\pi b}{aT_0} \cdot \frac{\pi b}{4l}. \qquad \text{.....................(40)}$$

The factor $2\pi b/aT_0$ is the value of μT_0 for a resonator with an opening of radius b in a thin flat plate or (practically) for a Helmholtz resonator with an opening of radius b. For a given T_0, we can diminish μT_0, as given by (40), by diminishing b/l. If energy were "lost" only by radiation, we could obtain sharper tuning with a resonator having a tubular neck than with a Helmholtz resonator. It will, however, appear in § 229 that the effect of viscosity is to cause so much dissipation of energy in the neck that the rate of decay is greater for the tubular neck than for the opening in a thin plate.

226. Effect of viscosity. When air passes steadily through an opening, the air in immediate contact with the walls is at rest. Unless the stream be so violent that the motion is turbulent, the velocity increases with the distance from the wall. When the motion is oscillatory, as in the neck of a sounding resonator, and is sufficiently gentle to be non-turbulent, the air in contact with the wall is at rest. It appears, however, that, when the distance from the wall exceeds a small length depending upon the frequency, the motion is practically the same as if the air were not viscous.

The essentials of the problem for a tubular neck may be gathered from the simple case in which air moves parallel to Ox, the axis of x; the horizontal plane Oxy is the face of a solid substance, and Oz points upwards. Let u be the velocity in the direction of x at time t in a plane S at distance z above Oxy. Then the rate of shearing is du/dz, and the force per unit area which the air above S exerts upon the air below S is $\eta\, du/dz$, where η is the viscosity. If a second plane T be at distance $z + dz$ from Oxy, the viscous stress across that plane is

$$\eta \{du/dz + (d^2u/dz^2)\, dz\}.$$

The resultant force per unit area on the layer ST is $\eta\, (d^2u/dz^2)\, dz$ in the positive direction.

Let the fall of pressure per unit length parallel to Ox be $F \cos \omega t$, where F is constant. If n be the frequency, $\omega = 2\pi n$. If the density of the air be ρ, the part of the layer ST, which has unit length, parallel to Ox, and unit width, has mass $\rho\, dz$. The force on it due to the difference of pressure between its ends is $F \cos \omega t \,.\, dz$. The force required to give it acceleration du/dt is $\rho\,(du/dt)\,dz$, and this equals the sum of the two other forces. Hence, on dividing by dz, we have

$$F \cos \omega t = \rho\, du/dt - \eta\, d^2u/dz^2. \qquad \text{.........(41)}$$

The solution of (41), which is periodic in t, is finite for all values of z and makes $u = 0$ when $z = 0$, is

$$u = (F/\rho\omega)\{\sin \omega t - e^{-kz} \sin (\omega t - kz)\}, \qquad \text{.........(42)}$$

where $k^2 = \frac{1}{2}\rho\omega/\eta = \pi\rho n/\eta$. Since kz is a number, k is the reciprocal of a length, and

$$1/k = \sqrt{\eta}/\sqrt{(\pi\rho n)}. \qquad \text{.........(43)}$$

Hence $1/k$ is proportional to $n^{-\frac{1}{2}}$.

The general periodic solution of (41) includes a term with Ae^{kz} as a factor, where A is an arbitrary constant. We put $A = 0$. If A were not zero, the amplitude of u would increase indefinitely as z increases.

If $\eta = 0$, k is infinite and the expression for u reduces to the first term, which gives the velocity when there is no viscosity. The second term diminishes rapidly as z increases. For air at normal pressure and at 0° C., $\rho = 1{\cdot}29 \times 10^{-3}$, $\eta = 1{\cdot}70 \times 10^{-4}$. Hence $k = 4{\cdot}88\sqrt{n}$. If $n = 400$, $k = 97{\cdot}6$ cm.$^{-1}$.

When $kz = \pi$, 2π, ... the exponential term in (42) has the following values:

kz	$e^{-kz} \sin (\omega t - kz)$	z cm.
π	$-4{\cdot}32 \times 10^{-2} \sin \omega t$	0·032
2π	$1{\cdot}87 \times 10^{-3} \sin \omega t$	0·064
3π	$-8{\cdot}07 \times 10^{-5} \sin \omega t$	0·097
4π	$3{\cdot}49 \times 10^{-6} \sin \omega t$	0·129

The Table gives the corresponding values of z for $n = 400$. For distances from the fixed plane greater than 0·064 cm., the motion may be regarded as unaffected by viscosity. If the *amplitude* of the velocity for these distances be u_0, we have

$$F/\rho\omega = u_0. \qquad \text{.........(44)}$$

227. Distribution of velocity in circular opening in thin plate. The flow of air in the mouth of a resonator has been examined by E. G. Richardson* by means of a hot-wire anemo-

* *Proc. Phys. Soc.* Vol. XL, p.206 (1928).

meter. A platinum or nickel wire, about 0·025 mm. in diameter and a few mm. long, is stretched between two narrow supports and carries an electric current which raises it to red heat. The resistance is measured while the current flows. When air streams past it, the wire is cooled and its resistance is diminished. The apparatus is calibrated by moving the wire at known speeds through air. The hot wire was placed in different positions in the mouth of the resonator. When the mouth was a circular opening of radius b in a plane plate, the curve showing v/v_0 in terms of r/b was similar to ABD in Fig. 122; here v_0 and v are the velocities at

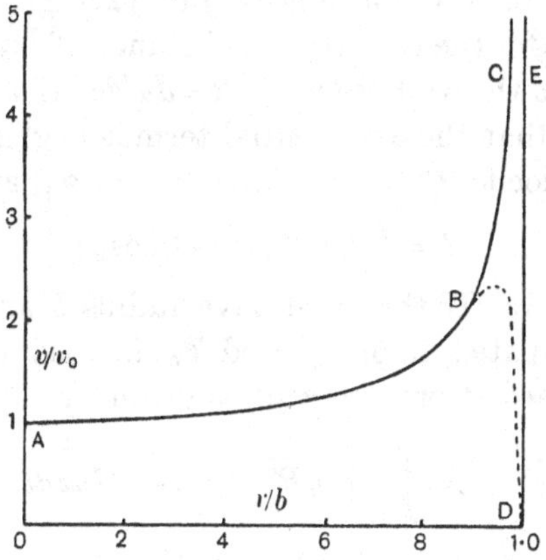

Fig. 122

the centre and at distance r from the centre in the plane of the opening. The amplitude had its maximum value at a distance from the edge which was found to be proportional to $n^{-\frac{1}{2}}$, as was to be expected from (43).

For a non-viscous liquid or gas, theory gives

$$v/v_0=(1-r^2/b^2)^{-\frac{1}{2}}, \quad \text{..................(45)}$$

and thus v increases from v_0 at $r=0$ to infinity at $r=b$. If, at any time, the total flow be f, we have

$$f=\int_0^b 2\pi rv\,dr=2\pi b^2 v_0. \quad \text{..................(46)}$$

On account of the viscosity of the air, the velocity at $r=b$ is not infinite but zero. The continuous curve ABC in Fig. 122 shows how v/v_0 would depend upon r/b if the air were not viscous. Actually v/v_0 increases from unity as r/b increases from zero, until, at a value of r/b a little less than unity, it reaches a maximum; it then decreases to zero at $r/b=1$, as shown by the dotted curve.

228. Dissipation of energy in resonator with tubular neck. On account of viscosity, the air is sheared near the wall of the tube, and energy is dissipated in the form of heat. Heat is produced at the rate h ergs per c.c. per sec., and, by (3), § 146, $h=\eta R^2$, where R radians/sec. is the rate of shearing. Unless the tube be narrow or the frequency be low, formula (42) of § 226 gives a sufficiently accurate value for the velocity u at distance z from the wall. If we measure z from the wall, we have $R=du/dz$. If, at such distances from the wall that the exponential term is negligible, the *amplitude* of the velocity be u_0, we have $F/\rho\omega=u_0$, and thus

$$R=ku_0e^{-kz}(\sin g+\cos g),$$

where $g=\omega t-kz$. If the tube have radius b and length l, the energy J_z dissipated, in one period T_0, in a layer of thickness z, measured inwards from the wall, is given by

$$J_z=\int_0^{T_0}\int_0^z \eta R^2 \,.\, 2\pi(b-z)\,l\,dz\,dt.$$

Now $(\sin g+\cos g)^2=1+\sin 2g$, and the integral of $\sin 2g$ from 0 to T_0 vanishes. Hence

$$J_z=2\pi T_0\eta u_0^2k^2l\int_0^z e^{-2kz}(b-z)\,dz$$

$$=\tfrac{1}{2}\pi T_0\eta u_0^2 l\{2kb-1-e^{-2kz}(2kb-2kz-1)\}.$$

For any but low frequencies, e^{-2kz} is very small when z exceeds a fraction of a millimetre. We may therefore neglect the exponential term and write for the energy dissipated in one period

$$J=\pi T_0k\eta lu_0^2(b-1/2k).$$

By § 226, $1/2k=(195{\cdot}2)^{-1}=0{\cdot}00512$ cm. for $n=400$. We may neglect $1/2k$ in comparison with b, unless the tube be very narrow. Then, since $k=\sqrt{(\pi\rho/\eta T_0)}$, we have

$$J=\pi T_0lbu_0^2\sqrt{(\pi\rho\eta/T_0)}. \quad \text{..............(47)}$$

If q be the amplitude of the flow, $q=\pi b^2 u_0$, and hence, by (35), § 208, if G be the energy carried off by sound waves during one period, we have, with $\Omega=4\pi$,

$$G=WT_0=\pi^3\rho b^4 u_0^2/2aT_0. \quad \text{...............(48)}$$

The maximum kinetic energy is $\frac{1}{2}\rho u_0^2\pi b^2 l$, if we neglect the energy due to the open ends of the tube. If the damping factor of the amplitude be $e^{-\mu t}$, we have, by (36), § 223, for small μ,

$$J+G=2\mu T_0 \cdot \tfrac{1}{2}\rho u_0^2 \pi b^2 l.$$

Hence $$\mu T_0=\frac{1}{b}\sqrt{\left(\frac{\pi\eta T_0}{\rho}\right)}+\frac{\pi^2 b^2}{2aT_0 l}. \quad \text{...............(49)}$$

229. Comparison of Helmholtz with tube-neck resonator. We will compare the values of μT_0 for the two resonators. In each case the frequency is 400 sec.$^{-1}$; then $T_0=400^{-1}$ sec. The radius, b, of the mouth is 0·5 cm. We take $a=3{\cdot}313\times10^4$, $\eta=1{\cdot}70\times10^{-4}$, $\rho=1{\cdot}293\times10^{-3}$. For the Helmholtz resonator, by § 224,

$$\mu T_0=2\pi b/aT_0=0{\cdot}0379.$$

By (35), § 223,

$$T_0=(2\pi/a)\sqrt{(S/2b)},$$

since, by § 218, $c=2b$. Hence we have $S=174$ cm.^{3}

The neck of the other resonator is of length l. If $l=5$ cm., we find, by (49),

$$\mu T_0=(1/b)(\pi\eta T_0/\rho)^{\frac{1}{2}}+\pi^2 b^2/(2aT_0 l)=0{\cdot}0642+0{\cdot}0030=0{\cdot}0672.$$

It will be seen that, of the two terms which make up μT_0, the term 0·0642, which is due to viscosity, is much larger than the term 0·0030, which is due to radiation of sound.

The damping due to both causes is greater for the tube-neck resonator than the damping due to radiation for the Helmholtz resonator. The damping due to viscosity in the Helmholtz resonator is probably small, for the serious shearing is confined to a small region near the edge of the opening.

230. Selectivity of a resonator. In some cases, the frequency of a vibrating system can be measured directly, as when a phonic wheel is driven by an electrically maintained tuning fork. In other cases, we compare the frequency of the system with the known frequency of a second system. One method employs beats.

In another method, one system is tuned until there is maximum response of one to the other, as in EXPERIMENT 31. The *accuracy* of the adjustment depends upon the selectivity of the responding system. The greater the selectivity, the more rapidly will the response decrease as the frequency of the exciting system departs from that corresponding to maximum response. The amplitude of the exciting system is, of course, supposed to remain constant.

The damping of the vibrations for a resonator of the Helmholtz type is less than for one with a tubular neck. Hence the Helmholtz resonator is preferred when it is desired that the response of the resonator to sound waves, whose frequency differs slightly from that of the resonator, should fall off as rapidly as possible when the discrepancy increases.

CHAPTER XI

EXPERIMENTS IN SOUND

EXPERIMENT 30. **Effect of temperature upon velocity of sound.**

231. Method. It is shown in § 204 that, if U be the velocity of sound in a gas at absolute temperature θ,

$$U = \sqrt{(\gamma R \theta)}, \quad \text{.........................(1)}$$

where γ is the ratio of the specific heat of the gas at constant pressure to the specific heat at constant volume, and R is the constant in the gas equation $pv = R\theta$, in which v is the volume of unit mass of the gas. The velocity is independent of the pressure.

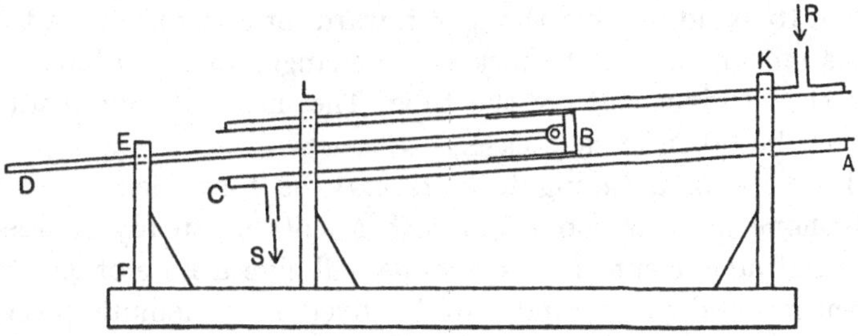

Fig. 123

We can test the relation between U and $\sqrt{\theta}$, if we can measure the velocity of sound in air at different temperatures.

The apparatus consists of a brass tube AC (Fig. 123), about 92 cm. in length and 2·5 cm. in internal diameter, provided with a movable piston B which slides within it. The tube is held in suitable supports K, L. A steam jacket is fitted to the tube; the steam enters the jacket at R and leaves at S. The end A is raised above the end C in order that the condensed water may drain away.

If good resonance is to be obtained, the piston must fit the tube closely. At the same time it must slide easily. The tube must therefore be of high quality, drawn with special care so as to be straight and of very nearly uniform section throughout its length. The piston is formed of a piece of high quality tube which just fits the resonance tube. It may be "ground in" with very fine grinding powder. The length (8 cm.) of the piston ensures that there is very little leakage.

The long piston rod *BD* passes through a hole *E* in the upright *EF* and is thus kept parallel to the axis of the tube. The piston rod is attached to the piston by a hinged joint. In order that the observer may be able to adjust the piston while his ear is near *A*, a distant control, in the form of a rod passing through bearings and sliding parallel to the tube, is provided. This control rod can be clamped to the piston rod when an approximate position of the piston has been found, and can then be used in making the final adjustment for best resonance.

The position of the piston for greatest resonance is best found by moving the piston while the fork is sounding. Care must be taken to avoid moving the fork towards or away from the tube, since movements of the fork cause changes in the intensity of the sound given out by the tube. The mean of four readings should be taken in each case.

In place of a tuning fork, it may be convenient to use a telephone receiver connected with a suitable steady source of alternating current. The telephone will give a note of constant intensity and pitch, and can be fixed in a definite position relative to the resonance tube. There may, however, be some difficulty in ascertaining the frequency of the fundamental note emitted by the telephone.

A vibrating tuning fork is held outside the tube near the end *A*, and the piston is moved along the tube until the column of air *AB* gives maximum resonance to the fork. The position can be found to 2 or 3 mm. When maximum resonance occurs, the air in the tube and near *A* vibrates with the same frequency as the fork. There is then a node at *B* and an anti-node at a point near *A*; the anti-node is not exactly in the plane of the end *A*, since there is some effect due to the open end of the tube. The condition for resonance can be satisfied if the distance from *B* to the anti-node near *A* be $(2m+1)\lambda/4$, where m is an integer and λ is the wave-length of the note of the fork.

If the tube be long enough to allow *AB* to be 60 cm., positions of *B* can be found for which $m=0$ and $m=1$, for a frequency of

512 vib. sec.$^{-1}$, when the temperature of the air column is 100° C. or 373° absolute. With greater frequencies, higher values of m will be possible.

Let the anti-node near A be outside the tube and at a distance x from the plane of A; we assume that x is independent of m. Then

$$AB = \tfrac{1}{4}(2m+1)\lambda - x. \quad \text{.................(2)}$$

If $B_0, B_1, \ldots$ be the positions of B for $m = 0, 1, \ldots$, we have, by (2),

$$AB_0 = \tfrac{1}{4}\lambda - x, \qquad AB_1 = \tfrac{3}{4}\lambda - x, \qquad AB_2 = \tfrac{5}{4}\lambda - x,$$

from which $\lambda = 2(AB_1 - AB_0)$, ...(3)

and $\lambda = AB_2 - AB_0$. ....(4)

Generally, $\lambda = (2/m)(AB_m - AB_0)$. ....(5)

If we use AB_2 and AB_1 in place of AB_2 and AB_0, we have $\lambda = 2(AB_2 - AB_1)$, but, since the resonance is more definite with AB_0 than with either AB_1 or AB_2, it seems better to pass over this equation in favour of (4).

We also find $x = \tfrac{1}{2}(AB_1 - 3AB_0)$, ...(6)

and $x = \tfrac{1}{4}(AB_2 - 5AB_0)$. ....(7)

Generally, $x = \{AB_m - (2m+1)AB_0\}/2m$. ....(8)

If the tube be wide enough, AB may be measured directly by a metre rule. In any case, since AB differs by a constant length from ED, the part of the rod projecting beyond EF, the value of AB can be found if ED be known for one value of AB.

Two sets of measurements are taken. In the first set, the jacket is open to the air and the tube is at the atmospheric temperature, which can be measured by a thermometer placed inside the tube. In the second set, steam from a boiler is passed through the jacket. The temperature of the steam is found from the barometric height. At points in the tube whose distances from A exceed two or three diameters of the tube, the temperature of the air will not differ appreciably from that of the steam.

When the temperature of the jacket, and so of the main part of the vibrating column of air, differs from that of the air outside the tube, there will be differences of temperature in the air near A and, consequently, the correction for the open end will not be the same as when the air in the tube has the same temperature as the air outside the tube. Thus the value of x will depend upon the temperature of the jacket. But, for any given temperature, λ is given independently of x by (3) and (4).

From the values obtained for AB, the wave-length λ is found in the case of each temperature. If the fork make n complete vibrations per second, the velocity of sound is given by

$$U = n\lambda. \qquad \text{(9)}$$

If U_1, U_2 be the velocities at the low and high absolute temperatures θ_1, θ_2, we have, by (1),

$$U_2/U_1 = \sqrt{(\theta_2/\theta_1)}. \qquad \text{(10)}$$

We may test the theory by comparing the value found for U_2/U_1 with that of $\sqrt{(\theta_2/\theta_1)}$.

For a given fork, n is constant, and then, by (9),

$$\lambda_2/\lambda_1 = \sqrt{(\theta_2/\theta_1)}. \qquad \text{(11)}$$

232. Practical example.

J. L. Hinton used a C fork of 512 and a G fork of 768 vib. sec.$^{-1}$.

At room temperature of 19·6° C., when $\theta_1 = 273° + 19·6° = 292·6°$, the readings for C fork were

AB_0: 15·55, 15·60, 15·65, 15·65. Mean $AB_0 = 15·61$ cm.
AB_1: 49·00, 48·95, 49·05, 48·95. Mean $AB_1 = 48·99$ cm.
AB_2: 82·70, 82·50, 82·20, 82·30. Mean $AB_2 = 82·42$ cm.

By (3), $\lambda = 2(AB_1 - AB_0) = 66·76$ cm.

By (4), $\lambda = AB_2 - AB_0 = 66·81$ cm.

Mean $\lambda = 66·78$ cm. Hence $U_1 = 512 \times 66·78 = 3·419 \times 10^4$ cm. sec.$^{-1}$.

Steam was passed through the jacket. Barometric height was 76·55 cm. and temperature of steam was 100·2° C. Then $\theta_2 = 273° + 100·2° = 373·2°$. The new readings, for the C fork, were

AB_0: 17·30, 17·25, 17·30, 17·30. Mean $AB_0 = 17·29$ cm.
AB_1: 54·75, 54·90, 54·80, 54·75. Mean $AB_1 = 54·80$ cm.

The tube was not long enough for AB_2.

By (3), $\lambda = 2(54·80 - 17·29) = 75·02$ cm.

Hence $U_2 = 512 \times 75·02 = 3·841 \times 10^4$ cm. sec.$^{-1}$.

We have $U_2/U_1 = 3·841/3·419 = 1·123$.

Similar observations were made with G fork. At $\theta_1 = 292·6°$,

$AB_0 = 10·32$, $AB_1 = 32·50$, $AB_2 = 54·54$ cm.

$\lambda = 2(AB_1 - AB_0) = 44·36$, $\lambda = AB_2 - AB_0 = 44·22$. Mean $\lambda = 44·29$ cm.

Hence $U_1 = 768 \times 44·29 = 3·401$ cm. sec.$^{-1}$.

At $\theta_2 = 373 \cdot 2°$, with G fork,

$$AB_0 = 11 \cdot 42, \quad AB_1 = 36 \cdot 30, \quad AB_2 = 61 \cdot 36 \text{ cm.}$$

$\lambda = 2(AB_1 - AB_0) = 49 \cdot 76$, $\lambda = AB_2 - AB_0 = 49 \cdot 94$. Mean $\lambda = 49 \cdot 85$ cm.

Hence $U_2 = 768 \times 49 \cdot 85 = 3 \cdot 828 \times 10^4$ cm. sec.$^{-1}$.

We have $U_2/U_1 = 3 \cdot 828/3 \cdot 401 = 1 \cdot 126$.

Mean of two values of U_2/U_1 is $1 \cdot 124$.

The ratio derived from the temperatures is

$$\sqrt{(\theta_2/\theta_1)} = \sqrt{(373 \cdot 2/292 \cdot 6)} = 1 \cdot 129.$$

EXPERIMENT 31. **Resonance with a bottle.**

233. Introduction. The formula (9) of § 211 gives the frequency of a resonator in terms of the velocity of sound, the internal volume of the resonator and the length c which depends upon the form of the neck. If we keep to air, the velocity cannot be changed unless we can vary the temperature of the resonator and of the air in its immediate neighbourhood. The value of c can be changed by varying the neck, if suitable apparatus be available. We can, however, very easily vary the volume by pouring water into the resonator, and then we can readily investigate the relation between S, the volume of air in the resonator, and n, the frequency of the note to which the resonator responds.

234. Method. An ordinary bottle with a capacity of about 500 c.c. forms a convenient resonator. The bottle is first filled with water up to the top of the neck, and is weighed when full. Water is then poured out until resonance occurs when a tuning fork of frequency n is sounded near the opening. The observer will be guided towards the correct adjustment if he blow across the opening of the bottle and so cause the resonator to give out a note of its natural frequency. By comparing this note with that of the fork, he can decide whether the frequency of the resonator is too high or too low. If the frequency of the resonator be too high, water is poured out of the bottle. It is best to go by small stages and to pour out only a small quantity of water at a time; a comparison between the resonator and the fork is made at each stage.

The final adjustment is best made by finding the point of maximum resonance when the fork is held near the opening. The fork should not be nearer to the opening than 3 or 4 cm. If it be brought too near, it will change the effective form of the opening by an appreciable amount. In this adjustment, it is convenient to have a small beaker (10 to 20 c.c.) from which water can be poured into the bottle; the same beaker receives water poured out of the bottle. In this way, it is easy to keep close to the volume giving the best resonance. A little care is needed to get the *maximum* resonance and not merely a moderately good response. At the maximum resonance, the note emitted by the resonator is clear and distinct. When the adjustment is complete, the bottle is weighed again. The loss of mass in grammes may be taken as giving S, the volume of air in the bottle. For accurate work, we may use the formula

$$S = (m/\rho)(1-\sigma/\rho')(1-\sigma/\rho)^{-1},$$

where the mass lost is balanced by a mass m of metal weights of density ρ', and ρ, σ are the densities of water and air.

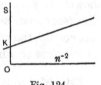

The observations are repeated with a number of forks, and the volume of the air in the resonator is plotted, as in Fig. 124, against $1/n^2$, where n is the frequency. The straight line which lies most evenly among the plotted points passes nearly, but not quite, through the origin. It cuts the axis OS in K; we denote OK by S_0. We thus find that $1/n^2$ is proportional to $S - S_0$. It will be found that S_0 is positive and approximately equal to half the volume of the (roughly) cylindrical part of the neck (see § 213).

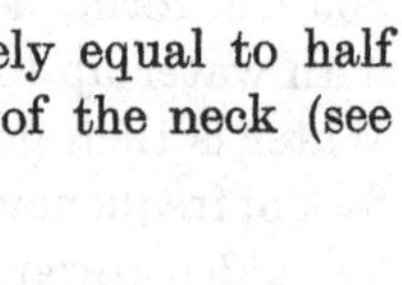

Fig. 124

235. Practical example.

Mr C. Chaffer made the observations; it would have been better to record the masses to tenths of a gramme and the volumes to tenths of a cubic centimetre.

Fork	n sec.$^{-1}$	$n^{-2} \times 10^6$ sec.^{2}	S c.c.	Fork	n sec.$^{-1}$	$n^{-2} \times 10^6$ sec.^{2}	S c.c.
C	512	3·815	37	F	341·33	8·583	73
B	480	4·340	41	E	320	9·766	83
A	426·67	5·493	49	D	288	12·056	102
G	384	6·782	59	C	256	15·259	125

When S was plotted against $n^{-2} \times 10^6$, the points lay nearly on a straight line giving $S_0 = 6$ c.c. approximately.

If we put $n^{-2}\times 10^6 = x$ and $S = y$ and apply Awbery's method (NOTE I) to find the straight line $y = mx + c$, we find $X = 8{\cdot}262$, $Y = 71{\cdot}125$. We see that X lies between the first four and last four values of x. Thus

$$4a = 20{\cdot}430, \qquad 4b = 186, \qquad 4h = 45{\cdot}664, \qquad 4k = 383.$$

Hence $$m = (k-b)/(h-a) = 7{\cdot}807.$$

Then $$S_0 = c = Y - mX = 71{\cdot}125 - 64{\cdot}501 = 6{\cdot}624 \text{ c.c.},$$

and $$S = 7{\cdot}807 \times n^{-2} \times 10^6 + 6{\cdot}624.$$

The eight values of S given by this formula are, to one place of decimals,

36·4   40·5   49·5   59·6   73·6   82·9   100·7   125·8.

EXPERIMENT 32. **Resonator with multiple openings.**

236. Method. A deflagrating jar, with a ground flange at the top, is convenient for the experiment. On the flange rests a plate of lead (Fig. 118, § 214) pierced by a number of equal circular holes. In drilling the holes, care must be taken that they are not deformed and that they are true to size. The lead is flattened on an iron "surface plate"; a flat piece of wood is placed on the lead and the wood (*not* the lead) is hit by a hammer. The lead is turned over from time to time. A tap with the finger nail will indicate at once whether the lead be flat or whether it be still curved. With a little care, it is possible so to flatten the lead that, when it is laid on the flange of the jar, it makes a nearly perfect joint. If the flange be wetted, the joint will be air-tight against the feeble differences of pressure which occur. If there be a leak in this joint, it will act as an additional opening. Care should be exercised to avoid bending the flattened plate by rough handling; the plate should not be lifted by one corner.

A plate of brass thick enough to resist bending may be used, if the under-side be ground or otherwise finished to a plane surface.

One or two small plates of glass are used to cover the holes. They make air-tight joints if wetted. A plate closing one hole should not be placed so close to an open hole as to cause any appreciable effect upon the lines of flow of the air which passes in and out of the open hole.

All the holes except one are closed; the volume of water in the jar is adjusted so that there is resonance to a fork of frequency n, about 256 sec.$^{-1}$, and the distance z cm. between the surface of the water and the lower surface of the perforated plate is measured.

It is assumed that the plate is horizontal. If the cross-section of the jar be constant and equal to A cm.^{2} and if S be the volume of air, $S = Az$ c.c. The observations are repeated with 2, 3, ... openings. Points plotted to show how z depends upon the number of holes will lie close to a straight line passing through the origin. The experiment thus confirms the theoretical result of § 214, that the volume is proportional to the number of holes.

We have assumed A to be constant. If A be not constant, the jar can be calibrated by pouring into it known volumes of water.

237. Practical example.

Mr E. C. Jubb found the values of z for five holes.

Number of holes $= r$	1	2	3	4	5
z	3·45	7·15	10·95	14·65	18·25 cm.
z/r	3·450	3·575	3·650	3·662	3·650 cm.

When plotted, the points lie nearly on a straight line. The value of z/r for one hole is too small as judged by the results for three, four and five holes. The jar was not quite cylindrical but opened out a little, after the fashion of a bell, near the flange. Hence A is larger near the flange than at a distance from it. If the jar had been calibrated and S had been found, S/r would have been more nearly constant than z/r.

Experiment 33. Resonator with variable cylindrical neck.

238. Introduction. The expression for the kinetic energy of the air in and near the mouth of a resonator involves a length c. If the neck of the resonator be a tube of circular section, c will depend upon the length l and the internal radius r of the tube in a manner which we will now determine. Fig. 125 represents a longitudinal section of the neck AD. The end A opens into the atmosphere. Near the ends A, D of the tube, the lines of flow of the air are curved. When the length of the tube is considerable compared with its diameter, the lines of flow in the central part of the tube are nearly straight and parallel to the axis of the tube. Let BC be the part where the lines of flow may be considered straight, and let $AB = x_1$, $CD = x_2$. We may assume that the velocity at time t at a point between B and C depends only upon the flow f, where

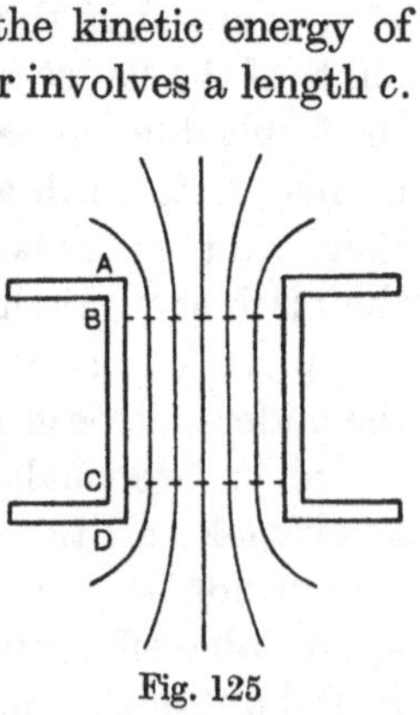

Fig. 125

$f = dv/dt$, and not upon the position of the point (see § 211). Hence, for a given flow f, the kinetic energy of the air in BC will be proportional to $l - x_1 - x_2$, the length of BC. Hence, if ρ_0 be the density of the undisturbed air, and if T be the whole kinetic energy, we have

$$T = \tfrac{1}{2}\rho_0 \{F(l - x_1 - x_2) + G_1 + G_2\} f^2, \quad \ldots\ldots\ldots(1)$$

where F, G_1, G_2 are constants. The constant G_1 is such that the kinetic energy of the air above the section B is $\tfrac{1}{2}\rho_0 G_1 f^2$. Similarly, $\tfrac{1}{2}\rho_0 G_2 f^2$ is the kinetic energy of the air below the section C. Equation (1) may be expected to correspond with facts, provided the length of the tube be not less than $x_1 + x_2$. The equation can be written

$$T = \tfrac{1}{2}\rho_0 \{Fl + H\} f^2, \quad \ldots\ldots\ldots\ldots\ldots\ldots\ldots\ldots(2)$$

where $$H = G_1 + G_2 - F(x_1 + x_2),$$

and H, on our assumptions, is constant. Comparing (2) with equation (3) of § 211, we have

$$c = (Fl + H)^{-1}.$$

By (10), § 211, if a denote the velocity of sound, and S be the "effective" volume (see § 241), the frequency n is given by

$$n = \frac{a}{2\pi}\left(\frac{c}{S}\right)^{\frac{1}{2}} = \frac{a}{2\pi}\left\{\frac{1}{S(Fl + H)}\right\}^{\frac{1}{2}}. \quad \ldots\ldots\ldots\ldots(3)$$

Thus, for a given frequency, $Fl + H$ is proportional to $1/S$. Hence, if we plot the observed value of $1/S$ against l, as in Fig. 126, we may expect that the resulting curve will differ little from a straight line cutting the negative part of the axis of l in K. Then $OK = H/F$ approximately.

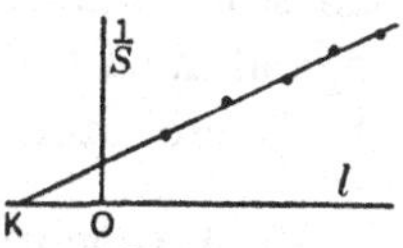

Fig. 126

239. Estimation of *F* and *H*. Let the internal radius of the tube be r. Then, if there were no motion of the air except in the tube and if the velocity were uniform throughout the tube, the velocity would be $f/\pi r^2$. Hence the kinetic energy would be

$$\tfrac{1}{2}\rho_0 (f/\pi r^2)^2 \pi r^2 l,$$

or $$\tfrac{1}{2}\rho_0 (l/\pi r^2) f^2.$$

Hence $$F = 1/\pi r^2. \quad \ldots\ldots\ldots\ldots\ldots\ldots\ldots\ldots\ldots\ldots(4)$$

If (2) hold for all lengths of tube—of course, a large assumption—it holds when the tube is of zero length. The channel between the air in the resonator and the atmosphere is then simply a circular opening of radius r in a thin plane plate. For such an opening, the kinetic energy is, by § 211 and § 218,

$$\tfrac{1}{2}\rho_0 (2r)^{-1} f^2.$$

Hence, on this assumption,

$$H = 1/2r. \qquad \text{.........................(5)}$$

From (4) and (5), $H/F = \frac{1}{2}\pi r$. This value of H/F may be compared with that deduced graphically, as in Fig. 126.

240. Correction for open ends. The curve found on plotting $1/S$ against l is nearly a straight line; this indicates that the kinetic energy of air streaming through a cylindrical neck of length l is, for a given flow, nearly the same as if the air were at rest everywhere except in a neck of length $l + H/F$, where its velocity is uniform. When the conditions at the two ends of the neck are similar, i.e. when the upper end has a flange, the correction for each end is $\frac{1}{2}H/F$. If, as in § 239,

$$F = 1/\pi r^2, \quad H = 1/2r,$$

we have

$$\tfrac{1}{2}H/F = \tfrac{1}{4}\pi r = 0{\cdot}785r.$$

The correction for an open end with a flange must lie between $\pi r/4$ or $0{\cdot}785r$ and $8r/3\pi$ or $0{\cdot}849r$. Lord Rayleigh* showed that the true value is probably not far off $0{\cdot}82r$. These estimates depend upon the assumption that the air, in and near the mouth of the resonator, flows as a non-viscous incompressible fluid.

241. Experimental details. A number of necks N (Fig. 127) of brass tube of different lengths, but of the same diameter, can be fitted to a cylindrical vessel J in the manner indicated in the figure. A large flange F can be fitted to the top of the neck. The ends of the tube N are slightly reduced in diameter to suit the hole in F and the hole in the upper plate of J. The length of the reduced part equals the thickness of the plate into which it fits. The volume of the air space in J is varied by means of water. The adjustment is facilitated if a tap be fitted to the vessel. For each

* *Theory of Sound*, Vol. II, § 307.

length of neck, the volume of air is adjusted to give resonance to a fork of known frequency. The neck is removed, and the vessel J is filled with water. It is then weighed. The neck is now put into place, and water is allowed to flow out by the tap until resonance to the fork is obtained. If too much water have been allowed to escape, a little is poured in by the neck. When the best adjustment for resonance has been found, the resonator—without the neck—is again weighed. The volume V of air between the surface of the water and the top plate of J is deduced from the loss of weight. The volume of the air in the neck is W, and $W = \pi r^2 l$, where r is the internal radius of the neck and l is its length. If S_0 be the whole volume of the resonator, $S_0 = V + W$.

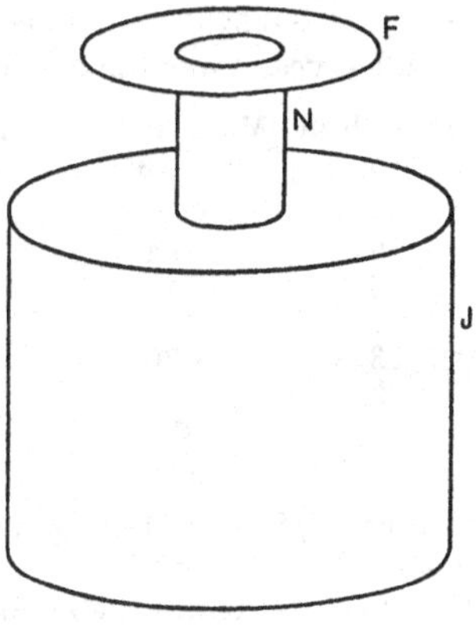

Fig. 127

The "effective" volume S of the resonator is greater than V and less than S_0. By § 213, we have approximately*

$$S = S_0 - \tfrac{2}{3}W = V + \tfrac{1}{3}W = V + \tfrac{1}{3}\pi r^2 l.$$

One observation may be made with the circular opening in the vessel without any neck. The radius r' will be slightly greater than the internal radius r of the necks. Let V_1' be the volume giving resonance to the fork for the opening of radius r', and let V_1 be the volume which would resound to the same fork for an opening of radius r. By (3), S is proportional to c, and $c = 2r$, the diameter of the opening. Since there is no neck, S is identical with V. Thus, for $l = 0$,

$$S = V_1' r / r'.$$

The results are exhibited graphically by plotting $1/S$ against l.

* The correction $\frac{1}{3}\pi r^2 l$ for the air in the neck may be tested by experiments with a single neck. For these observations, the volume, V, of the vessel (without the neck) and also the frequency, n, are varied. One of the longer necks should be used. When $1/n^2$ is plotted against V, the resulting straight line will cut the axis of V at a short distance from the origin on the negative side; this distance represents the volume correction due to the neck.

242. Practical example.

Mr T. P. Nicholas used fork of frequency $n=256$ vib. sec.$^{-1}$.

Internal radius of necks $=r=0\cdot71$ cm.; for neck of length l cm., correction $\frac{1}{3}\pi r^2 l=0\cdot528l$ was added to volume found by change of mass.

When vessel was full of water, total mass was 850 grm.

Length of neck, l	Mass, vessel and water	Uncorrected volume, V	Corrected volume, S	$\frac{1}{S}$
cm.	grm.	c.c.	c.c.	c.c.$^{-1}$
6	754	96	99·2	0·01008
5	741	109	111·6	0·00896
4	715	135	137·1	0·00729
3	670	180	181·6	0·00551
2	602	248	249·0	0·00402
1	491	359	359·5	0·00278
0	280	570	570·0	0·00175

When $1/S$ is plotted against l, the points lie nearly on a straight line cutting the axis of l at about $l=-0\cdot9$ cm. The correction for each end is therefore 0·45 cm., which equals $0\cdot634r$. Lord Rayleigh's correction $0\cdot82r$ gives 0·58 cm.

Experiment 34. Resonator with thin-plate opening.

243. Method. By § 211, the frequency n of a resonator of volume S is given by

$$n=ac^{\frac{1}{2}}/(2\pi S^{\frac{1}{2}}), \qquad \text{.......................(1)}$$

where c is a length determined by the mouth of the resonator and a is the velocity of sound. When the mouth is an opening of area A in a thin plane plate, c is necessarily proportional to $A^{\frac{1}{2}}$ and we can write $c=\mu A^{\frac{1}{2}}$, where μ is constant as long as the openings are geometrically similar. Thus (1) becomes

$$n=\mu^{\frac{1}{2}}aA^{\frac{1}{4}}/(2\pi S^{\frac{1}{2}}). \qquad \text{....................(2)}$$

In any case, μ is approximately equal to $2/\pi^{\frac{1}{2}}$ or 1·12838, the value for circles, unless the width of the opening be small compared with the length. To test the constancy of μ, we might take a series of non-adjustable openings and, for each opening, adjust the volume S to resound to a fork of frequency n. This takes much time, unless S be readily adjustable. An alternative is to give S a series of known values and to adjust A in each case to obtain resonance to the fork.

On the top of a deflagrating jar is placed a lead plate L (Fig. 128), in which is a rectangular slot D, of width h. A scale EE is fixed to the plate so as to be parallel to the slot. The area of opening can be varied by a flat metal plate F, which slides along the scale. The length of the opening can be found from the scale reading of F. To avoid confusion, one end of the plate F should be of irregular form. The lead plate L is made flat (see § 236) and should be carefully handled. If the top of the jar be wetted, the plate L will make an air-tight joint with the jar. Only "grubby" students smear the jar and plate with vaseline. If F be wetted, it will make good contact with L. The scale may be held at one end by a small bolt (fitted with a nut) which passes through the scale and the lead plate. A short slit, as at G, is made in L and a small portion K is turned up, so as to provide a stop against which the scale EE can rest.

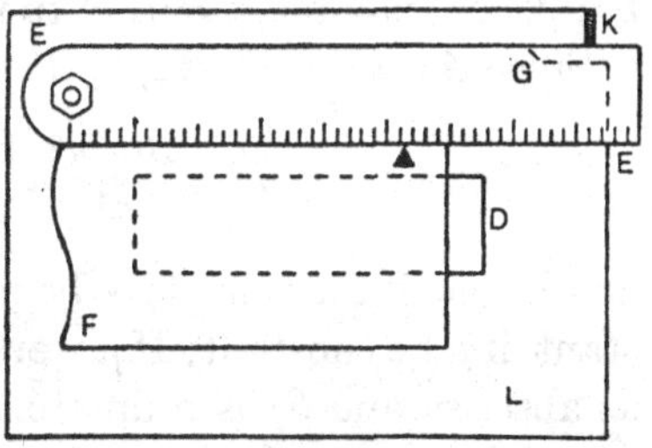

Fig. 128

The jar is first filled with water, so that the surface of the water is level with the under-side of the plate L. Then, by a 50 c.c. pipette, 100 c.c. of water are taken out, and the sliding plate F is adjusted so that resonance to a tuning fork of known pitch is obtained. The length, x, of the opening can be found from the scale reading of the index mark on the plate F, if the reading when the opening is just closed be known. Three independent settings of F should be made for each value of S. The area of the opening is $A = hx$.

Water is taken from the jar by steps of 50 c.c., and the sliding plate is adjusted for resonance at each stage; the process is continued till the full length of the slot is open. In each case the volume S of the space between the water and the plate L is recorded, and also the corresponding value of x.

When S is small, the area A, which gives resonance to the fundamental frequency (marked on the stem) of the fork, will be small, and the sound emitted at resonance will be feeble and may escape notice. If, for the same S, the area A be considerably increased, vigorous resonance to the first overtone of the fork may be obtained. It is, perhaps, not surprising that students should keep to the overtone, and disregard the fundamental, for a series of values of S, but it is strange that they should, as has often happened, pay no attention to this warning.

Since $A=hx$, we have, by (2),

$$\mu^2=\frac{16\pi^4n^4}{a^4}\cdot\frac{S^2}{A}=\frac{16\pi^4n^4}{a^4h}\cdot\frac{S^2}{x}, \quad\text{.................(3)}$$

and hence, if the same fork be used throughout, S^2/x will be constant, if μ be constant. If μ were constant, a curve plotted with x as abscissa and S^2 as ordinate, as in Fig. 129, would be a straight line through the origin. Since, however, μ increases as the ratio of the length to the width of the opening increases, we may expect that S^2/x will increase as x increases from h, the width of the slit, or as x diminishes from h, and that, in consequence, the curve will take the form PQR. The straight line OQ, which touches the curve, corresponds to square openings. The curve lies above OQ, for there are two positions of the sliding plate for which the ratio of length to width of the opening has equal values, and for each of these S^2/x is greater than for a square.

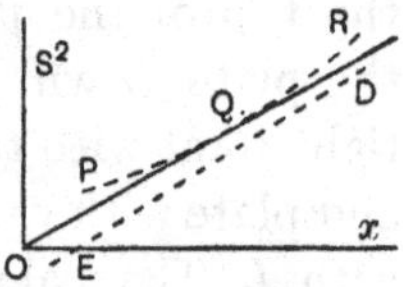

Fig. 129

If S^2/x be found for the point Q, an approximate value of a, the velocity of sound at the temperature of the room, will be found if we put $\mu=2/\pi^{\frac{1}{2}}=1{\cdot}12838$ in (3). Then we have

$$a=\left[\frac{4\pi^5n^4}{h}\cdot\frac{S^2}{x}\right]^{\frac{1}{4}}. \quad\text{.....................(4)}$$

If we use Maxwell's estimate $C=h\times 0{\cdot}3607$ for the capacity of a square of side h, the value of c for the square opening is πC or $\pi h\times 0{\cdot}3607$. Then $\mu=1{\cdot}13317$, which differs little from $1{\cdot}12838$, the value for a circle (see § 220).

244. Correction for edge effect. On account of viscosity, the air along a narrow strip of width ϵ round the perimeter of the opening is practically at rest; the width ϵ depends upon the frequency. Hence the effective area is not xh but B, where

$$B=(x-2\epsilon)(h-2\epsilon). \quad\text{.....................(5)}$$

We suppose that the frequency with viscous air and an area xh

equals that with non-viscous air and area B, but the assumption is—of necessity—very crude.

If we neglect the effect of the *form* of a rectangular opening of given area, we have, by (4),

$$S^2 = \frac{a^4(h-2\epsilon)}{4\pi^5 n^4}(x-2\epsilon). \quad \text{..................(6)}$$

Hence, if we plot S^2 against x, we shall obtain a straight line, such as ED (Fig. 129), cutting Ox at $x=2\epsilon$, and thus we can find 2ϵ. The slope of the line gives us

$$a^4(h-2\epsilon)/(4\pi^5 n^4),$$

and hence, as h, 2ϵ and n are known, we can find a, the velocity of sound at the temperature of the room.

245. Practical example.

The plate used by Mr R. K. Kerkham had thickness $b=0{\cdot}2$ cm. Mean width of slit $=h=1{\cdot}114$ cm. Column "p" shows volume as a multiple of 98·36 c.c., and $S=p\times 98{\cdot}36$ cm.^{2} Pipette used contained 49·18 c.c. and not 50 c.c. Column "x obs." gives observed length of opening, the mean of four observations. Frequency of fork $=n=288$ sec.$^{-1}$. Mean temperature 16·25° C. or 289·25° on absolute scale.

p	p^2	x obs. cm.	x cal. cm.	p	p^2	x obs. cm.	x cal. cm.
1	1	0·12	0·340	6	36	2·48	2·423
2	4	0·31	0·518	7	49	3·24	3·197
3	9	0·82	0·816	8	64	4·09	4·090
4	16	1·24	1·232	9	81	5·07	5·102
5	25	1·83	1·768				

When p^2 was plotted against x, points lay close to straight line

$$x = 0{\cdot}28 + 0{\cdot}05953p^2, \quad \text{.......................(7)}$$

such as ED, Fig. 129, except for $p=1$ and $p=2$, where observations were difficult and influence of form of opening is great. Column "x cal." gives x as calculated from (7). Term 0·28 equals 2ϵ. Since $h=1{\cdot}114$, $h-2\epsilon=0{\cdot}834$ cm. By (7),

$$p^2/(x-0{\cdot}28) = (0{\cdot}05953)^{-1} \text{ cm.}^{-1},$$

and thus

$$S^2/(x-0{\cdot}28) = (0{\cdot}05953)^{-1}\times 98{\cdot}36^2$$
$$= 1{\cdot}625\times 10^5 \text{ cm.}^5$$

By (6), velocity of sound at 289·25° K. is given by

$$a = \left[\frac{4\pi^5 n^4}{h-2\epsilon}\cdot\frac{S^2}{x-2\epsilon}\right]^{\frac{1}{4}} = \left[\frac{4\times\pi^5\times 288^4\times 1{\cdot}625\times 10^5}{0{\cdot}834}\right]^{\frac{1}{4}}$$
$$= 3{\cdot}579\times 10^4 \text{ cm. sec.}^{-1}.$$

This gives, at 0° C.,

$$a_0 = 3{\cdot}579\times 10^4\sqrt{(273/289{\cdot}25)} = 3{\cdot}477\times 10^4 \text{ cm. sec.}^{-1}.$$

NOTE I

AWBERY'S METHOD FOR COMBINATION OF OBSERVATIONS

In many cases, the variables x and y satisfy, according to theory, a linear relation of the form

$$y = mx + c. \qquad \text{.............................(1)}$$

In an experiment it may happen that a considerable number, n, of values, $x_1, x_2, \ldots, x_n$, of x and the corresponding values, $y_1, y_2, \ldots, y_n$, of y are observed. These values of x and y are, in general, affected by errors of observation. If only two observations be made, we have

$$m = (y_2 - y_1)/(x_2 - x_1), \qquad c = (x_2 y_1 - x_1 y_2)/(x_2 - x_1).$$

When more than two observations are made, errors will, in general, cause the values of m and c derived from the pairs x_p, x_q and y_p, y_q to differ from the values given by x_1, x_2 and y_1, y_2. Thus, a method of combining a set of slightly discrepant observations, so as to give a single value for m and a single value for c, is essential, unless we reject all but two of the observations. Such a method, if simple, may take the place of the elementary graphical method, in which m and c are found from the straight line which, in the opinion of the computer, lies most evenly among the points representing (x_1, y_1), (x_2, y_2), ... The graphical method cannot deal effectively with values of x and y having several significant figures, and, of course, introduces errors due to faulty plotting, to inaccuracies in the ruling of the squared paper or to non-uniform expansion or contraction of the paper.

In Awbery's* method, the algebraical means X, Y are calculated. Thus

$$X = (x_1 + x_2 + \ldots + x_n)/n, \quad Y = (y_1 + y_2 + \ldots + y_n)/n. \qquad \text{......(2)}$$

The point (X, Y) is the centroid G of the n points. We assume that the ideal straight line, $y = mx + c$, which we seek, passes through G. The straight line given by the method of least squares also passes through G.

We next determine the direction of the line. We divide the n points into two groups by a straight line through G parallel to the axis of y, and find (i) the coordinates a, b of the centroid E of the points 1, 2, ..., i for which $x < X$ and (ii) the coordinates h, k of the centroid F of the points $i+1, i+2, \ldots, n$ for which $x > X$. Thus

$$a = (x_1 + x_2 + \ldots + x_i)/i, \qquad b = (y_1 + y_2 + \ldots + y_i)/i,$$

$$h = (x_{i+1} + x_{i+2} + \ldots + x_n)/(n-i), \qquad k = (y_{i+1} + y_{i+2} + \ldots + y_n)/(n-i).$$

The straight line EF passes through G, since G is the centroid of the whole system of n points. We assume that EGF is the ideal line we seek. We then have

$$m = (k - b)/(h - a). \qquad \text{..........................(3)}$$

* J. H. Awbery, *Proc. Phys. Soc.* Vol. XLI, p. 384 (1929). See also W. E. Denning, *Proc. Phys. Soc.* Vol. XLII, p. 100 (1930). The method was proposed by Cauchy in 1847; Mr Awbery discovered it independently.

Using this value of m and the values of X and Y already found, we have

$$c = Y - mX. \qquad \text{.............................(4)}$$

If one of the x's equal X, we may put the corresponding point into either the first or the second group. Our choice will have little effect upon m, since the y of the point nearly equals Y.

As an example, we will fit a straight line to the values of x and y given in columns 2 and 3 of the following Table. The corresponding points, numbered 1, 2, ..., do not lie on one straight line. The values of y in column "True" are the exact values of y found by putting $x = 0{\cdot}31$, $1{\cdot}03$, ... in $y = 3{\cdot}02x - 5{\cdot}19$. The inexact values of y in column 3 were found by rounding off the "True" values to two decimal places.

	x	y	True	Awbery
1	0·31	−4·25	−4·2538	−4·2513
2	1·03	−2·08	−2·0794	−2·0773
3	1·54	−0·54	−0·5392	−0·5374
4	1·81	0·28	0·2762	0·2778
5	2·32	1·82	1·8164	1·8177
6	2·91	3·60	3·5982	3·5991
7	3·48	5·32	5·3196	5·3202
8	4·71	9·03	9·0342	9·0340
9	5·34	10·94	10·9368	10·9362

The nine points do not lie on a single straight line. We find, for example,

$$\frac{y_2 - y_1}{x_2 - x_1} = 3{\cdot}0139, \qquad \frac{y_3 - y_2}{x_3 - x_2} = 3{\cdot}0196, \qquad \frac{y_4 - y_3}{x_4 - x_3} = 3{\cdot}0370.$$

For the centroid of the nine points, we have

$$X = \tfrac{1}{9}\sum_1^9 x = \tfrac{1}{9} \times 23{\cdot}45 = 2{\cdot}6056, \qquad Y = \tfrac{1}{9}\sum_1^9 y = \tfrac{1}{9}(30{\cdot}99 - 6{\cdot}87) = 2{\cdot}6800.$$

The first group of points contains the first five, since for these $x < X$. We thus obtain

$$a = \tfrac{1}{5}\sum_1^5 x = \tfrac{1}{5} \times 7{\cdot}01 = 1{\cdot}402, \qquad b = \tfrac{1}{5}\sum_1^5 y = -\tfrac{1}{5} \times 4{\cdot}77 = -0{\cdot}9540,$$

$$h = \tfrac{1}{4}\sum_6^9 x = \tfrac{1}{4} \times 16{\cdot}44 = 4{\cdot}110, \quad k = \tfrac{1}{4}\sum_6^9 y = \tfrac{1}{4} \times 28{\cdot}89 = 7{\cdot}2225.$$

Hence $$m = \frac{k - b}{h - a} = \frac{8{\cdot}1765}{2{\cdot}708} = 3{\cdot}01939,$$

and $$c = Y - mX = 2{\cdot}6800 - 3{\cdot}01939 \times 2{\cdot}6056 = -5{\cdot}1873.$$

The values of y in column "Awbery" are, to four decimal places, those given by putting $x = 0{\cdot}31$, $1{\cdot}03$, ... in

$$y = 3{\cdot}01939x - 5{\cdot}1873.$$

The greatest difference of "Awbery" from "True" is 0·0025, and the mean difference, without regard to sign, is 0·0013.

Other methods.

(A) The method of least squares gives

$$m=\frac{n\Sigma xy-\Sigma x\,.\,\Sigma y}{n\Sigma x^2-(\Sigma x)^2},\quad c=\frac{\Sigma y\,.\,\Sigma x^2-\Sigma x\,.\,\Sigma xy}{n\Sigma x^2-(\Sigma x)^2},$$

where n is the number of points. In the example $n=9$, and

$$\Sigma x=23{\cdot}45,\quad \Sigma y=24{\cdot}12,\quad \Sigma x^2=83{\cdot}4653,\quad \Sigma xy=130{\cdot}3782.$$

Hence, to seven decimal places,

$$m=3{\cdot}0195454,\quad c=-5{\cdot}1875933.$$

If we put $x-X=\xi$, $y-Y=\eta$, and $m=\tan\theta$, we have

$$\tan\theta=\frac{\Sigma\xi\eta}{\Sigma\xi^2},\quad \tan 2\theta=\frac{2\Sigma\xi^2\,.\,\Sigma\xi\eta}{(\Sigma\xi^2)^2-(\Sigma\xi\eta)^2}.$$

(B) The method of least moment of inertia was brought to my notice by Dr G. T. Bennett. The "best" straight line is that about which the moment of inertia of the n points, each of unit "mass," is least; of necessity, it passes through (X, Y). The method gives

$$\tan 2\phi=\frac{2\Sigma\xi\eta}{\Sigma\xi^2-\Sigma\eta^2},$$

where $\tan\phi$ is the gradient of the line. When η/ξ is constant,

$$\tan 2\phi=\tan 2\theta.$$

When η/ξ has nearly the same value for each of the n points, $\tan 2\phi$ nearly equals $\tan 2\theta$. For the example,

$$\tan 2\phi=-0{\cdot}7439450,\quad \tan\phi=3{\cdot}0195463.$$

NOTE II

EXACT EQUATION FOR SPHERICAL WAVES*

Equation (13), of § 205, can be obtained, in an alternative manner, by using the standard equations of hydrodynamics. By (3), § 205, we have, for the adiabatic changes concerned, $dp/d\rho=\gamma R\theta$. The dynamical equation is

$$\rho\left(\frac{du}{dt}+u\frac{du}{dr}\right)=-\frac{dp}{dr}=-\frac{dp}{d\rho}\frac{d\rho}{dr}=-\gamma R\theta\frac{d\rho}{dr}. \quad \text{.........(1)}$$

As in (6), § 205, the equation of continuity is

$$r^2 d\rho/dt=-d\,(\rho ur^2)/dr. \quad \text{.........(2)}$$

Hence
$$\frac{d\rho}{dt}=-u\frac{d\rho}{dr}-\frac{\rho}{r^2}\frac{d}{dr}(ur^2). \quad \text{.........(3)}$$

Since r is independent of t, we have, by (1),

$$\rho\frac{d}{dt}(ur^2)=-\gamma R\theta r^2\frac{d\rho}{dr}-\rho ur^2\frac{du}{dr}. \quad \text{.........(4)}$$

* I owe the method of this NOTE to Mr W. R. Dean.

By (3), $$\rho \frac{d}{dr}(ur^2) = -r^2 \frac{d\rho}{dt} - ur^2 \frac{d\rho}{dr}. \qquad \text{.....................(5)}$$

If we differentiate (4) with respect to r, and (5) with respect to t, and use the result

$$d^2(ur^2)/dr\,dt = d^2(ur^2)/dt\,dr,$$

we find, on subtraction, that

$$\frac{d}{dt}\left\{r^2 \frac{d\rho}{dt} + ur^2 \frac{d\rho}{dr}\right\} - \frac{d}{dr}\left\{\gamma R\theta r^2 \frac{d\rho}{dr} + \rho ur^2 \frac{du}{dr}\right\}$$

$$= \frac{d\rho}{dr}\frac{d(ur^2)}{dt} - \frac{d\rho}{dt}\frac{d(ur^2)}{dr}. \qquad \text{........................(6)}$$

If we use the result

$$\frac{d\rho}{dr}\frac{d(ur^2)}{dt} - \frac{d}{dt}\left(ur^2 \frac{d\rho}{dr}\right) = -ur^2 \frac{d^2\rho}{dr\,dt},$$

equation (6) becomes

$$r\frac{d^2(r\rho)}{dt^2} - \frac{d}{dr}\left(\gamma R\theta r^2 \frac{d\rho}{dr}\right) = \frac{d}{dr}\left(\rho ur^2 \frac{du}{dr}\right) - ur^2 \frac{d^2\rho}{dr\,dt} - \frac{d\rho}{dt}\frac{d(ur^2)}{dr}$$

$$= \frac{d}{dr}\left(\rho ur^2 \frac{du}{dr}\right) - \frac{d}{dr}\left(ur^2 \frac{d\rho}{dt}\right)$$

$$= \frac{d}{dr}\left(\rho ur^2 \frac{du}{dr}\right) + \frac{d}{dr}\left\{u \frac{d(\rho ur^2)}{dr}\right\}, \quad \text{by (2)},$$

$$= \frac{d^2}{dr^2}(\rho ur^2 . u). \qquad \text{..........................(7)}$$

But $$\frac{1}{r}\frac{d}{dr}\left(r^2 \frac{d\rho}{dr}\right) = \frac{d^2(r\rho)}{dr^2},$$

and thus (7) becomes, on division by r,

$$\frac{d^2(r\rho)}{dt^2} = \gamma R\theta \frac{d^2(r\rho)}{dr^2} + \gamma Rr \frac{d\rho}{dr}\frac{d\theta}{dr} + \frac{1}{r}\frac{d^2(\rho u^2 r^2)}{dr^2}. \qquad \text{.........(8)}$$

This equation is identical with (13), § 205.

Printed in the United States
By Bookmasters